HOW TO CONTROL COSTS IN YOUR POLLUTION PREVENTION PROGRAM

HOW TO CONTROL COSTS IN YOUR POLLUTION PREVENTION PROGRAM

Judith A. Cichowicz

John Wiley & Sons, Inc.
New York ■ Chichester ■ Weinheim ■ Toronto ■ Singapore ■ Brisbane

This text is printed on acid-free paper.

Library of Congress Cataloging-in-Publication Data

Cichowicz, Judith A.

How to control costs in your pollution prevention program / Judith A. Cichowicz.

p. cm.

Includes bibliographical references and index.

ISBN 0-471-18015-7 (alk. paper)

1. Pollution prevention—Economic aspects. I. Title.

HD30.255.C535 1997

658.15′53—dc21 95-51065

This book is dedicated to my parents
Florence G. Cichowicz
and
Zigmund T. Cichowicz
who taught me how to think.

ACKNOWLEDGMENTS

No good book is written in isolation. I would like to especially thank the following individuals for their own unique forms of support. Special thanks go to:

R. Larry Lynch, P.E., for reading the manuscript, giving useful comments and acting interested when I would call during a small crisis.

Barbara Kirchem, who made sure I made all my airplanes and kept me pumped up with the elixir of Seattle, coffee.

Myral Bernstein, who organized the footnotes, ran the spell checker, and kept encouraging me.

Rick Arrasmith, who let me pick his brain in order to use more effective real-world examples.

Michael Jaffe, who taught me how to play.

And Vladislav Vishnevetski.

CONTENTS

HOW TO CONTROL COSTS IN YOUR POLLUTION PREVENTION PROGRAM

CHAPTER 1

INTRODUCTION

To achieve competitive status in the global marketplace, businesses cannot afford to use limited resources on reactive or "shoot from the hip" answers to the operational questions they face every day. Profitable businesses thoughtfully adopt changes in management that result in incremental improvements in operating systems. The result is production efficiencies and reduced costs.

This book views environmental issues as business conditions, made manageable through the implementation of appropriate, cost-effective pollution prevention strategies. Its purpose is to present the reader with a range of practical, immediately usable tools with which to do this.

This book has been written specifically to help small and medium-sized businesses become more adept at controlling costs associated with environmental issues in general, and pollution prevention in particular.

While many readers come from production functions in manufacturing organizations, this book also provides support for pollution prevention personnel working for service industries or providing a service function within a manufacturing company. Even government agencies that own contaminated property or conduct hazardous activities can benefit from the ideas

Hazards of Government Service

While federal, state, and local governmental organizations are traditionally thought of as enforcement authorities, many of their service activities have the potential to create a pollution event, incurring expensive disposal and remediation costs for the taxpayer.

Recognizing that the total costs of these services can be more cost-effectively controlled, the state of Wisconsin operates an environmental compliance program within its Bureau of State Risk Management of the Department of Administration. Among other pollution prevention tasks, this program facilitates management of the 700 million pounds of hazardous waste generated annually by the state.

Where does the waste come from? Waste streams are generated by such activities as the manufacture of license plates, vehicle maintenance, and laboratory residues from the state's 26 college campuses.

Think about the community in which you live, work, and pay taxes. Where can the pollution risks be reduced?

presented here. Thus, this book is equally useful to the manager of an environmental testing laboratory, the operator of a government motor pool, or the owner of a dry cleaning establishment.

It presents:

- A discussion of environmental management systems and their application to pollution prevention issues
- Processes for identifying and characterizing the most cost-effective pollution prevention opportunities
- A decision tree for determining whether a source creates a waste stream
- An analysis of performance measurements, from statistical process control to environmental cost accounting
- Methods for enhancing cost control, from purchasing to payment of invoices
- A process for selecting functional pollution prevention techniques, with an emphasis on cost control criteria

At the end of each chapter, there is an exercise designed to begin real-world practice of the ideas presented. Some of these exercises lead the reader to a series of suitable tools and references found in the appendices, including:

- Checklists and process model worksheets to facilitate identification of pollution prevention opportunities
- Worksheets for calculating the financial impacts of pollution prevention alternatives and for costing specific pollution prevention activities
- Reference materials from the federal government and other agencies
- Sources of additional information

This chapter introduces the regulatory framework that influences pollution prevention in the United States. The concept of cost control is presented with a glossary of financial terms. Capital access and the inertia of the status quo are considered.

THE EVOLUTION OF POLLUTION PREVENTION

From the advent of the Industrial Revolution, most businesses operated under the maxim that "the solution to pollution is dilution." Only since the late 1960s has industrialized society at large recognized the failure of that system to eliminate environmental risks or reduce environmental impacts. As we began to understand that the supposed solution was not working, we responded in several ways.

Historically, governments at the federal, state, and local levels operated reactively with a command-and-control mentality. If a problem occurred, specific rules, regulations, and enforcement mechanisms were designed. It is the nature of free societies that these rules or reactions are developed in response to a problem.

Using the command-and-control scenario, regulatory inspections were conducted in the spirit of finding fault. If a company followed the rules, it could avoid a fine and no one was thrown in jail. This didn't mean that the source of the pollution was eliminated or the overuse of a natural resource reduced. But the company bought some time and was able to turn back to its "real work" until the next visit.

Industry, when motivated by the legal and financial liabilities and costs associated with environmental pollution, either actual or potential, used a variety of mechanisms to respond in an effort to reduce compliance costs.

Bubbling Over with Conservation

Mesa, Arizona, is a rapidly expanding community in the desert Southwest. Because of its location, the efficient use of water resources is critical to its long-term quality of life. Unfortunately, the local government's zoning and planning authorities don't always seem to understand this in practice. One owner of a parcel of land being developed as a small industrial park was required by these authorities to install irrigation bubblers for all the naturally occurring cacti being left on the property as part of the landscape plan. The bubblers were installed, but never used, since they would lead to overwatering and weakening the cactus plants. In this example, implementation of an admirable goal (efficient delivery of water to vegetation) failed to take into account that certain native plants do not need irrigation. The local business paid money to install a system it could not use without jeopardizing the survivability of a native species.

Depending on how sophisticated the business decision makers were, new "state of the art" plants were constructed. Old plants were closed or sold to unsuspecting buyers. At others, the facility operator walked away, leaving a landlord or financial institution to face the enforcement music.

With the advent of the federally funded Superfund program, cleanup or remediation became an environmental mantra. But first, a company had to study the situation and take lawyer-intensive enforcement actions; maybe, someday, permission to actually clean up would be awarded.

In the interim, since societal perceptions were that industry could not be trusted, accidents were expected to happen. In these cases, industry and government itself mobilized to ship in the "first responders," decked out in fancy gear to save us from ourselves.

While this was all very heroic and made for exciting news coverage, it missed the environmental point—that is, determine where the problem originates and take action to eliminate it and mitigate its impacts.

To be effective, pollution prevention extends beyond this limited scope of "fire fighting." It recognizes that, with a focused effort, environmental problems can be prevented or controlled. It acknowledges that pollution is not merely the result of catastrophic releases. It can be the result of many small acts or careless past practices that accumulate over time and overwhelm the capacity of the environment to cleanse and regenerate itself. Further, pollution prevention recognizes the significance of resource conservation and optimal resource utilization in reducing environmental degradation.

But to accomplish this requires a management effort that is as consistent and systematic as financial management. Changing a gasket on a regular schedule may not win any accolades from the White House, or the news media, but it can prevent a pollution event.

Regulatory Classifications

The U.S. regulatory structure has adopted several methodologies for identifying chemicals and chemical products (Table 1.1). These approaches have been derived from the application of both physical and chemical parameters of the materials, as well as the regulatory missions of the agency organizing the classification system.

It is important to recognize these regulatory authorities and their classifications when evaluating a workplace for pollution prevention (P^2) opportunities. This knowledge also becomes useful when managing responses to enforcement actions or other legal activities.

Some states have adopted their own unique vocabulary to describe chemicals, with parameters that can exceed federal requirements. The smart pollution prevention professional is knowledgeable in the language of the regulatory community (state, federal, local) where a work process takes place.

For international matters, such as the international transportation of chemicals, additional descriptors (dangerous goods, marine pollutant) are used.

COST CONTROL

It is a misnomer to consider the function of cost control merely as cost containment. While cost containment can be an element, the work process of cost control has a much broader scope. Cost control is the process that provides direction and oversight to the commitment and payout of expenditures.

In application, it does not mean selecting the lowest-priced vendor or product, but, rather, making a rational choice that integrates various decision-making factors:

■ TABLE 1.1 Regulatory Classifications for Chemicals

Federal Agency	Regulatory Concern	Classification Name	Regulatory Citation
U.S. Dept. of Transportation	Protection of the public during transportation of chemicals	Hazardous Materials (for international transport purposes, these are classified as Dangerous Goods)	49 CFR Subchapter C Parts 171-180
U.S. Environmental Protection Agency	Protection of all environmental media (air, land, water) from storage and releases of designated chemicals	Hazardous Substances	40 CFR Subchapter J Parts 300-399
U.S. Environmental Protection Agency	Protection of all environmental media (air, land, water) from storage and releases of designated chemicals	Extremely Hazardous Substances	40 CFR Subchapter J Parts 300-399
U.S. Environmental Protection Agency	Protection of the environment and the public from releases to ambient air of designated chemicals	Regulated Substances Hazardous Air Pollutants	40 CFR Subchapter C Parts 50-79
U.S. Environmental Protection Agency	Protection of the environment during treatment, storage, and disposal of waste streams with hazardous characteristics	Hazardous Wastes	40 CFR Subchapter I Parts 240-299
U.S. Occupational Safety & Health Administration	Protection of employees from chemicals with the capability to cause physical or health hazards to employees	Hazardous Chemicals	29 CFR 1910.1200
U.S. Occupational Safety & Health Administration	Protection of employees from the consequences of chemical accidents involving highly hazardous chemicals	Highly Hazardous Chemicals	29 CFR 1910.119

- Desired quality of the work (service) or material purchased
- Schedule for obtaining the work (service) or material
- Availability of the work (service) or material
- Perceived capacity of the vendor to provide the work (service) or material in accordance with the work specifications
- Terms for providing the work (service) or material, such as payment schedule, risk transference criteria, assignment of particular personnel, and responsibilities for cleanup and/or wastes generated

Through integrated decision making, the true costs of a work (service) or material are identified and understood up front. This has the further benefit of reducing the number of surprises that can occur and lead to change orders and cost overruns.

In later chapters, as methods for cost containment are described, the reader may notice that many are simple, distinctly low-tech solutions. They are common sense applications and improvements that can bring serious money to the pollution prevention bottom line.

THE P^2 FINANCIAL GLOSSARY

Throughout this book, reference is made to financial terminology not normally used in the everyday lexicon of the P^2 professional. The following is a handy reference list of words and definitions that can improve communication of P^2 issues to both internal and external audiences.

Activity-Based Costing

Through this accounting technique, all operating expenditures (including overhead) are allocated to the goods and services for which the cost is incurred.

Asset/Fixed Asset

When a business owns something with a monetary value, this item is viewed as an asset. A fixed asset is tangible property, such as a piece of equipment, that is not readily consumed in the business process. For depreciation purposes, fixed assets have defined life spans.

Capital Costs

These are the costs incurred in order to raise capital or money. They can be interest, fees, or related expenses.

Depreciation

This is an accounting mechanism by which the reporting of the total costs of a fixed asset are spread out over a defined time period. The dollar amount taken or depreciated is charged in accounting records against revenue of the defined time period. Although the time period selected can be based on the estimated life of the fixed asset, the time selected may be driven by tax code requirements.

Externalities (Societal Costs)

In economic terms, externalities are societal costs for which an organization is not legally held accountable. They can include factors that are difficult to track, such as the costs of environmental impacts for the disposal of packaging after a product has been shipped to and received by the end user. The parameters of legal accountability vary from jurisdiction to jurisdiction. For example, Seattle, Washington, adopted a Street Utility Business Charge, whose proceeds are dedicated for use in the maintenance and repair of streets and related transportation infrastructure. While this would be an externality in most communities, in Seattle it is more appropriately classified as a private cost.

G&A

General & Administrative (G&A) costs are associated with overhead. These costs are not directly assigned to the costs of goods and services provided.

Life-Cycle Accounting

This term describes " . . . the assignment and analysis of product-specific costs within a life cycle framework including unusual, hidden, liability, and less tangible costs."[1]

Private Costs

These costs directly affect the bottom line and represent those things for which a business can be held legally responsible. They can be referred to as the "cost of doing business." Included are such expenditures as those for

1. U.S. Environmental Protection Agency, *An Introduction to Environmental Accounting as a Business Management Tool: Key Concepts And Terms*, EPA 742-R-95-001, June 1995, p. 28.

equipment, payroll, advertising, and so on. Private costs are routinely addressed in accounting systems.

Return on Investment (ROI)

In traditional accounting, ROI is represented as a ratio between earnings and investment. It expresses how much return is attained from a certain monetary investment.

$$\text{ROI} = \text{Earnings/Investment}$$

When applied to pollution prevention activities, this ratio continues to relate earnings to investment. To be meaningful, however, the elements that comprise earnings do change, depending on the nature of the projected activity. For example, if the investment is to purchase a more efficient dust collector for a bagging operation, earnings can be defined to include such items as:

- Costs saved from reduced maintenance
- Reduced workers' compensation claims
- Improved production due to a less hazardous work environment

When computed, an ROI that is greater than one indicates a favorable return. An ROI of less than one is an unfavorable return.

Risk Transfer

With regard to risk management, the risk transfer process results in transferring a financial or legal risk and/or liability to another party. For example, when a contractor is hired, evidence of general liability insurance coverage may be required, naming the organization purchasing the contractor's services as an additional insured. In the event of a liability event tied to the contractor, the purchasing organization can turn to the insurance provider for the contractor to recover financial costs through a process called subrogation. A purchasing organization may still be sued, but the costs of defense are transferred to the contractor.

Time Value of Money

This economic concept recognizes that the value of money changes with time. For example, if you invest a dollar today at 5 percent interest, it will be worth $1.05 in one year. For a true picture of the costs associated with a pollution prevention project, the time value of money must be taken into account.

Total Cost Assessment (TCA)

This accounting methodology integrates environmental costs into capital budget analyses. It can address both private and societal costs relative to the following intended applications:

- Cost accounting
- Capital budgeting
- Process/product design.

Opportunity Cost

This is the cost, or value, of what is given up in order to do something else. In evaluating pollution prevention projects, this can include the costs incurred when a less-hazardous raw material is substituted into a process.

Pollution Credits

When an organization obtains a permit for discharges and emissions, there are normally parameters for discharge (both by composition and quantity) that are applied to the geographic scope of the permit. If these credits are "banked" and not used, they can be traded later in regulatory negotiation for alternative credits. In the event of the possible sale of that piece of property, pollution credits may be assets if they can be transferred to a new owner, who then has purchases the permits to expand capacity.

IS CAPITAL ACCESS A BARRIER TO POLLUTION PREVENTION?

There is a business adage that banks won't lend you money unless you don't need it. For the small business considering pollution prevention, it has been suggested that the lack of access to capital created a barrier to pollution prevention implementation. In practice, though, the barrier is commonly internal to the organization, and not created by the capital markets.

Well-managed small businesses that have the foresight to consider pollution prevention are generally privately held and often loan-averse. There is a general unwillingness to implement an improvement unless it can be financed from current cash flow. So, although the organization may qualify for a competitive loan, and the project payback may be less than two years, these small business owners are reluctant to sign personal guarantees or assign other assets to potential lenders in order to pay for the money.

Where the Money Is

Operating since 1987, the Cascadia Revolving Fund (Seattle, Washington) is a nonprofit community loan fund that provides financing to disadvantaged businesses, including those that are small and owned by minorities or women. In a traditional lending sense, these businesses may also be disadvantaged because they:

- Are young (less than 2 years old);
- Have a small sales volume;
- Have insufficient collateral to satisfy bank security requirements; or
- Have a loan need of less than $150,000.

Since 1994, the fund has operated a special program focused on meeting P^2 capital requirements for these businesses. It has experienced limited interest in obtaining capital from this targeted group. Among factors influencing this are:

- Greater ability to access capital from traditional lenders with commercial banks;
- Lack of inertia to make P^2 improvements unless required to by a regulation;
- Not every P^2 solution requires capitalization.

Additionally, many of these business owners are reluctant to provide the detailed financial information, such as business and personal tax returns, required to secure a pollution prevention loan. These business owners come from a culture of self-sufficiency and independence that has trained them to respond, "Mind your own business." The preparation of this type of loan application paperwork can seem onerous to someone who is skilled at producing widgets but unfamiliar with the communications scenarios of loan officers.

While the rate of small business ownership is high, so is the rate of small business failure. The owners who have survived see the risks associated with a getting a loan as too great. If the organization is a service business, the risk is of even greater concern since these organizations typically have few capital assets and rely heavily on the production and output of a few key individuals. If these individuals leave the organization, contracts are cancelled, or payments defaulted on, there may be no way for the small business to repay the loan principal and interest.

When the service business is a government agency, it must get taxpayer approval for an investment. If the voters say no, there are few financial alternatives available.

THE INERTIA OF THE STATUS QUO

In most organizations, the threat of change represents an enormous psychological barrier to the implementation of any type of new activity.

Even though change is all around us, as human beings, we instinctively resist. This is especially apparent if we feel there is no choice. The lack of, or limitations in, organizational buy-off of a change throughout the decision-making process can condemn even the most critical project to failure.

In mature, medium-to large-sized organizations, the momentum for retaining the status quo can is substantial. The P^2 challenge is to get leverage on the organization in order to overcome that inertia.

Who applies the leverage necessary for these risk-averse organizations to take on debt and implement P^2 capital improvements? External influences are:

- Regulatory agencies
- Landlords/property owners
- Customers

The leverage of a regulatory enforcement action is easy to recognize. Do it or you go to jail, face fines, and spend lots of time doing anything but whatever you work at.

In other cases, regulatory effects are more subtle. For example, a building permit for an expanded warehouse may be withheld unless the entire structure is brought up to code, including special containment areas for chemical storage. This pressure is market driven only to the point where an organization can decide to not expand or move to another jurisdiction.

The interests of landlords and property owners are distinctly market driven, but it is their market, not the market of the tenant. For example, a closed-loop dry cleaning system can effectively reduce the amount of solvents used and/or waste produced in the dry cleaning process. If the cost of the system is $50,000, the small operator is resistant to taking on debt in order to make the change. But the landlord has a vested interest in ensuring that tenants operate in an environmentally responsible manner.

First, the new system protects the resale and leasing value of the property by preventing contamination. But resale is in some distant future. The landlord or property owner makes money today from rents and the ability to draw on the equity value of the property to refinance debt. If there is environmental contamination at a piece of property, the ability of the property owner to refinance or gain access to that equity is seriously handicapped.

However, if the property owner requires the tenant to implement a P^2 capital improvement, such as the closed-loop system, the environmental risk to the property owner is reduced. The property can be refinanced or sold. There is greater access to insurance protection (risk financing), so the property owner is happy. Further, with the pollution risk reduced, the public has been served.

In some types of businesses, outside vendors (or products or services) create the pressure for change. For instance failure to provide environmentally responsible warehousing and handling for products can cause a company to cancel use of an outside distribution service. At the same time, these same companies can be supporters of pollution prevention efforts and serve as an alternate source of capital.

There are two types of customers: internal and external. Internal customers are organizational employees in other departments for whom various products (manufacturing component) or services (maintenance of a production line) are provided by an internal supplier. External customers include not only the purchaser (or, in the case of a governmental authority, the taxpayer/recipient of the service) of the product or service, but also other stakeholders, such as regulatory agencies.

Customers are the final category of participants who create leverage. If the customer is a large, publicly traded company, that company frequently has its own environmental management systems. It doesn't want to be surprised years later with litigation because of a pollution event relative to its activities with its suppliers.

In order to continue doing business with these types of customers, organizations establish and enforce their own pollution prevention requirements. A major public marine terminal recently completed a $4 million retrofit to an existing chemical tank farm in order to meet the pollution prevention requirements of its major customer. It can now handle more volatile products with less direct risk to the public due to systems engineering features it has installed. Without that investment, it would have lost existing business and the growth potential of new product lines.

POLLUTION PREVENTION

The essence of pollution prevention is simple:

- Use fewer natural resources.
- Create less waste.

The process begins by identifying pollution prevention opportunities. This can be done either formal or informally, depending on the time, resources, and objectives of the organization.

A simple opportunity identification technique is the "no-brainer" approach. Some people would call it common sense, but in the realm of pollution prevention, common sense is common only if you possess it and use it.

In the "no-brainer," an individual, such as a warehouse worker, observes a potential pollution event, decides what action to take, and takes it. For example, the warehouse worker observes a container of a chemical material balanced precariously on a shelf and moves it to a more stable position.

Commonsense Spill Control

For an inexpensive and entertaining way to teach basic spill containment/response concepts, try this:

1. Determine exercise goals (e.g., spill containment concepts you want to stress).
2. Make several water balloons of different sizes (filled, partially filled).
3. If you have different colored balloons, for instructional purposes designate them as different hazard classes (e.g., yellow/oxidizers; pink/flammable liquid; blue/corrosive liquid).
4. Have participants stationed at each exercise location and deliver a balloon (or several, depending on how complex you want the exercise to be) to each location.
5. Walk to each exercise location and prick each balloon with a pin. Some will break; others will just leak.
6. Watch the results as participants "respond" to the release.
7. Have participants review spill containment concepts applied or misapplied.
8. Repeat the exercise for additional reinforcement of the ideas if desired.

In many cases, this is all that is needed. But at other times, a team approach and more elaborate planning is required, due to:

- The complexity of the work process
- Regulatory enforcement actions
- The need for organizational buy-off

SUMMARY

This chapter presented a primer to:

- The regulatory framework that influences pollution prevention in the United States
- Financial terms associated with cost control and economic analysis
- Barriers such as capital access and status quo

The next chapter will provide an introduction to management as an overall business function. It considers the theory and practice of continuous improvement in business operations.

Presented are typical industrial operating practices and applications which create environmental risks. As environmental management systems evolve into institutionalized methodologies for identifying, quantifying, and costing environmental exposures, the financial groundwork for defining and implementing cost-effective pollution prevention activities becomes firmer.

Exercise

Objectives:

- Select a work process to be analyzed for potential pollution prevention implementation.
- Begin to use financial language to communicate the value of a project.
- Practice applying the concepts and tools presented in this chapter to real-world situations.

Instructions:
The following exercise is most beneficial if the reader considers actual situations faced in the workplace. If the reader is a student or is not yet familiar with the business activities of the organization, select activities associated with home life. For example, the home activity could include automobile maintenance.

1. Identify one work process in your organization in which there is a potential for a pollution event (e.g., production of toothpaste, maintenance of equipment, etc.).
2. Brainstorm, either individually or in a group, in order to:
   a. Identify a list of potential environmental effects (i.e., air, groundwater, surface water, soil).
   b. For each environmental effect, list at least one source of a potential pollution event (e.g., hazardous materials transferred between containers, wastewater stream produced, etc.).
3. Take the results and rank the potential pollution events in order of concern (1 is greatest concern, 10 is least).
4. For the top three items ranked:
   a. List at least two types of information useful to consider in determining the financial impacts (costs) of the potential pollution events.
   b. Research your organization in order to clearly identify where this type of information could be obtained within your organization.

CHAPTER

2

ENVIRONMENTAL MANAGEMENT SYSTEMS AND POLLUTION PREVENTION

MANAGEMENT DEMYSTIFIED

Management is a planned, coherent response to a set of circumstances. As such, it is not reactive, but proactive in focus.

Within the socioeconomic and political context of the United States, the profitability and survivability of a business are linked to whether and how well the business and its component activities are managed. Management utilizes a frame of reference that anticipates business needs and requirements, then plans and implements a course of action that, presumably, causes these needs and requirements to be met.

Although many aspects of business management are viewed as unique functional specialties (e.g., financial, quality, operations, information systems, sales, and so on), with their own special vocabulary and tools, the basic techniques for managing are similar. They include:

- Issue or problem identification
- Planning work activities
- Budgeting
- Implementation
- Evaluation of results

This chapter provides an overview of basic management, including policy making, issue identification, planning, budgeting, and implementation. It discusses the influence of management systems such as Responsible Care® or and ISO 9001 (Quality Management). From this structure, it then considers the application of these approaches to successful pollution prevention.

THE PURPOSE OF POLICY

Within the business world, it is common to hear discussions about corporate policy. From human resources to production, policy and the statements that comprise it cover a broad range of functional topics. Normally issued by senior management, policy states the overall intent and direction for an organization.

While the appearance of policy statements can be deceiving, they are significant to an effective organization. Useful policy statements are simple. They define and communicate a specific functional intent. Policy is a foundation. As an organization moves through its day-to-day business activities, it enhances its performance by designing and implementing objectives to support its policy.

Policies can be misused if they are released with limited forethought with regard to their internal and external consequences. Additionally, if not periodically reviewed and revalidated for their continuing suitability, they can create legal liabilities. More than one corporate officer has had the sting of lawsuits, depositions, and trials imposed due to a failure to revise corporate policy to reflect contemporary needs.

Procedures and work practices are not policy statements. They are the detailed second-tier documents that explain the "what, where, and how" a business activity is conducted.

Organizations that are successful in implementing work activities or projects achieve their success, in part, because the other elements of business management—from problem identification to evaluation of results—have been designed to support an organizational policy.

Identify the Issues

Management relates needs and wants in order to identify problems and issues suitable for further consideration. It then develops alternative methods of action, described as work activities. It is through completion of the work activity that a targeted result or objective is achieved.

Environmental issues derive from such sources as:

- Inspections by an adversarial third party, such as a regulatory enforcement agency
- Inspections by a business partner, such as a financial institution, insurance company, or customer
- Internal inspections and/or audits
- Thoughtful implementation of regulations
- Customer or supplier complaints or criticisms
- Accident/incident investigation
- Changes in operating practices
- Changes in products produced and/or services rendered by an organization
- Mergers/acquisitions

Plan Work Activities

During the planning stage, research is conducted on the issue or problem. Alternative courses of action are proposed to resolve and/or respond to

Define the Product of a Service Industry with Care

Defining a product for service industries can be complex. Misdefining can create conflicts with customers. One Responsible Care® chemical company made a decision to develop baseline data on employee inhalation exposures to chemicals during packaging activities. While employee inhalation exposures were not anticipated to be excessive, the company understood that, based on the actual results, corrective action planning could be required.

The company contracted with a reputable national firm with appropriate credentials in industrial hygiene to perform specific employee inhalation exposure monitoring at its facilities. The certified industrial hygienist managing the project expected the service provider to send back raw data. Instead, the company got reams of paper filled with subjective observations, lengthy conclusions, recommendations for corrective action, and, yes, the requested data. In this example, the service company saw its product as being a lengthy report. Its customer saw the product as being objective data. By providing more than was asked for, the service provider created a minefield for future plaintiffs' attorneys.

a matter. The professional manager seldom conducts planning in a vacuum. Planning commonly involves dialogue with internal and external groups that may represent competing viewpoints. Through this approach, there is greater likelihood of devising a plan of action that has a realistic chance of being both approved and completed.

Through this research and collaboration period, a manager begins to get organizational "buy-off" on whatever the final recommendation may be. This period is also when the manager further defines and acknowledges certain "unknowns" that can affect a project, but cannot be satisfactorily identified or resolved until the project is in progress.

Budget

Budgets serve to describe and quantify the cost factors associated with a proposal. A budget is a professional "best guess" with respect to the time, money, and related resources needed to accomplish a project or task.

After a budget is developed, it is typically approved by a higher management authority. In many cases, this is more than a simple justification of costs and benefits. The specific project must be internally sold or marketed to that approval authority. In this respect, all managers must be salespeople.

The selling process requires not only designing potential solutions, but also taking that descriptive information or project schedule and expressing it in dollars, since dollars are the language of any budget.

Each organization has its own system or methodology for preparing and submitting budgets for approval. Smaller organizations may develop their budgets month to month, depending on cash flow to determine whether they can make an expenditure at all. In these cases, some budget requests may be deferred or moved to another month when the financial picture has changed.

Large organizations have longer budgeting lead times because of their complexity. They consider large dollar amounts, along with numerous locations, work processes, and proposed solutions. Typically, they collect budget "wants" from many sources, each with its own approval authority. These budget "wants" are then "rolled up" into a generalized operating budget, which is finally adopted by senior management.

Whether it means using unfamiliar spreadsheet software or preparing detailed piping and instrumentation diagrams for an industrial operation that may never be built, the manager must learn what the budgeting process is in the organization and work within that process to facilitate obtaining budget authorization.

While internal politics can be a hindrance to budget approval, those who have worked with organizations facing financial problems (e.g., bankruptcy or merger) appreciate that even the most highly justified project can be denied due to lack of funds. The professional manager adapts to the realities associated with the lack of capital.

One of the vivid illustrations of the power of politics over common sense was the so-called U.S. government shutdown during 1995 and 1996. Because Congress had not approved operating budgets for selected agencies, including the Environmental Protection Agency, employees identified as "nonessential" were not permitted to work for a period of time. "Essential" employees found themselves unable to return long distance phone calls or use copy machines. Eventually, the budget (i.e., capital) became available and personnel were again permitted to return to work, receive salaries, and use copy machines.

Implement the Project

After budget dollars have been approved, the project or activity can be scheduled for implementation. Implementation is where seasoned managers expect to have the most difficulty.

Implementation becomes the true management test because it involves using technical and creative skills, as well as people, to accomplish an objective. While, in some cases, this is as easy as calling a vendor and ordering materials, it can often be much more difficult than that.

This is especially true when the people who need to be managed or motivated to action do not have direct reporting responsibility to the manager charged with making the activity happen. It is also the stage when that early "buy-off" considered during planning can become most useful.

When personnel involved in implementation are not "direct reports," management must break away from the narrow scope of command-and-control and move into a phase of greater coordination and collaboration. The manager may assume the role of facilitator between conflicting or confusing interests, as each step is taken to complete the activity.

Evaluate Results

Following implementation, the results of a project are evaluated to validate whether the project choice was suitable, stayed within budget, and used optimal methods for the organization. The benefit of evaluating results comes from the assembly of an organizational history.

EMERGENCE OF MANAGEMENT SYSTEMS

The coherent, coordinated organization of various business activities and their supporting tools and resources comprise a management system. Certain "tests" can be applied to verify that the coordination is coherent and functions as a system. Systems verification requires validating that:

- Objectives (policy) of the system are identified
- Objectives of the system are understood throughout the organization
- The system is documented
- System implementation can be independently verified

If management were as simple in practice as described here, the world would be perfect. But reality teaches that even well managed, properly executed activities contain flaws. Where there are flaws, mistakes can occur, resulting in additional, frequently unbudgeted costs.

In response to these mistakes or nonconformities, various applied management systems approaches have been adopted by business organizations.

Total Quality Management

The traditional quality control model centers itself on inspecting work products to reduce the frequency of flaws. Its fallacies result from the assumption that life is static and checking things results in perfection.

Total Quality Management, or TQM (Figure 2.1) evolved beyond that inspection objective, centering itself not on perfection, but rather on continuous improvement in work processes in order to achieve customer satisfaction. TQM assumes that all work, whether production of a good or a service, is a process. Processes are composed of interrelated phenomena (inputs and outputs) that, when organized together, comprise the dynamic system.

If a mistake or nonconformity occurs, the work process is objectively reviewed to determine the root cause of the event. This evaluation process, or root cause analysis, seeks to remove the knee-jerk response of finger-pointing and assigning blame.

Through objective analysis, the organization learns that, although John opened the wrong valve and spilled a hazardous material, the root cause was that the valve labels had been blown away several days earlier in a windstorm, and valve labels had not been included in the formal pre-startup safety review.

The system is fundamental to Total Quality Management. In order to achieve quality in a work process, the system must be organized to:

- Be consistent;
- Be verifiable; and
- Promote incremental change (i.e., continuous improvement).

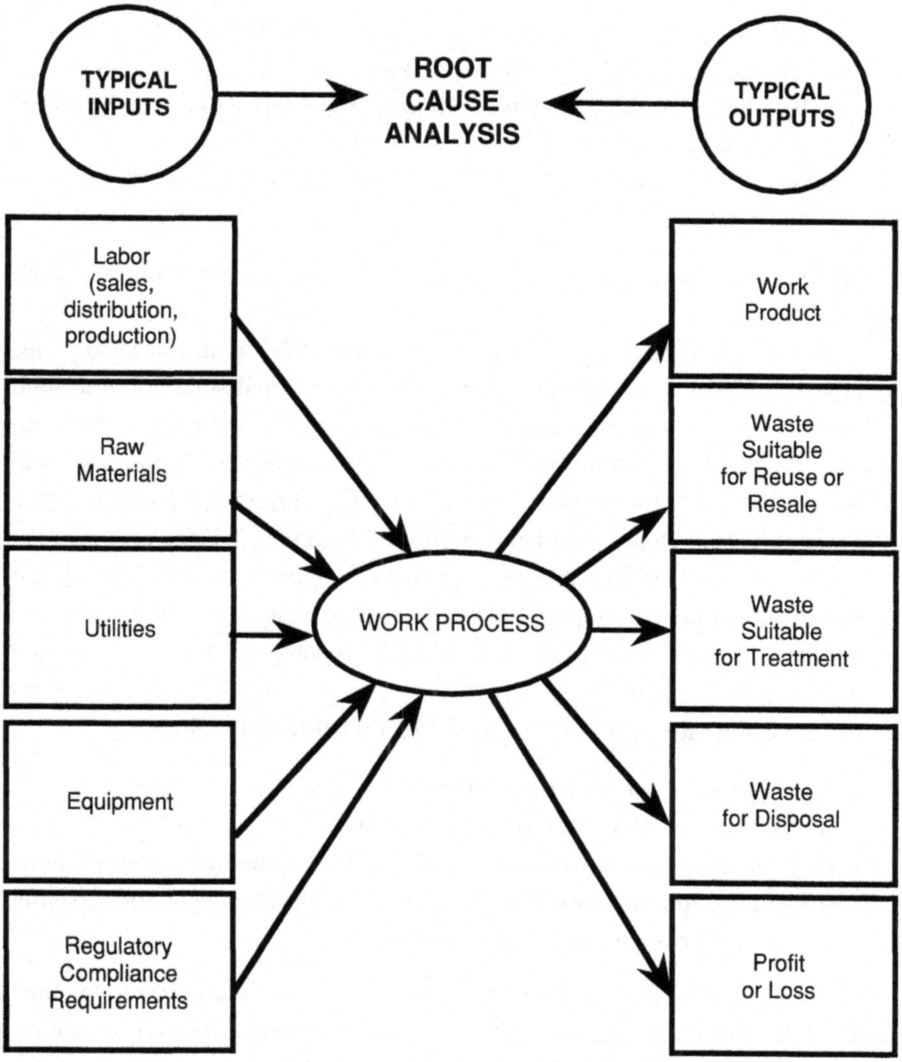

■ **FIGURE 2.1 Components of Total Quality Management**

ISO 9000 Quality Management Systems

The ISO 9000 series, adopted by the International Standards Organization in Geneva, Switzerland, consists of international consensus standards and guidance documents adopted to promote quality management. The series includes:

- ISO 9001 *Quality Systems—Model for Quality Assurance in Design, Development, Production, Installation, and Servicing*
- ISO 9002 *Quality Systems—Model for Quality Assurance in Production, Installation, and Servicing*
- ISO 9003 *Quality Systems—Model for Quality Assurance in Final Inspection and Test*
- ISO 9004-1 *Quality Management and Quality System Elements—Guidelines*

ISO 9001 and ISO 9002 are the most commonly implemented standards in the series.

While it is possible for a company to "self-certify" to an ISO 9000 series standard, this is not considered acceptable to traditional stakeholders, such as customers and vendors. Therefore, many companies retain a registrar to provide an independent third-party evaluation of compliance to the standard. The accreditation authority in the United States for registrars is the Registration Accreditation Board, headquartered in Milwaukee, Wisconsin.

The ISO 9000 Quality Management System is based on the concept that organizations (known as suppliers), by adopting consistent work practices as they relate to quality, are able to ensure their customers that a predefined level of quality is attained.

The systematic components of the ISO 9000 series include:

- Identifying customer requirements
- Planning for meeting those requirements
- Identifying nonconformities to that plan and customer requirements
- Defining and implementing corrective and preventive actions to eliminate recurrence of nonconformities

The growing adoption of the ISO 9000 series has been market-driven. In many parts of the world and in many industries, attaining registration to an appropriate ISO 9000 standard is mandatory for conducting business.

ISO 9000 gained its initial economic stimulus as the European Union (EU) moved toward a single market in the 1980s. Many EU members represented countries with long histories of wars between and invasions of one another, so the level of trust that existed in commerce was quite unlike the trust within the United States, with its tradition of free interstate commerce.

As the EU countries looked for a logical mechanism by which to verify the quality of products and services purchased from suppliers abroad, ISO 9000 registration became a mark of credibility.

With the exception of key multinational businesses, companies in the United States have been much slower to jump on the ISO 9000 bandwagon than their European counterparts. In some industries, there is sufficient business activity to satisfy supply within the United States. In others, it rubs against the American spirit of fierce independence, free thinking, and solving problems individually. And there is a cost involved.

But a growing need to trade internationally, pressures from European markets, recognition of the cost savings inherent in preventing mistakes, and, finally, changes in the auto industry, have accelerated the rate at which U.S. based businesses are moving to seek ISO 9000 series registration.

The automotive Big Three—Chrysler, Ford, and General Motors—have developed a similar, consensus approach to quality, known as QS-9000. Its objective is to promote continuous improvement in the auto industry by enhancing quality systems, eliminating redundant requirements, and reducing costs. First issued in 1994, the basis for its requirements are:

- ISO 9000
- Sector-specific (automotive)
- Customer-specific (Chrysler, Ford, General Motors)

QS-9000 can be characterized as "ISO 9000 plus." Registration to QS-9000 is a mandate for entities planning to do business with the Big Three. It is projected to become a mandate for organizations doing business with direct suppliers to the Big Three.

ENVIRONMENTAL MANAGEMENT SYSTEMS AND THE POLLUTION PREVENTION IMPERATIVE

On December 4, 1984, a toxic release occurred at a pesticide production facility in Bhopal, India, resulting in the deaths of more than 2,500 people and injuries to more than 200,000. That event has been classified as the worst industrial accident in history. It was a wake-up call for an entire industry. The worst that could be expected actually happened.

In Bhopal's aftermath, the Canadian Chemical Producers Association developed a voluntary industry initiative for continuous improvement in chemical manufacturing, naming it Responsible Care. Soon afterward, the initiative was embraced in the United States by the Chemical Manufacturers Association (CMA).

Responsible Care®

CMA began by adopting a policy known as Guiding Principles (Appendix A.1) . Using an institutional consensus process, the association prepared "Codes of Management Practices" concerning:

- Pollution prevention (Appendix A.2)
- Process safety
- Community awareness and emergency response
- Employee health and safety
- Distribution
- Product stewardship

Each code details management practices to which CMA member companies must subscribe. While there is considerable flexibility in the details of how a company implements each management practice, implementation results are captured for reporting to the CMA.

One management practice of the Pollution Prevention Code requires "a Quantitative Inventory at each facility of wastes generated and releases to the air, water, and land, measured or estimated at the point of generation or release."[1]

Performance to codes is measured in several ways. The Pollution Prevention Code requires submittal of data, which are assembled to track industry trends. This includes data on:

- SARA 313 release reports
- Waste generated, as defined in the CMA annual waste survey (hazardous and nonhazardous)

Performance is further measured through the annual submittal by association members of a self-evaluation, which shows the member company's status relative to each management practice. Results are compiled and submitted for all company facilities based on six descriptive categories:

- NA—No Action.
- EV—Evaluating existing company practices against the management practice

1. "Pollution Prevention Code of Management Practices" Chemical Manufacturers Association, Arlington, VA, (no publication date), p. 2.

- DP—Developing plan to implement management practice
- IA—Implementing action plan
- PP—Management practice in place
- RI—Reassessing management practice implementation

Member companies with multiple facilities face an organizational challenge in defining what constitutes having the "management practice in place." Management practices can be specified:

- Commonly for all facilities by the corporate level of management;
- At facility-specific level by each facility; or
- With a combination of corporate and facility-specific implementation requirements.

Whatever the tasks or actions are the comprise "management practice in place," they need to be verifiable in order to promote accountability.

Responsible Distribution Process®

In response to subtle market pressures created within the chemical industry by Responsible Care®, the Canadian Chemical Distributors Association (CACD) and the National Association of Chemical Distributors (NACD) in the United States adopted their own voluntary approach for continuous improvement in chemical distribution, the Responsible Distribution Process® (RPD). Similar to Responsible Care®, RDP® defines codes and elements while establishing targets for performance improvement in environmental areas such as waste reduction, emergency response, and handling & storage.

Participation in Responsible Distribution® is a condition of membership in the NACD. This requires regular reporting of progress toward element compliance. In 1995, for example, member companies were subject to an initial program verification by an independent outside third party conducted by Underwriters Laboratories. Companies submitted procedures supporting four of nine codes. They were reviewed and verified for their capacity to address all elements of the four codes they supported.

In order to improve RDP® implementation and verification, the NACD expanded its management systems requirements in 1996 to mandate:

- Document and data control;
- Internal audit; and
- Corrective/preventive action.

The following RDP Code elements link directly with pollution prevention opportunities:

- Ensure that all self-generated waste and empty containers are disposed of in a responsible manner, and in accordance with existing regulations.
- Encourage waste reduction programs or procedures, giving preference first to source reduction, second to recycle/reuse, third to treatment, and fourth to disposal.[2]

The impact of Responsible Care® as a management system has expanded beyond CMA's traditional membership, primarily producers of commodity chemicals. During the early 1990s, CMA introduced a partnering option for other trade associations and their membership. The Synthetic Organic Chemical Manufacturers Association (SOCMA), for instance, represents members of small chemical companies, including chemical processors, repackagers, and chemical industry service providers. It has become a Responsible Care® partner, but has taken the initiative to develop implementation tools and reporting strategies more appropriate to the performance improvement of its own membership. SOCMA has produced a variety of pollution prevention materials useful to this group (Table 2.1).

■ TABLE 2.1 P^2 Resources from SOCMA

The following materials can be ordered from:
Synthetic Organic Chemical Manufacturers Association
1100 New York Avenue NW, Suite 1090
Washington, D.C. 20005
phone (202) 414-4100
fax (202) 289-8584
World Wide Web: http://www.socma.com

- *Pollution Prevention in the Dye Industry*, Catalog #ET01
- *SOCMA's Guide to Pollution Prevention*, 1996, Catolog #PI12
- *Case Studies in Pollution Prevention*, Catalog #PI13
- *SOCMA's Guide to Hosting a Pollution Prevention Open House*, 1996, Catalog #PI14
- *Pollution Prevention Study*, Catalog #GR11
- *Step-By-Step Guide on Calculating Potential Emissions from Batch Processes*, Catalog #GR06

2. *NACD Responsible Distribution Process (R) Implementation Guide*, National Association of Chemical Distributors, Arlington, VA, 1991, p. 11.

Chemical Specialties Manufacturers Association

The Chemical Specialties Manufacturers Association represents producers of specialty chemicals for sale as consumer products. The trade names and brands of these products are common household words. This trade association has taken a different track in the interests of improving the performance of its membership. Operating under the assumption that the employees of its members are an essential asset, it began the voluntary Chem-Life Program.

Chem-Life is a communications effort designed to upgrade employee knowledge about a company, its operations, and the role of the employee in making those operations function appropriately. It has excellent applicability to pollution prevention implementation.

Chem-Life consists of a series of educational tool kits, which explain complex chemical processing issues in lay terms. Each tool kit contains the following:

- Discussion leader background sheet
- Background questions for discussion
- Leader discussion guide
- Prototype copy for newsletter article
- Artwork for overhead reproduction
- Study guides
- Experiment sheets
- Suggestions for customizing company-specific study guides

From an organizational perspective, these tool kits are suitable for upgrading overall communication between management and line workers (Appendix B.2.). This then stimulates the synergy needed for effective design and implementation of pollution prevention and other management plans.

Initial study guides available from the association address:

- Measurement and dosage
- Natural versus manmade
- Risks and benefits

ISO 14000 Environmental Management Systems

With a recognition that organizational performance is improved through consistent application of common practices and principles, the International Standards Organization, in 1992, began to prepare standards on environmental management—the ISO 14000 series. By its nature, the process of adopting international consensus standards is slow, and the final adoption of the first of the series, ISO 14001, did not occur until the fall of 1996.

As an interim measure and concurrent with this international standards writing effort, several standards writing organizations adopted their own mechanisms for fostering environmental improvements. Notable are BS 7750 (British standard, adopted in 1994) and NSF 110 (U.S. Standard, adopted in 1995).

BS 7750, on environmental management systems, has been influential in the structuring of ISO 14001. It is, however, considered to be more prescriptive than ISO 14001, providing less flexibility in framing an environmental management system (Table 2.2).

NSF 110 is the only U.S. standard for environmental management systems. The U.S. EPA participated as a member of a joint committee of regulatory, public health officials, users, consumers, and industry that prepared this consensus document. It is anticipated that this standard will be superseded by ISO 14001.

■ **TABLE 2.2 Types of Environmental Management Systems**

Standard	Environmental Management System Definition	Comments
BS 7750 (British)	The part of the overall management system that includes organizational structure, planning activities, responsibilities, practices, procedures, processes, and resources for developing, implementing, achieving, reviewing, and maintaining the environmental policy	Requires an overall management system in other functional areas
NSF 110 (United States)	The part of the overall management system that implements the environmental policy, objectives, and targets and includes the organization structure, responsibilities, practices, procedures, processes, and resources for implementing the policy	Requires that overall management system supports environmental policy
ISO 14001 (International)	The organizational structure, responsibilities, practices, procedures, processes, and resources for implementing environmental management	Enables EMS to take plan independent of whether there is an overall management system

The ISO 14000 series, as an international consensus standard, is directed at requirements and guidance for environmental management systems, dividing environmental management into two broad evaluation areas:

- Organization
- Product and process

The specification standard for organization is ISO 14001. Under this scheme, the EMS addresses specific functional areas. An environmental policy is required and, within that requirement, a commitment to P^2, or prevention of pollution, is mandated.

The international phrase "prevention of pollution" is defined within ISO 14001 as the "use of processes, practices, materials or products that avoid, reduce or control pollution, which may include recycling, treatment, process changes, control mechanisms, efficient use of resources and material substitution." It is further noted that its benefits are ". . . the reduction of adverse environmental impacts, improved efficiency and reduced costs."[3]

In contrast to the ISO 9000 series, which measures consistency in work process performance, the ISO 14000 series emphasizes implementing systems that are directed toward improvement in environmental performance. The latter requires the identification of measurable performance goals or metrics to verify improvement, rather than maintenance of the status quo.

The planning requirements of ISO 14001 are more comprehensive than merely setting environmental objectives and targets. Organizations subscribing to this EMS approach must also review their "activities, products, or services" in order to define environmental aspects of those activities. Environmental aspects are elements of the organization that interact with the environment. These aspects are considered significant if they have either an adverse or a beneficial change to the environment resulting from the activity, product, or service.

Thus, a company can achieve greater productivity and cost efficiency through introduction of an EMS than it may have experienced through introduction of a quality management system (QMS). While large corporations have significant staff and financial resources to apply toward achieving compliance with ISO 14001, small to medium-sized companies have to take more innovative approaches.

3. ISO 14001. *Environmental management systems—Specification with guidance for use.* First edition 1996–09–01, p. 2.

Innovate or Fail to Compete: ISO 14001

The standard process for compliance with international standards has been to hire someone, familiar with your industry and the standard and have that individual provide consultancy service to tell you how to set things up. The weakness in this approach is that when the consultant leaves, there has been little internal buy-off and perhaps few, if any, champions nurtured.

But ISO 14001 is a new standard, and, despite all the high-level intellectual debates going on, it presents more opportunity for innovation because it is new. Preconceived, biased notions of what should be done have not been ingrained into an audit community.

Because the ISO 14000 series consists of international standards subject to application throughout the world in countries with varying degrees of regulatory, infrastructural, and managerial development, the ISO 14000 series offers substantial latitude for a company to define facility-specific baselines from which improvement is measured.

While the draft ISO 14000 standards on product and process evaluation have not advanced through the consensus approval process as far as the ISO 14001 specification on EMS, familiarity with these draft standards also offers a significant opportunity for the P^2 practioner to gain credibility within the company, while responding to stakeholder demands on environmental management. There are three components:

1. Life-cycle assessment
2. Environmental labeling
3. Environmental aspects in product standards

Life-Cycle Assessment Life-Cycle Assessment operates as both a management system and a tool. As a management system, its objective is to promote development of environmental impact statements, also known as environmental consequences. These are product-specific. A product may be a tangible item or material (e.g., paint, pharmaceutical, etc.) or a less tangible service (e.g., laboratory testing, garbage collection, etc.). The impact/consequences research covers the product or service from its design to its final disposal by an end user. Although it may be supported by the data elements of life-cycle accounting, the traditional life-cycle assessment concerns itself with impacts rather than costs.

Service organizations adopting life-cycle assessment as part of an environmental management system can choose to consider various elements of their service as products or deliverables. They are manufactured or produced in a different manner than a tangible product, but they are products none the less.

Although life-cycle analysis is a synonym , the U.S. Environmental Protection Agency usage is "life-cycle assessment." The assessment consists of three elements:

- Inventory
- Impact assessment
- Improvement analysis

EPA assigns four stages to a product life cycle:

- Raw materials acquisition
- Manufacturing/production
- Consumer use/reuse/maintenance
- Recycle/waste management

As a tool, life-cycle assessment is a systematic approach for identifying and analyzing environmental consequences. The pollution prevention professional should be aware that this tool is at an immature stage of development in its own life cycle. As such, each academic or consultant carries his or her own bias with respect to what comprises the tool and how it can be used. There is no one formula that fits all situations.

Collaborate/Coordinate

While it is now a fundamental tool for most environmental professionals, it was only on November 20, 1985, that the OSHA Standard on Hazard Communication (29 CFR 1910.1200) became effective. It meant that all manufacturers, distributors, and importers of hazardous chemicals had to make available to their customers in the manufacturing Standard Industrial Classifications (SICs) material safety data sheets (MSDSs) and product labeling in accordance with the standard for each chemical.

This represented a monumental task for one industry group, chemical distributors. They had to collect MSDSs prior to the effective date and organize them to achieve compliance and get the right information to the the correct customers.

McKesson Chemical Company was, at that time, one of the top three chemical distributors in the United States. It had more than

100,000 customers on its customer list and handled more than 10,000 products. While some of these products were not hazardous, and many were merely the same material in different sized containers, the amount of information to be gathered, organized, and redistributed to its customers was substantial.

Managing that process required both the assembly of a technical team capable of evaluating the products for their hazards and coordination with a centralized computer department and McKesson Chemical's employees in 60-plus facilities throughout the United States. The company kept its compliance deadline because it managed the work process, taking small steps over a compressed period of time, working with diverse departments to achieve its goals.

Department	**Hazard Communication Resource Capability**
Marketing	Contact and clout with suppliers of hazardous chemicals
Operations	Development and maintenance of content: ■ MSDS database ■ product labels ■ employee training
Computer Services	Information systems support to organize and distribute MSDS database
Sales	Customer contacts
Service Centers (field locations)	Customer contacts Dissemination of MSDSs and product labels Employee training

Only after the initial implementation could the company sit back and evaluate what it could then do to improve the system it had put into place. A testament to its success occurred less than one year later, when the company was acquired by a competitor. Within several months of that acquisition, the approach developed by McKesson Chemical Company was adopted and implemented by its new owners.

Why would an organization decide to undertake a life-cycle analysis? In one case from Clorox Corporation (Oakland, California), a life-cycle analysis was conducted on its liquid bleach product. The intent of the study was to:

- Identify best environmental practices
- Obtain a baseline and ability to compare plant to plant, year to year, formulation to formulation
- Lay groundwork (baseline) in the event community outreach was conducted
- Identify any new angles on eliminating waste

Among the study results were that, in its baseline plant in Chicago, 62 percent of energy costs were for heating. Because electricity is six times more expensive than gas in that area, the greatest opportunity for return lay in reducing electrical consumption. At a 10 percent reduction in energy, the plant could save $35 million a year.[4]

The Recycling of a Negative Life Cycle

Is there life—and product potential—after a negative product life cycle? In the 1960s, the horrors of "thalidomide babies" born with birth defects were splashed across newspapers and televisions throughout the Western industrialized world. The drug, thalidomide, had been consumed by women for nausea and other symptoms of pregnancy. It was withdrawn from the market when its side effects became known. But, more recently, it has been resurrected because of its viability for treating symptoms of AIDS, cancer, rheumatoid arthritis, and tuberculosis. It inhibits the growth of blood vessels and, thus, can contribute to the starvation of cancer tumors. Will thalidomide be reborn? The jury is still out. But it offers an object lesson for pollution prevention professionals. With creative analysis, products and services have the potential to be regenerated to add value to an organization.

Environmental Labeling Product labeling is becoming an increasingly complex issue. When the U.S. Department of Agriculture revised its nutritional labeling scheme a few years ago, food producers howled in protest. But after the required testing was done, some found out that their claims and products were not as pure as originally thought.

4. "Life Cycle Assessment Case Studies Highlighted," *Chemical Times and Trends*, October 1995, p. 29.

With the proliferation of regulatory requirements, the persuasiveness of marketing, and a customer's need to know certain information, product labels can contain a great deal of clutter.

The environmental labeling element of the ISO 14000 series addresses a process for the verification of environmental or "green" labeling claims. Using a third-party verification mechanism, consumers can make informed decisions with a higher level of confidence.

Environmental Aspects in Product Standards This component of the ISO 14000 series is directed at the people who write product standards. It consists of a systematic mechanism for addressing so-called environmental aspects, such as scale, severity, and the probability and permanency of impact. It offers the opportunity to improve access to information for the P^2 professional considering product substitution.

The ISO 14000 series is projected to become a significant force in environmental management, including pollution prevention. To improve the effectiveness of this book for its readers, the language of the 14000 series standards will be used throughout.

THE POLLUTION PREVENTION CHALLENGE: USING THE OUTPUTS TO DETERMINE THE INPUTS AND ROOT CAUSES

The earlier discussion on TQM briefly presented the concepts of inputs, outputs, and root causes. By starting with identifiable environmental outputs, the pollution prevention practitioner can trace inputs to their source and determine the root causes of undesirable outputs. Armed with that information, pollution prevention opportunities can be designed, with a greater chance of budget authorization and implementation success.

While in real life, activities for defining inputs and outputs can be compressed because of a crisis that demands a quick-fix response, the methodology that follows is useful in defining and planning for solutions with a longer lead time.

Table 2.3 illustrates a methodology for defining and measuring or valuing an output. From the information given in this example, it is unknown why 2,000 pounds of hazardous substance are unaccounted for. It may have been lost, stolen, spilled, or misrecorded. The direct output, measured at 4,000 pounds, is expressed at its purchase price of $13,000.

■ TABLE 2.3 Defining Outputs
Case: Printed Circuit Board Manufacturer Using a Hazardous Substance

Direct Output Valuation

Hazardous Substance Activity Calendar Year			Quantities
Initial Inventory, January 1			5,000 lbs.
Additional purchased throughout year			25,000 lbs.
	Subtotal		**30,000 lbs.**
Final inventory, December 31			2,000 lbs.
Throughput recorded in inventory records for use in manufacturing			24,000 lbs.
	Subtotal		**26,000 lbs.**
Direct output: potential product loss		30,000 − 26,000 =	**4,000 lbs.**
Value of potential product loss: 4,000 lbs. @ $3.25/lb			**$13,000**

Indirect Output Valuation

Indirect Output	Est. Cost/Unit	Number of Units	Cost
Labor: track/reconcile inventory loss	$50/ hour (includes benefits)	8 hours	$400
Emergency response due to product release	$2,500/event	3 events	$7,500
Labor productivity loss due to "fast track" maintenance in area of concern	$65/hour (includes benefits)	16 hours	$1,040
Waste collection/ disposal	$20/lb.	1,000 lbs.	$20,000
"Fast track" regulatory compliance costs	$65/hour (includes benefits)	25 hours	$1,625
Regulatory fines	$10,000/event	1 event	$10,000
Increased insurance premium	$15,000/ enforcement action	2 actions	$30,000
Total indirect outputs			**$70,565**

Direct Outputs + Indirect Outputs = Total Value
$13,000 + $70,565 = $83,565

Indirect outputs can also be factored into the equation. The indirect outputs of each pollution prevention opportunity will be unique to each process situation. In the example shown, indirect outputs equate to 84 percent of the costs associated with unaccounted-for product.

In identifying inputs to the work process(es) that created the outputs, the pollution prevention practitioner can begin research with internal audit and/or inventory control to determine whether the unaccounted-for material is within the range of a normal inventory adjustment due to variations in inventory collection. For example, if the material is stored in a bulk tank, tank readings vary depending on the time of day, ambient temperatures, operator skill, and the method the operator uses for taking the measurement.

A simple, quick observation of the operation that receives and stores the material prior to use may indicate poor maintenance (e.g., a leaky gasket, stained surface at hose connections) or sloppy work practices.

In reacting to the outputs, the P^2 practitioner becomes Sherlock Holmes. Logic and deductive reasoning are applied to identify all inputs involved in the processes of receiving, storing and using the hazardous substance. These include, but are not limited to, materials, equipment, personnel, utilities, and site conditions (weather, visibility, and the like).

The assembling of this input/output information promotes a systematic determination of the true root cause(s) of the unaccounted-for product. Alternative solutions for correcting the matter, similarly quantified in dollars, are compared with the direct output cost in assembling a project justification.

INFLUENCE OF REGULATIONS ON MANAGEMENT OPTIONS

There is no question that local, state, and federal regulations in the United States are a major force in directing and controlling approaches to EMS. From the viewpoint of each individual governmental entity, the rationale for any environmental regulation is a response to some real or perceived need to address a public good.

Environmental regulation is, in essence, market-driven. The public demands, and the government entity supplies. The costs are borne directly by the organizations subject to the regulation and indirectly by their customers, as costs are passed on.

Stated in 1995 dollars, total regulatory costs increased only 2.1 percent from 1977 to 1995 in the United States. There was a real reduction in many regulatory requirements and costs, such as the elimination of price controls because of the deregulation of trucking. However, that portion of regulatory costs directed at the environment and risk reduction expanded in both real dollars and as a percent of total regulatory costs. In 1995, these environmental and risk reduction regulatory costs comprised more than 35 percent of total regulatory costs.[5]

Since the 1970s, the U.S. environmental regulatory structure, particularly at the national level, has developed into an environmental management system with a relatively common language and tools for addressing environmental requirements.

Consider the OSHA Process Safety Standard (29 CFR 1910.119) and the EPA Risk Management Program (40 CFR 68). If a company handles certain chemicals in certain quantities, compliance with these requirements is mandated.

While the OSHA focus is on-site consequences of a chemical release and employee health and safety, EPA's is off-site consequences and the impacts of a release on a broader population and the environment. In both cases, though, the concern is a chemical product release, which, by default, is an environmental concern and source of pollution.

A common element of both sets of regulations relates to the use of outside contractors working at a facility on or near a covered process. "The owner or operator, when selecting a contractor, shall obtain and evaluate information regarding the contract owner or operator's safety performance and programs."[6]

In practice, organizations covered under these programs have taken other federal requirements and used them as criteria for selecting outside contractors. They are not necessarily required by law to do so, but this has become a de facto cost of doing business.

For example, gasoline terminals and petroleum refineries commonly contract work such as tank farm maintenance, welding, or tank cleaning. By internal fiat, they commonly require that any employee working for such a contractor at their facilities have a current 40-hour certification per OSHA Hazardous Waste Operations and Emergency Response (HAZWOPER, 29

5. Thomas Hopkins, Rochester Institute of Technology for U.S. Small Business Administration, *Profiles of Regulatory Costs*, November 1995 (Report # PB96128038).

6. OSHA Process Safety Standard, 29 CFR 1910.119.

CFR 1910.120). Having this type of certification for the contractor implies a management system that ensures that employees are trained in accordance with HAZWOPER requirements, records are maintained and there is recurrent training. HAZWOPER has become a component of the management systems for both contractors and the customers who purchase their services.

Even when an organization is not required to be in compliance with certain regulations, it can adopt elements of those regulations to provide structure to its management system. To the extent that these regulations are used as models, particularly when not directly subject to regulatory enforcement inspections, they can improve productivity while addressing environmental concerns.

THE FINANCIAL GROUNDWORK FOR IMPROVEMENTS

This chapter has introduced management and management systems to the P^2 practitioner. Properly used, these systems enable complex technical requirements to be communicated to decision-making authorities within the organization.

True communication is a two-way street, which requires the P^2 practitioner to both send messages and receive them. Learning and using financial language complements this communication process.

Keep in mind that not all costs are easily identifiable, or necessarily need to be reported with precision. In these cases, the manager makes reasonable assumptions, which are expressed in order to enable a logical decision to be reached. The level of precision needed in reporting costs should be related to organizational needs and to how much time there is to collect data, rather than to some personal bias for perfection and absolutes. Very few managers have the luxury of dealing in absolutes in all cases.

How can information relating to spills and releases be effectively used to influence an expenditure decision, and ultimately reduce the pollution exposures? Take as an example an organization that operates in multiple states at multiple sites, using its own fleet to move raw materials between selected facilities and product (hazardous substances) to customers. The following approach can also be used in assessing multiple functions taking place under one roof.

The environmental manager of this organization periodically gets reports of chemical releases involving these facilities' operations and wants to plan preventive action for the future. Using data collected from internal

■ **TABLE 2.4 Estimating Financial Impacts of Chemical Releases**

Type of Event	Assumption: Minimum Cost to Company/Event	Number of Events 6/1/95-5/30/96	Estimated Minimum Cost to Company
Release contained on-site	$2,500	35	$87,500
Release with off-site consequence	$7,500	1	$7,500
Transportation emergency on highway	$8,500	22	$187,000
Transportation emergency at customer location	$8,600	5	$43,000

reports and regulatory records, the manager identifies 63 events and categorizes them for costing (Table 2.4).

The cost assumptions were not pulled from a hat; they reflect a professional "best guess" based largely on estimated labor and disposal costs. In the example given, the manager may conclude that determining pollution prevention strategies for transportation may bring the largest benefit to the organization. When pollution prevention alternatives are presented for budget approval, a completed tool (Table 2.4, Appendix C.1) becomes a part of the justification.

Exercise

Objective: Initiate a comprehensive evaluation of a work process through the compilation of life-cycle assessment information.

1. Using the work process (an activity, product, or service) identified in Exercise 1, or an alternative work process:
   - **a.** Review the ranked potential pollution events.
   - **b.** List the raw materials inputs to the activity, product or service.
   - **c.** Identify any special features of the activity, product, or service that may contribute to the potential pollution event.
   - **d.** Identify the waste outputs of the activity, product, or service.

2. Describe how customers (internal or external) use the output of the activity, product, or service.
3. List the potential pollution events (environmental consequences or effects) that may apply to this customer use.
4. From the perspective of your organization, and also considering customer usage of the activity, product, or service, identify one process change to the activity, product, or service that can be made that can result in:
    a. Reduction of raw materials (including energy) used in the activity, product, or service
    b. Substitution of a less hazardous raw material
    c. Production of less waste output (at the organization or the customer)
    d. Creation of fewer environmental consequences at the customer level.

CHAPTER

3

CONDUCTING A POLLUTION PREVENTION ASSESSMENT

Pollution prevention is the total of actions applied to processes, practices, materials, products, or services that reduce, eliminate, or mitigate pollution. Its function is to implement changes that alter or modify the outputs and produce benefits. Pollution prevention opportunities are linked to specific business problems or issues.

This chapter looks at the process of pollution prevention assessment. It presents an overview of the benefits of pollution prevention and relates it to the need to communicate with both regulators and the financial community. It describes the tools and information already available for a cost-effective assessment process, finally addressing the need to compare alternative solutions to problems.

IDENTIFY THE ISSUES

To conduct a pollution prevention assessment, it is necessary to clearly define the environmental problem or issue before developing solutions. This systematic issue identification results in designation of opportunities for improve-

Buckets of P^2 Progress

In the 1970s, a California-based chemical tank farm manager took the following steps to "prevent pollution." Where there was a leaky valve or gasket, a drip bucket was put in place. A union employee was given the task of emptying the bucket into the tank when the bucket was full. The cost to replace the gasket or change the valve was minimal, but not conceived of as being a necessary response.

Do these types of "past practices" still exist? Perhaps. If identified and corrected, they can have a dramatic positive impact on productivity, operating costs, and pollution prevention.

ment through pollution prevention actions. In this chapter, we consider how to:

- Identify and define business processes that may present pollution prevention opportunities
- Develop a profile of environmental characteristics for those opportunities
- Use the language of finance to communicate about pollution prevention opportunities

POLLUTION PREVENTION BENEFITS

The benefits or objectives of pollution prevention include, but are not limited to:

- Reduced adverse environmental impacts
- Improved operating efficiency
- Decreased operating costs
- Minimization of management involvement in non-value added, pollution-oriented litigation or regulatory enforcement

An analysis of pollution prevention opportunities, or P^2 assessment, requires defining what the organization wants to accomplish—that is, the desired benefits of pollution prevention.

The pollution prevention assessment has several components:

- Selection of activity, product, or service (ISO 14001) and definition of scope subject to evaluation for pollution prevention opportunities

- Designation of environmental aspects of that activity, product, or service
- Assembly of baseline performance characteristics
- Definition of pollution prevention objectives for the designated activity, product, or service
- Assignment of measurable pollution prevention targets that can be related to baseline performance characteristics

Only after the P^2 assessment is completed can realistic solutions be proposed and evaluated for implementation.

SPEAK THE LANGUAGE OF FINANCE

In support of ISO 14001, the starting point for environmental planning is to delineate environmental business issues as activities, products, or services. The choice is dependent on the organization and how it describes work processes subject to a pollution prevention and environmental management assessment.

The descriptor selected should be driven by accounting consequences and the capacity of the organization to collect financial data in a specific way. Financial records are usually organized by centers of responsibility for profit and loss. This approach can be restricted by the reporting constraints of accounting software packages in use, out-of-date accounting frameworks, or company bias. Traditionally, financial information is collected and reportable by product line, business function, geographic location, or similar appropriate divisions.

In the example shown in Table 3.1, the company is a manufacturer that produces windows for residential use. The environmental issue is the release of volatile organic compounds (VOCs) from a paint spray booth operation

■ **TABLE 3.1 Alternatives for Describing Activity, Product, Service***

	General Description	Specific Description
Activity	Paint manufacturing	Paint manufacturing in Production Area 3
Product	All customer branded paints	ABC Painting Supplies, Inc. branded products produced to ABC formulation
Service	**Custom formulation**	R&D lab activities. Developing samples of custom formulations to meet customer application requirements

*Company: XYZ Co.; Facility business: Manufacturing windows, including paints and coatings applied to protect surfaces.

that paints the frames. This work process can be defined by the activity or the product.

Because the manufacturer retains its financial records by product line, the pollution prevention assessment is classified by product. This permits any costs associated with the assessment or its results to be appropriately assigned as a cost of doing business for that product.

If the work process is a service or the organization is not a manufacturer, an alternative classification is more appropriate. The following example applies to a trucking company providing transportation services.

The trucker's environmental concern is hazardous waste generated during vehicle maintenance in its maintenance shop. This work process can be defined as either an activity or a service. Although the company has only one service, trucking, it keeps its financial records by business functions that support the provision of that service. Therefore, classifying the environmental issue by activity is more suitable.

Understanding the system of financial recordkeeping in an organization and adapting the pollution prevention assessment process to that system enhance the ability to communicate and track costs and benefits of a specific implemented solution.

WHERE'S THE PROBLEM?

While regulatory inspections, complaints, and accidents are three common sources for the initial identification of a business problem of environmental concern, it can be more cost-effective to discover environmental issues without invasive third-party interventions.

Pollution prevention assessment requires pinpointing actual or potential sources of waste production and resource consumption. As a systematic, scheduled methodology, it can be implemented concurrently with the operation of normal business functions, minimizing disruption. It is proactive, rather than reactive. The result is controlled costs for identifying problems and generating solutions.

In order for the pollution prevention assessment to be effective, an assessment scope is required. The "scope statement" specifies the parameters of the work process being evaluated. By periodically comparing or testing assessment tasks against the scope, the assessment is held to a specific path. Unproductive tasks are stopped or eliminated, enabling the organization to better focus its resources on the designated issue.

■ **TABLE 3.2 Alternatives for Classification & Scope**

Manufacturing Process

Classification (select one)	Scope
Activity	Paint spray booth (VOCs in use)
Product	Production of #1325 Windows (i.e., paint spray booth when VOCs are in use)
Service	Does not apply

Service Process

Classification (select one)	Scope
Activity	Diesel engine maintenance
Product	Does not apply
Service	Trucking

In the examples shown in Table 3.2, the scope statements set simple, but clear, limits for the pollution prevention assessment. The window manufacturer limits its analysis to windows produced when VOCs are present in the paint. The trucking company limits its analysis to maintenance of diesel, rather than gasoline, engines.

Effective cost control requires early decisions on the limits of pollution prevention assessments. It is impossible to do all things. The scope clarifies the parameters for study.

REGULATORY REPORTS AS TOOLS OF ANALYSIS

If an environmental regulation requires compilation of specific records or reports, these represent a management tool that can contribute to the establishment of priorities for pollution prevention action.

If hazardous wastes are produced, as a minimum federal requirement, organizations that are generators must prepare Uniform Hazardous Waste Manifests and biennial reports.

The Uniform Hazardous Waste Manifest (and its various state equivalents) provides detailed information about a specific waste shipment going to an off-site hazardous waste treatment, storage, and disposal facility (TSDF). The biennial report consolidates that information for analysis and tracking purposes.

Biennial reports are submitted to the Regional Administrator of the U.S. Environmental Protection Agency by March 1 of each even-numbered year. It includes such information as:

- EPA identification number, name, and address for each U.S.-based, off-site treatment, storage, or disposal facility to which waste was shipped
- Name and EPA identification number of each transporter used for these shipments
- Description, EPA hazardous waste number, DOT hazard class, and quantity of each hazardous waste shipped off-site
- Description of the efforts undertaken during the year to reduce the volume and toxicity of waste generated (i.e., pollution prevention)
- Description of the changes in volume and toxicity of waste actually achieved during the year in comparison to previous years

These types of data are especially helpful when conducting a pollution prevention assessment for a process occurring at multiple sites. From volume of wastes to variety of waste streams, they provide objective evidence of the magnitude of the situation. They also facilitate a comparison of performance between related work processes at different physical locations.

Table 3.3 lists other federal regulatory reports that can provide useful management information. Although not all of these reports are required by all industries, or for all chemicals, the report structures themselves are helpful. They can be adapted for use as internal management and/or data collection tools.

Even if the organization is exempted from filing the report, the act of collecting the data in a systematic manner enables the organization to identify areas of potential pollution risk that may otherwise have gone unnoticed until a pollution event occurred.

Agencies such as the EPA are introducing electronic data interchange (EDI), which enables certain reports to be filed electronically, rather than through hard copy. Reports currently accepted electronically by the EPA include:

- Discharge monitoring (per National Pollutant Discharge Elimination System [NPDES])
- Reformulated gasoline and anti-dumping reports

EPA is phasing in the use of EDI as it creates or revises rules.

■ **TABLE 3.3 Regulatory Records as Pollution Prevention Management Tools**

Report/Record Description	Regulatory Citation	Potential Uses
SARA Tier 2 Chemical Inventory Report	40 CFR 370.20	■ Emergency response planning ■ Establish targets for substitution with hazardous inventory
Toxics Release Inventory Report	40 CFR Part 372	■ Establish targets for release reduction
Unintentional Release Form (DOT 5800.1)	49 CFR 171.16	■ Analysis of causes of transportation releases
Emergency Release Notification Report	40 CFR 355.40	■ Analysis of causes of chemical releases
Pesticide Production Control Records	40 CFR Part 169	■ Emergency Response Planning
Log and Summary of Recordable Injuries and Illnesses	29 CFR 1904.2	■ Analyze injuries or illnesses to determine areas of concern in chemical usage or work processes
Incident Investigation Report	29 CFR 1910.119	■ Identify incidents that caused or could have reasonably resulted in a catastrophic release
Five-Year Accident History	40 CFR 68.42	■ Analyze history to identify patterns of deaths, injuries, environmental damage
Employee Exposure and Medical Records	29 CFR 1910.20	■ Identify processes with notable chemical exposure risk to employees

While some organizations can anticipate substantial cost savings in electronic filing, small and medium-sized companies will want to factor in the time and dollar costs associated with setting up, learning, and fully implementing such a system. Attractive as electronic filing may look in concept, software glitches, insufficient memory, and configuration issues can add significantly to the cost of its initial introduction and full use.

CHECKLISTS

Another simple, inexpensive tool for identifying and initiating control of pollution prevention risks and their consequences is the checklist. Checklists

■ TABLE 3.4 Intermediate Bulk Container Fill Checklist

Tank Serial #: ____________
Tank Size (gallons): ________
Order Date: __/__/__
Bill of Lading #: ___________
Customer Name: ___________
Delivery Date: __/__/__
Material Ordered: __________

Condition Report

Exterior: Check correct column

Item	OK	Needs Improvement: Describe
Mfg. Specification Plate		
DOT Inspection Date		
Pressure Test Date		
Serial Number		
DOT Shipping Name, ID number on two opposing sides, 1" high letters		
DOT placard on two opposing sides		
Product Label		

Interior: Check correct column

Item	OK	Needs Improvement: Describe
Clean/dry		
No residual contents		

If residual present, describe (1) Type: __________ (2) Amount: ________
Cleaned Out Prior to Loading: __ Yes __No

Item	OK	Bad	Replaced
Gaskets: Dome			
Gaskets: Dust Cap			
Valve Condition			
Camlock Fitting Condition			

Shipment Report

Item	Yes	No	Comments
Retain Sample Pulled			
Tote Tank Approved for Shipment			
Leaks Detected After Filling			
If leaks, explain corrective action			

Operator's Name: _________________ **Date:** __/__/__
Supervisor's Name: _________________ **Date:** __/__/__

have a dual purpose. They can be designed to eliminate some issues, while identifying additional areas of environmental concern.

The checklist shown as Table 3.4, for instance, is applicable when filling a small bulk tank (known in the chemical industry as an intermediate bulk container) with a hazardous material. While the existence of the checklist leads to the conclusion that filling the container is itself a pollution risk, the objective of this checklist is to ensure that the container being filled has the structural integrity that permits it to continue to contain the hazardous material during storage, transportation, and use.

If the operator makes a decision about the condition of a gasket or valve before filling, the chance of a leak decreases once the container is filled and in use. After filling is completed, the checklist further reminds the operator to inspect for leaks. If a leak occurs, it directs that corrective action be taken. The notation of corrective action becomes data available for a root cause analysis and identification of future opportunities for pollution prevention.

OTHER RISK FACTORS

When time and budgets permit, other risk factors are reviewed in order to better identify the need for pollution prevention (Table 3.5). These include:

- Known or observable hazards of the process
- Projected consequences in the event of the failure of engineering controls applicable to the hazards of the process
- Projected consequences in the event of the failure of administrative controls applicable to the hazards of the process
- Application of detection methodologies, such as process monitoring and control instrumentation with alarms, and detection hardware, such as hydrocarbon sensors
- Process siting, such as proximity to personnel or the public
- Human factors, including the training and education of personnel on all shifts and during holidays, when the most experienced workers may be off-duty

ENVIRONMENTAL ASPECTS

After clearly identifying the work process scheduled for pollution prevention assessment, environmental aspects must be considered. The goals of this step are to:

■ TABLE 3.5 Risk Factor Review

Activity/Product/Service: Receiving, Storing, and Dispensing of Flammable Liquids

Item	Risk Factor	Issues
Known hazard	Flammability	Is grounding system adequate?
Consequences of engineering control failure	Tank gauge malfunction	Overfill potential
Consequences of administrative control failure	Pre-load checklist not completed	Tanker may be out of test date
Detection methodology	Overflow	If audible alarm, can it be heard throughout the facility?
Siting	Edge of property away from structures, but next to highway	What is the possibility that accident on highway could end up hitting the bulk storage tanks?
Human Factors	Night shift with single loader on duty	Can emergency shutdowns and alarms be activated if employee is injured?

- Define environmental aspects
- Quantify characteristics of those environmental aspects
- Establish a baseline for future comparison purposes

A worksheet (Appendix C.2) can be used to assemble this type of environmental aspects characterization. As this tool is applied to a work situation, the pollution prevention professional will want to pay special attention to the presence of:

- Hazardous waste streams
- Special waste streams
- Commingled waste streams

The baseline information collected through this worksheet is later used to compare alternative solutions and their projected impacts. The baseline enables the P^2 practitioner to ask whether a proposed corrective action:

- Applies to the problem area
- Reduces or mitigates the environmental impact
- Enables progress toward meeting the objective to be measured

Assess the Pollution Risk from Vandalism

Large, complex manufacturing organizations are not the only places that face a pollution risk. One retail quick-lube facility sustained a 1,500-gallon lubricating oil spill when, after normal working hours, a vandal hooked up a garden hose, turned it on, and left in running into an oil storage tank with an unsecured cap.

Take a look at retail facilities in your community and see if you can identify any pollution risks. Look at such ubiquitous businesses as dry cleaners (solvents), grocery stores (corrosives, pesticides), and hardware stores (flammable liquids, corrosives, pesticides).

Then conduct a site security survey of your own facility. Evaluate risks during normal working hours and on weekends and holidays.

If the organization is evaluating pollution prevention opportunities for a proposed new activity, product, or service, a baseline is still important. This baseline represents the minimum parameter for an environmental aspect.

Sources of baseline data for new activities, products, or services can come from industry standards, OEM specifications, regulations, or company defined-requirements. A reference list of industry standards-writing and professional associations that can assist in this process appears in Appendix B.1.

OBJECTIVES AND TARGETS FOR P² IMPROVEMENT

After assembling and analyzing this background information, the organization is able to state specific pollution prevention objectives and targets for the activity, product, or service.

In a functional management system, policy statements express the broad objectives of an organization. Environmental management systems also contain policy statements. An environmental policy is mandatory in an ISO 14001 system. These policy statements present overall organizational objectives as they apply to environmental issues.

By necessity, these are generalized objectives applicable to cross-functional areas. When linked to a pollution prevention assessment specific to an activity, product, or service, these foundation objectives are expanded upon and expressed with more detail appropriate to the activity, product, or service.

The example shown in Table 3.6 applies to a general contractor who dispatches work crews to dispersed customer locations in order to provide

■ TABLE 3.6 Environmental Aspects Characterization

Facility: ABC Gasoline Co.

Scope: Clean and purge unleaded gas tank (10,000 gallon capacity); visual inspection; examine welds and repair as necessary

__ Activity __ Product X Service
__ New __ Modification X Existing (no modification)
EPA ID #: WAD 00032145

	Ongoing	Ongoing	New or Modification	New or Modification	Action
ENVIRONMENTAL ASPECT(S)	**Describe Baseline Environmental Parameters**	**Proposed Environmental Objective for Improvement**	**Describe Baseline Environmental Parameters**	**Proposed Environmental Objective for Improvement**	**Briefly Describe How Objective Will Be Achieved**
Emissions to atmosphere (controlled)	VOCs; quantity unknown	Released through vent			Run VOC venting through hose/carbon can
Emissions to atmosphere (uncontrolled)	VOCs released through manhole; quantity unknown	None			None
Discharge to water (controlled)	400–600 gallons of ignitable liquid (to waste water treatment)	Reduce volume of ignitable liquid			Dispose through gravity separator and re-distill water
Discharge to water (uncontrolled)	Not applicable, if done properly	Avoid spills			Temporary plugs on storm drains in vicinity
Produce solid waste (nonhazardous)	2 drums (55-gallon)	Reduce volume 20%			Reduce use of disposable PPE; run decon water through gravity separator

■ TABLE 3.6 *(Continued)*

Produce solid waste (hazardous, not commingled)	Tank bottoms (unpumpable solids) 1/2 drum—est. 200 lbs.	Reduce volume 70%			Run through sludge press to dewater
Produce solid waste (hazardous, commingled)	Not applicable	Not applicable			Not applicable
Produce special waste (state variations, radioactive, biohazard)	Not applicable	Not applicable			Not applicable
Discharge to land (controlled)	Dewatered tank bottoms 1/2 drum—est. 200#	None			None
Discharge to land (uncontrolled)	Not applicable	Not applicable			Not applicable
Use of land, fuels, water, energy, and other natural resources	15 gallons diesel fuel to operate equipment	None			None
Odor, noise, dust, vibration and/or visual impact	Noise, odor	None			None
Other ecosystem effects	Not applicable	Not applicable			Not applicable
Other ecosystem effects	Not applicable	Not applicable			Not applicable

Completed by: Signature on File
signature

____
print name

Date Completed: 4/26/95

construction services such as the repair and maintenance of aboveground storage tanks and pipelines. Welding is often performed, so the company expects to have compressed gases, as well as other hazardous materials used in maintenance, on-site. Additionally, because the work is performed at customer locations, site conditions vary. The environmental controls required must adapt to the characteristics of the site.

The company has an environmental policy that includes as an objective: "Work site controls are selected to reduce risks to people and the natural environment." But how can that be translated into something meaningful as it relates to pollution prevention?

In its operations, the company treats each work activity as "new." During its project planning for an activity, an activity-specific environmental objective is established in support of the broad company policy. In this case, the objective is to reduce the volume of hazardous waste produced during the project. An estimate is made with regard to the projected maximum volume of hazardous waste. This becomes the stated target against which the actual performance of selected work controls (i.e., pollution prevention alternatives) is measured.

CORRECTIVE AND PREVENTIVE ACTIONS

Management systems define two styles of action: corrective and preventive. When a problem already exists, actions taken to respond to the matter are corrective in nature. If there is only the potential for a problem, actions taken to eliminate or minimize the potential are considered preventive.

Corrective actions can be quick fixes that treat the symptom and are taken in response to an immediate need. They minimize or mitigate negative impacts, but do not eliminate recurrence of problems or conditions of a similar nature. They also include steps taken to eliminate the root cause of an existing problem. This second type of corrective action typically has a longer implementation cycle.

Preventive actions are directed at stopping potential problems from occurring in the first place. Their mission is to eliminate root causes before the fact. Preventive actions are planned responses to probable consequences. They result in more permanent solutions and incremental changes to the way in which business is conducted.

Preventive actions are more cost-effective than corrective actions because they are initiated outside of a crisis mode and eliminate the costs associated with mistakes.

Consider the situation of a 55-gallon drum of acetone presented in damaged condition by a trucking company for delivery at a loading dock. The quick-fix corrective action can include:

- Turning the container so the flow of product is minimized or stopped;
- Patching the container and placing it in a salvage drum; or
- Pumping the container out to an undamaged container.

After the immediate response, a longer-term corrective action associated with damaged freight (as opposed to one damaged acetone drum) may address:

- Establishing a qualification process for hazardous materials freight haulers to validate the safe operating practices of the carrier;
- Eliminating the use of the lowest-cost carrier.

However, presuming the damaged freight had never been delivered, a preventive action would consider having liquid hazardous materials shipped on pallets and shrink-wrapped or otherwise banded.

While corrective and preventive action alternatives may present an additional cost of doing business in the short run, as new practices are developed and introduced, they have the long-term impact of reducing costs associated with an operation. In the case of the damaged freight example, this can include costs associated with:

- Managing damaged-freight claims;
- Conducting emergency response actions;
- Managing raw materials shortages created by inability to use the damaged freight.

These definitions do not mean that pollution prevention can be carried out only before there is a pollution event. They are included to assist pollution prevention practitioners in understanding management systems terminology in order to be better communicators to the decision makers they face.

Whether the specific pollution prevention project being assessed is considered a corrective or preventive action is not necessarily significant with regard to a specific end result. The P^2 professional should be aware of the subtle differences in definition, however, particularly if working with an organization seeking or having achieved registration to an ISO 9000 or ISO 14000 management systems standard. There is always the threat that poor communications and semantics can become roadblocks to implementing cost-effective, timely business decisions.

A COST APPROACH TO COMPARING POLLUTION PREVENTION ALTERNATIVES

In order to compare the costs of pollution prevention alternatives, numbers are calculated to relate costs over time.

When the work process exists and is ongoing, cost calculation is based on a specified time interval that makes sense to the organization's financial requirements. In some cases, this compares last year's costs with estimated costs in five years. In others, a one-year range is more useful. The key is to anticipate what the internal audience for this information wants or finds useful. Dollar amounts are then calculated to communicate in that manner.

The Decision to Do Nothing

No matter what the business issue, it is always appropriate to consider a "no action" option. Relative to pollution prevention, this consists of a series of simple computations (Table 3.7). A determination is made about which costs are assigned to the activity, product, or service when there is no implementation of pollution prevention alternatives.

This calculation is equally useful for objectively assessing new types of projects, since it minimizes the human bias factor present in all business requests.

Specific assumptions plugged into these "do-nothing" calculations vary. However, the assumptions should be listed. With reference to Table 3.7, assumptions include:

- Time frame for comparison (one year, three years, etc.)
- Changes in business climate (e.g., more business, fewer "changeovers," etc.)
- Variations in the frequency of waste stream generation (increase or decrease)
- Changes in type of waste stream generation

Assign Values to "Soft" Costs

There are other, less readily quantifiable factors that also affect cost calculations. They include such situations as:

- New or modified regulatory requirements (state, local, or federal) mandating phaseout of selected raw materials
- Process shutdowns or production curtailments
- Discovery of new characteristics in waste streams

■ TABLE 3.7 Calculating the No-Action Option

Company: XYZ Paint Co.
Facility Business: Manufacturing paints and coatings
Product: All customer-branded product
Environmental Effect: Produce solid waste (hazardous)
Objective: Reduce volume 25% in one year

Assessment Alternative: No additional pollution prevention effort
100 pounds waste/run of D001, Ignitable Sludge
Frequency: 8 runs/day
Current Daily Volume: 800 pounds/day

Cost Projection (Five years)
Assumptions:

- No reduction in volume of waste streams created per production run
- 25% increase in volume (size) of production runs
- 25% increase in number of "changeovers" between production runs because runs will be of different (customized) formulations, resulting in 10 runs/day

1. Compute anticipated daily volume in five years if no pollution prevention alternatives are implemented.

Current volume/ run	25% volume increase	Total volume @ 8 runs/day	25% increase in no. of runs	Est. total volume/day in five years
100 lbs	+25 lbs	1,000 lbs	×1.25	1,250 lbs

2. Compute daily disposal costs if no pollution prevention alternatives are implemented. This calculation assumes a 10 percent increase in disposal prices over five years.

Actual (as of 12/31/96)	**5 Years (projected)**
800 × $3.25 = $2,600	1,250 × $3.575 = $4,468.75

Current Price/lb for Disposal	**Estimated Price for Disposal in 5 Years**	**Projected Daily Volume (5 Yrs.)**	**Estimated Change in Disposal Costs (5 Yrs.)**
$3.25	$3.575	1,250 lbs	$1,868.75

3. Calculate other costs associated with the activity, such as those for:

- Maintaining employee HAZWOPER certification for hazardous waste handling
- Insurance premiums associated with hazardous waste shipments
- Workers' compensation premiums for employees packaging and handling hazardous wastes
- Emergency response associated with production line
- Regulatory compliance reporting

- Changes in disposal technology or accessibility by the disposal service vendor
- Contingency costs because the operation has been newly acquired and historical data are meager

While more professional judgment is necessary to assign value to these "soft" costs, determining these costs is an important step. First, it requires the organization to recognize the existence of the phenomena that comprise soft costs. Second, it causes the organization to communicate these soft costs as a hard dollar figure appropriate for comparisons.

Assessing the Alternatives

The organization is now ready to develop alternative techniques for achieving pollution prevention in the activity, product, or service. These alternatives are the pollution prevention opportunities.

Chapter 7 presents a catalog of techniques that may be suitable for accomplishing pollution prevention in a specific situation.

After alternatives have been listed, it is necessary to cost out each potential alternative project and compare costs to projected savings, or the P^2 return on investment (ROI).

Referring to Table 3.8, it is seen that if P^2 Alternative 3 is implemented, the company still anticipates a 31 percent increase in its daily waste disposal costs at the end of five years. However, taking no action would result in a 72 percent increase.

Assuming that the improvements that comprise Alternative 3 cost an estimated \$145,000 to implement, the following would be the return on investment after one year. The targeted savings are computed on the basis of a seven-day work week.

$$\text{Projected annual savings/cost} = \text{ROI}$$
$$\$391{,}462.50/\$145{,}000 = 2.7$$

■ TABLE 3.8 Waste Management Costs

Company: XYZ Paint Co.
Facility Business: Manufacturing paints and coatings
Product: All customer-branded product
Environmental Effect: Produce solid waste (hazardous)
Objective: Reduce volume 25% in one year

Assessment Alternative 3: Pump Efficiency Upgrade
100 pounds waste/run of D001, Ignitable Sludge
Frequency: 8 runs/day
Current daily volume: 800 pounds/day

Cost projection (five years) assumptions:

- 24% reduction in volume of waste streams created per production run
- 25% increase in volume (size) of production runs
- 25% increase in number of "changeovers" between production runs because runs will be of different (customized) formulations, resulting in 10 runs/day

1. Compute anticipated daily volume in five years by implementing Alternative 3.

Current volume/ run	25% volume increase	Total volume @ 8 runs/ day	25% increase in number of runs	Est. total volume/ day in five years (no reduction)	Est. Reduction with Alternative 3	Est. total volume/ day in five years with reduction alternative
100 lbs	+25 lbs	1,000 lbs	×1.25	1,250 lbs	300 lbs	950 lbs

2. Compute daily disposal costs if no pollution prevention alternative is implemented. This calculation assumes a 10% increase in disposal prices over 5 years.

Actual (as of 12/31/96)	**5 Years (projected, no alternative)**	**5 Years Projected with Alternative 3**
800 × $3.25 = $2,600	1,250 × $3.575 = $4,468.75	950 × $3.575 = $3396.25
Total Daily Costs	per day increase in costs	per day increase in costs
	$ 1868.00	$ 1072.50
	+ 2600.00	+ 2600.00
	Total Daily Costs $ 4468.00	Total Daily Costs $ 3672.50

Current Price/lb for Disposal	**Estimated Price for Disposal in 5 Years**	**Projected Daily Volume (5 Yrs.)**	**Estimated Change in Disposal Costs (5 Yrs.)**
$3.25	$3.575	950 lbs	$1072.50

3. Calculate other costs associated with the activity, such as those for:

- Maintaining employee HAZWOPER certification for hazardous waste handling
- Insurance premiums associated with hazardous waste shipments
- Workers' compensation premiums for employees packaging and handling hazardous wastes
- Emergency response associated with production line
- Regulatory compliance reporting

SUMMARY

A comprehensive pollution prevention assessment consists of preparing a suitable scope of the work activity, product, or service subject to assessment. This ensures successful delineation of areas requiring corrective and preventive action. Only then can alternative methods for implementing corrective and preventive action be identified and a thorough cost analysis conducted.

This chapter presented a process for such an assessment. It considered the benefits of P^2 within the framework of communicating with regulators and the financial community. It identified tools and information already available at many organizations that can be assembled to support such an assessment. Finally, it offered techniques for a financial comparison of alternative P^2 solutions.

Exercise

Objectives:

- Gain experience in preparing a scope statement
- Practice devising a baseline for measuring improvements

Instructions: Either continue with the work processes addressed in exercises 1 and 2, or select another activity, product, or service that presents a P^2 opportunity.

1. Write a scope statement that describes the characteristics of the activity, product, or service being evaluated for P^2. Make sure that the statement:
   a. Identifies the facility involved
   b. Identifies the location of the activity, product, or service within the facility
   c. Clarifies whether the activity, product, or service is:
      - Brand new
      - Existing
      - A modification (retrofit) to an existing activity, product, or service
      - The result of an acquisition
2. List the environmental effects of that activity, product, or service:
   a. By media (air, water, land)
   b. By regulatory requirement
3. Describe and enumerate the baseline performance measurement(s) from which improvements will be gauged.

CHAPTER

4

A REGULATORY FOUNDATION FOR POLLUTION PREVENTION OPPORTUNITIES

Many of the pollution problems faced in the United States are historic in nature, arising from past practices. The explosive growth of the 1950s created an atmosphere of consumption with few operating constraints, as long as product got out the door. If a site became contaminated, and the water polluted, it wasn't too difficult to pick up and move into a new facility. Avoiding pollution was neither a societal nor an organizational priority.

Since then, a national consensus has developed that places a high premium on a clean, protected environment. The natural environment is recognized, even by most school children, as being finite. Although the environment can cleanse itself and regenerate over time, expanding world populations continue to put relentless pressure on these resources.

As organizations began to understand that pollution costs money in lost materials, wasted time, injured workers, and inferior product, they began to make changes that placed an emphasis on minimizing pollution.

One of the earliest P^2 programs was a 1970s approach taken by 3M Corporation. Operating under the slogan "Pollution Prevention Pays," it was marketed to employees throughout 3M's facilities in an effort to reduce production costs, waste generation, and environmental contamination.

How a P² Brainstorming Session Works: A Team Approach

1. Assemble a team with skills and responsibilities useful to problem identification. Two to five members is workable. Larger groups rapidly become unwieldy.
2. Designate a "recorder," who records the team's ideas.
3. The team reviews a prepared summary problem statement (e.g., the amount of wastewater produced in manufacturing increased 25 percent) regarding the potential pollution event. To save time, the team leader or facilitator prepares this in advance.
4. During a specified time period (5 to 10 minutes), the team verbally relays "stream of consciousness" thoughts relating to the problem statement. The recorder listens and develops a list of thoughts and ideas.
5. The team reviews the list, and team members may be asked to clarify what a idea or phrase means.
6. In a shorter time period (1 to 2 minutes), each team member ranks the list (1 = best opportunity; 10 = worst opportunity).
7. The numerical rankings are totaled to determine which opportunities are rated most favorably by the team. In the ranking method used here, the lower the number, the more favorable.
8. Normally, only the top 2 or 3 ideas ever proceed further to the development of alternatives to address the issues raised.

This chapter addresses the decision-making process associated with reclassifying a process stream as a waste stream. It considers:

- The national regulatory framework for waste management
- Environmental auditing
- The comprehensive assessment of pollution prevention opportunities.

THE STRUCTURE OF WASTE MANAGEMENT REGULATIONS

There is no question that waste management in the United States is full of regulatory complexity (Table 4.1). No matter what the source, an organization must make careful evaluations to determine whether its waste stream meets any of the criteria that would classify it as a wastewater, Hazardous Air Pollutant, solid waste, hazardous waste, or even radioactive waste.

■ TABLE 4.1 Environmental Laws Controlling Hazardous Substances

Law	Description
Clean Air Act	Regulates emissions of hazardous air pollutants
Clean Water Act	Regulates discharge of hazardous pollutants into the nation's surface waters
Marine Protection, Research, and Sanctuaries Act	Regulates waste disposal at sea
Occupational Safety and Health Act	Regulates hazards in the workplace, including worker exposure to hazardous substances
Resource Conservation and Recovery Act	Regulates hazardous waste generation, storage, transportation, treatment, and disposal
Safe Drinking Water Act	Regulates contaminant levels in drinking water
Toxic Substances Control Act	Regulates the manufacture, use, and disposal of chemical substances
Hazardous Materials Transportation Act	Regulates the transportation of hazardous materials, substances, and wastes
Comprehensive Environmental Response, Compensation, and Liability Act	Provides for the cleanup of inactive and abandoned hazardous waste sites
Atomic Energy Act	Regulates nuclear energy production and nuclear waste disposal
Surface Mining Control and Reclamation Act	Regulates the environmental aspects of mining and reclamation
Domestic Chemical Diversion Control Act of 1993	Registration and recordkeeping requirements for manufacturers, distributors, importers, and exporters of drug precursors (Appendix G.2)

This complexity has given the traditional language of business operating practices precise legal meanings which, if not applied accurately and consistently in source evaluation, can result in enforcement actions, from fines to jail sentences.

Understanding these seemingly tangled requirements is important to the identification of pollution prevention opportunities. Understanding the characteristics of the process streams being evaluated may release an organization from some regulatory requirements because of an exemption.

For the purposes of this discussion, radioactive and biological wastes are not addressed. However, the technical issues surrounding these types of waste streams, while different, can appropriately make effective use of management systems similar to those described for chemical waste streams.

Who is Involved in Hazardous Waste?

Under the directives of the Resource Conservation and Recovery Act, the U.S. Environmental Protection Agency (and equivalent state agencies, where states have taken responsibility) administers a waste management system. The organizational structure of this system (Figure 4.1) is directed at

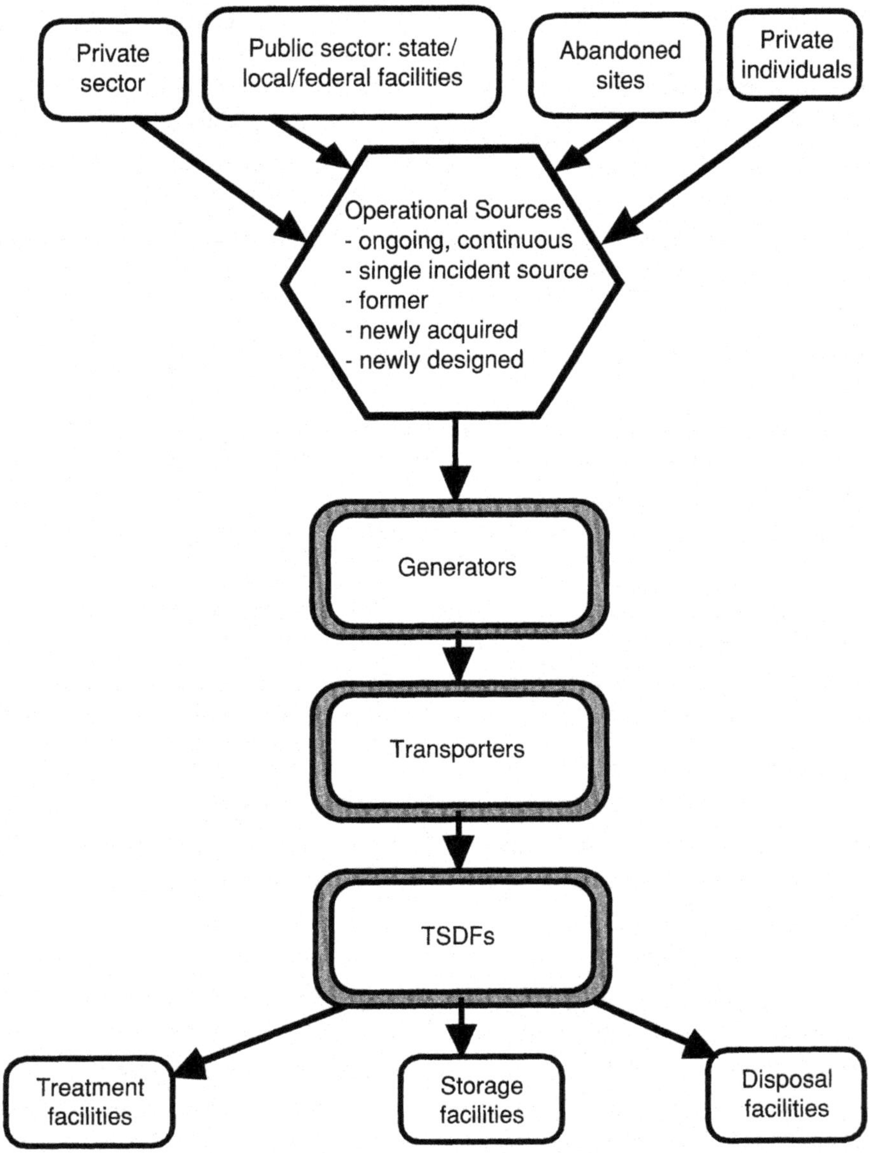

■ **FIGURE 4.1 The Players in Waste Management in the United States**

all of the parties that participate in the waste management loop, as well as the waste streams themselves.

Federal regulations specify three categories of hazardous waste generators (Table 4.2):

- Conditionally exempt small quantity generators
- Small quantity generators
- Large quantity generators

The category used is a function of how much regulated waste and how often regulated waste streams are generated. In concept, smaller quantity generators are subject to less federal regulation. For example, conditionally exempt small quantity generators are not required to obtain EPA identification numbers. In application, however, many of these generators obtain numbers, because the TSDFs they are sending their waste streams to require them due to the terms of the TSDF permit or TSDF operating procedure.

■ TABLE 4.2 Categories of Hazardous Waste Generators

Category	Monthly Generation Criteria	Summary of Federal Regulatory Requirements
Conditionally exempt small quantity generator	■ No more than 100 kilograms (about 220 pounds) of hazardous waste or ■ No more than 1 kilogram (about 2 pounds) of acutely hazardous waste	■ Identify all waste generated ■ Send waste to a hazardous waste facility ■ Never accumulate more than 1,000 kg of hazardous waste on the property
Small quantity generator	■ Generate more than 100 kg and less than 1,000 kg or hazardous waste or ■ Generate no more than 1 kg of acutely hazardous waste	■ Obtain U.S. EPA identification number ■ Comply with federal rules for managing hazardous waste, including those for accumulation, treatment, storage, and disposal
Large quantity generator	■ Generate more than 1,000 kg of hazardous waste or ■ Generate more than 1 kg of acutely hazardous waste	■ Obtain U.S. EPA identification number ■ Comply with all applicable hazardous waste management regulations

Source: U.S. Environmental Protection Agency.

Additionally, states have the option of developing criteria that are more stringent than federal criteria. If an organization thinks it falls into one of the less restrictive generator categories, it is important to check with appropriate state authorities to clarify whether the federal requirement or a more stringent state requirement is applicable.

Table 4.3 presents the requirements for determining whether the generator must additionally register with the U.S. Department of Transportation (DOT) because it offers hazardous materials or waste for shipment. If the generator must register, an annual filing with the U.S. Department of Transportation is required (Appendix D.1).

Transporters are also essential participants in the management system. They must meet registration requirements with the EPA and DOT and maintain a satisfactory level of safety fitness. This is helpful to the pollution prevention professional, because good transportation operators reduce the risk of a pollution event occurring during transportation of hazardous materials or wastes. Detailed criteria for evaluating transporters appear in Chapter 6.

TSDFs provide waste treatment, storage, or disposal. These facilities can provide one, two, or all three functions at any one location. Treatment facilities conduct such operations as wastewater treatment or liquid waste solidification. Storage facilities are permitted to keep hazardous wastes onsite for more than 90 days. These facilities frequently operate to consolidate waste streams from multiple sources for later treatment or disposal. Landfills, injection wells, and incinerators are all disposal activities, providing the terminus for a hazardous waste stream.

RCRA as a Management System

The RCRA waste management system is prescriptive and contains many performance measures. That is, it specifies, via regulations and permits, how various types of materials, waste streams, and potential waste streams are to be identified and handled. It includes a documentation process that facilitates tracking of the material or stream from its point of origin (cradle) to its point of final handling (grave).

While the initial thrust of this cradle-to-grave documentation process was to make accountability through enforcement easier to conduct, the results have provided performance measurements (Figure 4.2). Through the application of systematically designed tools, organizations can develop data

■ TABLE 4.3 U.S. Department of Transportation Hazardous Materials Registration Program Research and Special Programs Administration

You Are Required to Register if, Between July 1 of Any Year and June 30 of the Following Year, You Offered or Transported:

Amount	DOT Hazard Class	Class Description
Any quantity by highway	7	Radioactive material
More than 25 kg (55 pounds)	Division 1.1, 1.2, 1.3	Explosive material
More than one liter (1.06 quarts)	Division 2.3, Hazard Zone A Division 6.1, Packing Group I, Hazard Zone A	Material extremely toxic by inhalation
Liquid or gas bulk package with capacity: ≥ 13,248 liters (3,500 gallons)	Classes 1-9	1—Explosives 2—Gases 3—Flammable liquids 4—Flammable solids 5—Oxidizers and organic peroxides 6—Toxic materials and infectious substances 7—Radioactive materials 8—Corrosive materials 9—Miscellaneous dangerous goods
Solid materials bulk package with capacity > 13.24 meters (468 cubic feet)	Classes 1-9	1—Explosives 2—Gases 3—Flammable liquids 4—Flammable solids 5—Oxidizers and organic peroxides 6—Toxic materials and infectious substances 7—Radioactive materials 8—Corrosive materials 9—Miscellaneous dangerous goods
Single shipment of non-bulk package with a gross weight of 2,268 kg (5,000 pounds) or more of one hazardous materials class when placarding is required	Classes 1-9	1—Explosives 2—Gases 3—Flammable liquids 4—Flammable solids 5—Oxidizers and organic peroxides 6—Toxic materials and infectious substances 7—Radioactive materials 8—Corrosive materials 9—Miscellaneous dangerous goods

The registration form appears in Appendix D.1.

that can be used to establish performance targets and measure performance to those targets.

The process begins with the "maybe" generator identifying a material of concern and making a waste determination (Table 4.4). Consideration is given to whether a material or stream is discarded. Discarded materials are classified in the regulations as materials that are:

- Abandoned;
- Recycled; or
- Considered inherently wastelike.

Solid wastes are then defined as discarded materials. However, this solid waste definition makes no mention of the physical state of the material. Thus, a liquid can be considered a solid waste.

Abandonment of solid wastes includes:

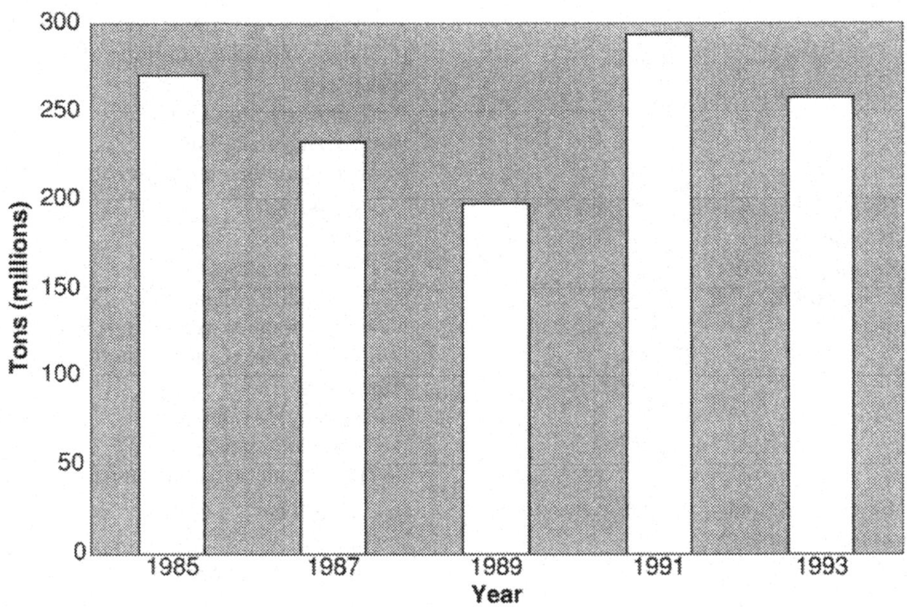

■ **FIGURE 4.2 RCRA Managed Hazardous Waste**

Note: The increase in 1991 is due to the addition in that reporting period of new waste codes and new criteria for classifying wastes using the Toxicity Characteristic Leaching Procedure.

Source: U.S. Environmental Protection Agency, *Biennial RCRA Hazardous Waste Reports for 1985, 1987, 1989, 1991* and *1993.*

- Disposal;
- Burning or incineration;
- Accumulation, storage, or treatment (but not recycling) before or in lieu of being abandoned by being disposed of, burned, or incinerated.

Is Your Waste Stream Exempted?

There are key exemptions to the solid waste definition that enable specific waste streams to fall off the solid waste regulatory chain. However, the stream can still be subject to other regulatory control. Some of the more notable exemptions are for:

- Domestic sewage;
- Industrial wastewater discharges that are point sources regulated under Section 402 of the Clean Water Act;
- Irrigation return flows.

A wide variety of specific industrial waste streams are also exempted. In conducting a stream evaluation, the reader is referred to 40 CFR 261.4 (Appendix E.1) for a comprehensive list of exempted waste streams. Meeting the criteria of an exempted waste stream significantly reduces the cost of waste stream management.

A major concern in solid waste management centers on whether a stream is hazardous. By legal definition, certain streams are not considered hazardous wastes, including:

- Household wastes;
- Commercial or industrial sources without hazardous waste;
- Some agricultural wastes, such as animal manure used as fertilizer.

Another major exemption from the definition of hazardous waste applies to samples. A solid waste sample or a sample of water, soil, or air collected solely for testing in order to determine its characteristics or composition is not classified as a hazardous waste. Specific performance criteria apply. The sample must be:

- Transported to a laboratory for the purpose of testing or;
- Transported back to the sample collector after testing or;

■ **TABLE 4.4 Small Quantity Generators: Typical Waste Streams**

Type of Business or Business Function	Typical Wastes Generated
Building cleaning and maintenance	Acids/bases Solvents
Chemical manufacturers	Acids/bases Cyanide wastes Heavy metals/inorganics Ignitable wastes Reactives Solvents
Cleaning agents and cosmetics	Acids/bases Heavy metals/inorganics Ignitable wastes Pesticides Solvents
Construction	Acids/bases Ignitable wastes Solvents Special wastes (asbestos, lead-contaminated debris)
Educational and vocational shops	Acids/bases Ignitable wastes Pesticides Reactives Solvents Radioactive wastes Infectious wastes
Equipment repair	Acids/bases Ignitable wastes Solvents Radioactive wastes
Food processing	Acids/bases Organic wastes Infectious wastes Radioactive wastes
Formulators	Acids/bases Cyanide wastes Heavy metals/inorganics Ignitable wastes Pesticides Reactives Solvents
Funeral services	Solvents Formaldehyde Infectious wastes

■ **TABLE 4.4** (*Continued*)

Type of Business or Business Function	Typical Wastes Generated
Furniture/wood manufacturing and refinishing	Ignitable wastes Solvents
Laboratories	Acids/bases Heavy metals/inorganics Ignitable wastes Reactives Solvents Radioactive wastes Infectious wastes
Laundries and dry cleaners	Dry cleaning filtration residues Solvents
Metal manufacturing	Acids/bases Cyanide wastes Heavy metals/inorganics Ignitable wastes Reactives Solvents Spent plating wastes
Motor freight terminals and railroad transportation	Acids/bases Heavy metals/inorganics Ignitable wastes Lead-acid batteries Solvents
Other manufacturing ■ Textiles ■ Plastics ■ Leather	Heavy metals/inorganics Solvents
Pesticide end users and application services	Heavy metals/inorganics Pesticides Solvents
Printing and allied industries	Acids/bases Heavy metals/inorganics Ink sludges Spent plating wastes Solvents
Vehicle maintenance	Acids/bases Heavy metals/inorganics Ignitable wastes Lead-acid batteries Solvents
Wood preserving	Preserving agents

Source: U.S. Environmental Protection Agency.

- Stored by the sample collector before transport to a laboratory for testing or;
- Stored in a laboratory before testing or;
- Stored in a laboratory after testing but before its return to the sample collector; or
- Stored temporarily in the laboratory after testing for a specific purpose (for example, until conclusion of a court case or enforcement action where further testing of the sample may be necessary).

Additionally, a sample collector shipping samples to a laboratory and a laboratory returning samples to a sample collector must comply with U.S. Department of Transportation, U.S. Postal Service (USPS), or any other applicable shipping requirements.

In cases in which the sample collector determines that DOT, USPS, or other shipping requirements do not apply to the shipment of the sample, the sample must be packaged so that it does not leak, spill, or vaporize from its packaging. Additionally, the following information must accompany the sample:

- Sample collector's name, mailing address, and telephone number;
- Laboratory's name, mailing address, and telephone number;
- Quantity of the sample;
- Date of shipment;
- Description of the sample.

In the workplace, people evolve a particular vocabulary that enables them to communicate within their job function. However, the legal, regulatory world discounts this common usage. Therefore, during a pollution prevention assessment, it is critical to avoid misunderstandings.

This is accomplished by recognizing and clarifying the differences between the common use of terminology at the process stream source and the regulatory definitions that are applied to direct the options for handling. The pollution prevention professional must become conversant in both systems in order to make things work at an optimal level of productivity. Linking the EPA definitions (Table 4.5) to workplace usage are essential to successful pollution prevention implementation.

Is the Solid Waste a Hazardous Waste?

If not otherwise exempted, waste streams need to be carefully classified by their chemical characteristics before any decisions can be made on the

■ TABLE 4.5 EPA Definitions of Common Operational Terms

Term	Definition
Sludge	Solid, semi-solid or liquid waste (exclusive of the treated effluent from a wastewater treatment plant) generated from a: ■ Municipal, commercial, or industrial wastewater treatment plant; ■ Water supply treatment plant; or ■ Air pollution control facility.
By-product	Material that ■ Is not one of the primary products of a production process and ■ Is not solely or separately produced by the production process. Examples are process residues, such as slag or distillation bottoms. The term does not include a co-product that is produced for the general public's use and is ordinarily used in the form it is produced in the process.
Reclaimed material	■ Processed to recover a usable product, or ■ Regeneration of a usable product. Examples are recovery of lead from spent batteries and regeneration of spent solvents.

Source: 40 CFR 261.

pollution prevention management options permitted under the regulatory structure. Hazardous wastes fall into two broad categories: (1) characteristic; (2) listed.

Characteristic Wastes Characteristic wastes possess specific physical parameters or characteristics that make them hazardous (Table 4.6). Streams being evaluated that meet any of these characteristics, and are not otherwise exempted, require management as hazardous wastes.

Listed Wastes Solid Waste identified as and meeting the criteria of listed wastes are hazardous unless specifically exempted. There are three categories of listed wastes:

- Hazardous waste from nonspecific sources (F-wastes)
- Hazardous waste from specific sources (K-wastes)
- Discarded commercial chemical products, off-specification species, and their containers and/or spill residues (U and P wastes).

■ TABLE 4.6 Characteristic Hazardous Wastes

Characteristic Name	Characteristic Criteria	EPA Hazardous Waste Number
Ignitability	■ Liquid, other than an aqueous solution of < 24% alcohol and flash point < 60°C (140°F); or ■ Not liquid but causes fire through friction, absorption of moisture or spontaneous chemical changes under standard pressure and temperature; burns vigorously and persistently when ignited; or ■ Ignitable compressed gas; or ■ Oxidizer per U.S. DOT regulations	D001
Corrosivity	■ Aqueous with pH ≤ 2 or ≥ 12.5; or ■ Liquid and corrodes steel per NACE Standard TM-01-69 and EPA SW-846	D002
Reactivity	■ Normally unstable and readily undergoes violent change without detonating ■ Reacts violently with water ■ Forms potentially explosive mixtures with water ■ When mixed with water, generates toxic gases, vapors, or fumes in sufficient quantity to present danger to human health or the environment ■ Cyanide or sulfide-bearing waste which, when exposed to pH conditions between 2 and 12.5, can generate toxic gases, vapors, or fumes in sufficient quantity to present a danger to human health or the environment ■ Is capable of detonation or explosive decomposition or reaction at standard temperature and pressure ■ Is an explosive or forbidden explosive per U.S. DOT regulations	D003
Toxicity	When Toxicity Characteristic Leaching Procedure (Test Method 1311 of EPA SW-846) used, extract contains contaminants listed in Table 1 of 40 CFR 261.24(b)	From D004 through D0043, depending on contaminant

Source: Based on 40 CFR Part 261.

The complete listing appears in Appendix E.2. Streams being evaluated that meet the criteria of these lists require management as hazardous wastes. They can, however, be exempted through petitioning the EPA under a process described in 40 CFR 260.20 and 260.22.

Appendix E.2 further includes the regulatory rationale for listing a chemical (e.g., toxic, reactive, etc.), introducing the subcategory of acute hazardous wastes. These waste streams are extremely toxic because they are fatal to humans in low doses. When human toxicity studies are not available, the acute hazardous waste classification is based on the following animal testing parameters:

- An oral LD 50 toxicity (rat) of less than 50 milligrams per kilogram;
- An inhalation LC 50 toxicity (rat) of less than 2 milligrams per liter;
- A dermal LD 50 toxicity (rabbit) of less than 200 milligrams per kilogram; or
- Is otherwise capable of causing or significantly contributing to an increase in serious irreversible, or incapacitating reversible, illness.

Residues

In applying the listed waste designation, special attention is warranted with regard to container residues. It is very difficult to truly empty most containers of all free liquid. But this residue content still represents a risk to human health and the environment.

In the EPA regulatory structure, container residues are intended for discard.[1] If the residue represents a listed waste, it must be managed as a hazardous waste unless specific qualifications are met. For example, if a container with residue is refilled with the same chemical product or chemical intermediate, this is a qualified legitimate reuse. However, if the container with residue is being reconditioned, the discarded residue is considered a hazardous waste and must be managed as such.

To assist in this residue management process, EPA has defined parameters that establish whether a container (or the inner liner removed from a container) is considered empty:

- Waste is not a compressed gas or identified as an acute hazardous waste in 40 CFR 261.31, 261.32, 261.33;
- All wastes are removed that can be removed using the common practices to remove materials from that type of container, e.g. pouring, pumping, and aspirating;
- No more than one inch (2.5 cm) of residue remains on the bottom or inner liner.

1. 40 CFR 261.33(d).

- If the container is less than or equal to 110 gallons in size, no more than 3 percent by weight of the total capacity remain
- If the container is greater than 110 gallons in size, no more than 0.3 by percent weight of the total capacity remain

Calculating Emptiness There are several issues in using these parameters to designate an empty container. The process of measuring for compliance to the so-called one-inch rule is messy and subject to a high degree of inaccuracy.

If the intent of the organization is to reduce residue streams considered as hazardous, it is more reliable to calculate weight limits (Table 4.7). However, three key pieces of information are needed:

- Gross weight of the container/and its contents
- Net weight of the previous contents
- Gross weight of the container after emptying

Typically, net weight appears on the container. Although gross weight may appear on a bill of lading accompanying a delivered product, this will usually be the gross weight of a shipment, rather than a single container. Some organizations have begun to request container-specific gross weight information to appear on all containers they purchase.

■ TABLE 4.7 Calculating Container Emptiness

Container Type: 55 Gallon Steel Drum
Previous Contents: Commercial Chemical Product (40 CFR §261.33)

1. Calculate weight of container	
Gross weight of container/contents	463 pounds
− Net weight of previous contents	− 440 pounds
Calculated weight of container	23 pounds
2. Calculate net weight of residue	
Gross weight of container/ residue	56 pounds
− Calculate weight of container	− 23 pounds
Calculated weight of residue	33 pounds
3. Calculate 3% of total capacity by weight	
Net weight of previous contents	440 pounds
× .03	× .03
Calculated 3% of total capacity by weight	13.2 pounds

Result: Residue weighs more than 3% of total capacity by weight. The container cannot be defined as empty.

While it may seem logical to take one weight and apply it to all containers of the same size and type, this is not always appropriate. Containers such as the ubiquitous 55-gallon steel drum may have been reconditioned and reused. Although this reuse is a worthwhile waste minimization effort, the practice of reconditioning these containers requires that the top, or "head," of the drum be cut off. That means every reconditioned container is of variable height and weight.

Whether an organization chooses to conduct calculations for emptiness is influenced by several factors:

- Number of "empty" containers
- Rate of potentially "empty" container production
- The previous contents of those containers
- The costs associated with managing container residues if they fall under the hazardous waste criteria

The value of the "unemptied residue" or remaining product can be included in any pollution prevention assessment that is directed at removing more material from that type of container.

Compressed Gases and Acute Hazardous Waste

Because of unique physical characteristics of compressed gases and acute hazardous waste, containers of these materials are managed differently. For example, when the pressure inside a compressed gas cylinder approaches atmospheric pressure, it is considered empty.

Specific cleaning is required when the container or its inner liner held a material listed as an acute hazardous waste. The cleaning options include:

- Selecting a solvent that can remove the specific residue in order to triple-rinse the container or inner liner;
- Cleaning by another method shown in scientific literature or by tests that removal equivalent to triple rinsing is achieved;
- Removal (and proper handling of) an inner liner, so the container is considered empty.

Construction, Demolition and Lead-Contaminated Waste Streams

Lead-contaminated wastes are considered hazardous (D008). While many organizations do not now work directly with lead in their processes, lead was used in paints and piping systems for many decades. Thus, activities associated with construction and demolition, particularly at facilities more than a few years old, can result in high costs for:

- Disposing of lead-contaminated waste streams
- Air monitoring, medical surveillance, mandatory medical removals, special personal protective equipment, and training of construction workers laboring at work areas with lead contamination

In some cases, the pollution risk associated with lead paints is managed most cost-effectively by leaving the surfaces alone and encapsulating them to minimize exposure to flaking chips and dust.

In the interest of construction worker safety, OSHA has identified the following types of construction activities that may involve lead contamination (29 CFR 1926.62):

- Demolition or salvage of structures where lead or materials containing lead are present
- Removal or encapsulation of materials containing lead
- New construction, alteration, repair, or renovation of structures, substrates, or portions thereof, that contain lead or materials containing lead
- Installation of products containing lead
- Lead contamination/emergency cleanup
- Transportation, disposal, storage, or containment of lead or materials containing lead on the site or location at which construction activities are performed
- Maintenance operations associated with the construction activities relating to the previously described activities

These construction activities include such typical tasks as:

- Manual demolition of structures (drywall), sanding, scraping
- Abrasive blasting, blasting enclosure movement and removal
- Rivet busting, welding, cutting, or burning of lead materials
- Power tool cleaning

When evaluating a pollution prevention alternative that involves demolition or maintenance of surfaces that may be contaminated with lead, consider the special OSHA requirements and their costs.

ENVIRONMENTAL AUDITING

Environmental auditing is a term used to describe a process of methodically evaluating factors with a potential negative environmental influence. They are called audits rather than inspections because they:

- Look at a sampling of conditions and responses to those conditions
- Are conducted by individuals without line responsibility for the activity being reviewed

Environmental audits are useful to the pollution prevention professional when designed to support the following objectives:

- Verification of the status of compliance with defined requirements, such as regulations, industry standards, or organizational targets
- Identification of areas where preventive improvements are feasible or necessary
- Establishment of a framework for corrective action planning and plan implementation

Environmental management systems auditing is a requirement of ISO 14001. In ISO 14001, the audit focuses on a determination of whether the management system is functional.

There are three types of environmental audits:

- Internal: conducted within an organization by organizational personnel
- External: conducted by the organization on external authorities such as suppliers, subcontractors, and customers
- Third-party: conducted on the organization by an outside third party, such as an environmental management system registrar (ISO 14001), a customer, or a regulatory agency

Selecting Auditors

If resources and time permit, using an audit team of at least two people provides a second set of eyes and less bias in the observation and collection of audit results. When employee interviews are included in the audit process, the team of two enables one person to write while the other asks appropriate questions.

It goes without saying that, to be objective, the auditor cannot have line responsibility for the activity, product, or service being audited. The following auditor characteristics and skills are also useful:

- Communication (i.e., listening, questioning, writing)
- Observation
- Flexibility
- Organization
- Fairness
- Non-intimidation
- Tenacity
- Focus

Internal Audits

Of the three environmental audit types, internal audits offer the most cost-effective mechanism for accomplishing audit objectives. Their timing can be controlled so they are not disruptive to ordinary business activities. They are not expected to result in regulatory fines or loss of business.

To be useful, internal audits need to be rigorous and unbiased in implementation. The design and use of a standard checklist ensures that critical information is not overlooked. Results are conveniently compiled in a manner easy to use for corrective action plan development.

The audit format provided in Appendix C.3 is generic in design, but covers many of the types of issues that affect pollution prevention. The activities it covers are applicable to a broad spectrum of organizations that handle inventory that has an environmental risk.

The format is easily customized in its last section by adding process criteria unique to the facility or activity being audited. Customization is beneficial because it causes the organization to identify its work processes and construct a performance statement for those processes. It helps to eliminate fuzzy thinking about what exists and how it should operate.

Many organizations successfully use internal audit activities as a way to improve employees' (i.e., auditors) knowledge of the organization while honing their auditing skills (e.g., communication, observation), which are readily transferable and useful to other work activities in the organization.

The results of environmental audits are not always favorable. Problems are found. Depending on the problem, organizations face issues concerning whether, when, and how to report problems to regulatory agencies.

EPA Policy on Self-Policing

In 1995, EPA adopted a final policy on "Incentives for Self-Policing." Its purpose is to encourage the elimination of environmental problems through promotion of voluntary actions on the part of regulated organizations.

It is accomplished through a sequential process, in which:

- An organization voluntarily develops and implements an environmental auditing scheme
- Deficiencies based on regulatory compliance requirements are promptly disclosed to the appropriate regulatory authority
- These same deficiencies are corrected
- The organization takes steps to prevent recurrence
- The organization takes steps to remediate any environmental damage.

Through this policy, EPA has the flexibility to:

- Eliminate or reduce civil penalties
- Not recommend cases for civil prosecution
- Not request audit reports as a trigger for enforcement investigations

Through this policy, environmental audits and their resultant corrective action plans are another element of the due diligence found in a well managed organization. An audit implies responsibility on the part of the organization.

COMPREHENSIVE POLLUTION PREVENTION ASSESSMENT

A process stream consists of material inputs that are commingled or worked in order to create a product. Products are either material goods (e.g., paint, canned fruit) or services (e.g., painting, restaurants). Knowing what the inputs are to a process stream is critical to the complete identification of potential waste stream outputs.

Table 4.4 identifies, by broad chemical description, the types of waste streams commonly associated with various types of business activities. A comprehensive process assessment or inventory takes into account specific sources for those wastes.

When a manufacturing process undergoes a pollution prevention assessment, the following potential waste sources should be looked for. This is especially important when evaluating a manufacturing process in a newly acquired facility. Potential sources of waste can include:

- Damaged raw materials
- Off-specification raw materials
- Out-of-date raw materials (past shelf life)
- Raw materials no longer used
- Samples of raw materials
- Damaged finished goods
- Off-specification finished goods
- Returned finished goods
- Out-of-date finished goods (old stock)
- Samples of finished products

Residual process streams can also be significant waste stream contributors. They consist of residuals contained in processing equipment such as hoses, piping systems, pumps, or processing vessels. Some of these streams, such as tank bottoms, are produced intermittently. But others are created routinely, as hoses, lines, and pumps are flushed and cleaned.

These types of manufacturing waste streams can be managed so that the total waste produced is reduced or eliminated. It requires assessing the identified stream and determining whether it can be:

- Put back into the process
- Reworked (raw material for another product)
- Sold, as is, to an alternative purchaser requiring a different grade
- Sold as another product

Many manufacturers have been able to take a waste by-product from one production operation and convert it into a new, profitable product line.

Process Safety and Risk Management

There are many factors that influence when a process stream crosses a line to become a waste stream or present a spill risk. The engineer continually analyzes production equipment in an effort to make alterations that optimize production return while keeping chemically hazardous components contained. This is a theoretically logical approach with the potential to control when, where, how and how much waste is produced in a process.

Work processes are run by people, however, and their individual responses within the cycle of a work process create the conditions that result in waste streams and the potential for a spill release. Therefore, it is necessary to analyze the roles of both people and machines in the pollution prevention equation.

Although their legislative mandates differ in focus, both the U.S. Occupational Safety and Health Administration and the U.S. Environmental Protection Agency have established requirements for comprehensive management systems directed at the prevention of chemical releases. They are:

- OSHA—29 CFR 1910.119—Process Safety Standard
- EPA—40 CFR Part 68—Chemical Accident Prevention

The OSHA Process Safety Standard, effective in 1992, establishes a series of prescriptive compliance requirements (Table 4.8) aimed at worker safety on-site. Its requirements are mandatory for:

- Processes involving certain chemicals (highly hazardous chemicals) at or above designated threshold quantities
- Processes involving flammable liquids or gases on site in any one location in quantities greater than 10,000 pounds

There are exceptions for:

- Hydrocarbons used solely for workplace consumption
- Flammable liquids, stored in atmospheric tanks or transferred, which are kept below their boiling point without benefit of chilling or refrigeration
- Retail facilities
- Oil or gas well drilling or servicing operations
- Normally unoccupied remote sites

While the EPA Risk Management Program requirements are similar to OSHA's, there are some variations necessitated by different legislative mandates (Table 4.9). The EPA focus is planning for the consequences of off-site releases. It also has three levels of compliance, referred to as Programs (Table 4.10). Depending on an organization's capacity to fall into Program areas 1 or 2, there can be fewer compliance requirements (Table 4.11). Initial compliance is required by August 20, 1999.

The process hazard analysis is a component of both OSHA and EPA (Program 3) requirements. However, even if a work process is not subject to these requirements by regulation, the process hazard analysis technique contributes to successful pollution prevention efforts.

■ **TABLE 4.8 Process Safety Standard: Pollution Prevention Applications**

Requirement	Summary Description (OSHA)	Pollution Prevention Application
Employee Participation	Employers must have a written plan and involve employees at an elemental level of process safety management.	Promotes the participation of employees knowledgeable in the hazardous work process in the evaluation of risks and development of corrective actions to reduce risks.
Process Safety Information	Written information (MSDSs, Piping-instrumentation diagrams etc.) required on process chemicals, technology, and equipment	These data must be assembled before conducting a process hazard analysis. They provide a road map for an effective PHA.
Process Hazard Analysis (PHA)	Systematic approach for identifying, evaluating, and controlling processes containing highly hazardous chemicals.	Although the regulation limits its application to specified covered processes and highly hazardous chemicals, effective PHAs result in the identification and evaluation of P^2 alternatives.
Training	Training to help employees and contractors understand the nature and cause of problems arising from process operations.	When personnel understand the risks associated with a work process, they are better motivated to reduce the risk and consistently implement actions to eliminate or minimize these risks.
Contractors	Systematic method of screening and communicating with contractors working in and around regulated processes.	Competent contractors with employees trained on job-specific safety issues represent less risk of a chemical release.
Pre-Startup Safety Review	Systematic safety review of new or modified process before operation.	Implementation reduces the unidentified contingencies inherent in any new or modified process.

■ **TABLE 4.8** (*Continued*)

Requirement	Summary Description (OSHA)	Pollution Prevention Application
Mechanical Integrity	Ensures that equipment used in a process is designed, constructed and maintained to minimize risk of releases.	Equipment that is properly designed, constructed, and maintained reduces the release risk.
Hot Work Permit	Administrative mechanism for communicating when hot work (e.g., welding, cutting) is performed.	Effective communication before starting hazardous activities reduces the risk of releases.
Management of Change	Process for managing all modifications to equipment, procedures, raw materials, processing conditions other than "replacement in kind."	Change inherently creates greater risk. By managing the change process, the risk of releases is decreased.
Incident Investigation	Administrative mechanism for initiating and collecting information on events that could have reasonably resulted in a catastrophic release.	Identify weaknesses in structural and nonstructural elements of a work process that has the potential to cause a release.
Emergency Planning and Response	Emergency action plans required.	Effective emergency response reduces the total size and impact of a release.
Compliance Audits	Self-evaluation of the effectiveness of the PSM.	A systematic administrative review process can uncover gaps in safety.
Trade Secrets	Limits protection of trade secret information if that information is needed to conduct a process hazard analysis.	Ensures that all information necessary to evaluate a hazardous process is available. If the process or chemical is not covered under the Standard, an organization will have to apply business judgment to determine whether it is necessary to include trade secret information in order to have an effective process hazard analysis.

■ TABLE 4.9 Comparison of Process Safety Standard (29 CFR 1910.119) and Risk Management Prevention Program (40 CFR Part 68) Requirements

OSHA (PSM)	EPA (RMPP)
Differences in Terminology	
■ Highly hazardous substance	■ Regulated substance
■ Employer	■ Owner or operator
■ Facility	■ Stationary source
■ Standard	■ Rule or part
Differences in Extent of Regulatory Implementation	
■ 1910.119(d)(2)(E) An evaluation of consequences of deviations, including those affecting the safety health of employees	■ 68.65 (c)(1)(v) An evaluation of consequences of deviations, including those affecting the public and the environment
■ 1910.119 (e)(3)(ii) The identification of any previous incident that had a likely potential catastrophic consequences in the workplace	■ 68.67(c)(2) The identification of any previous incident that had a likely potential for catastrophic consequences, including those on the public and the environment
■ 1910.119(e)(3)(vii) A qualitative evaluation of a range of the possible safety and health effects of failure of controls on employees in the workplace	■ 68.67(c)(7) A qualitative evaluation of a range of possible safety and health effects of failure of controls, (including those on the public and the environment
■ 1910.119 (m)(1) The employer shall investigate each incident that resulted in, or could reasonably have resulted in, a catastrophic release of a highly hazardous chemical in the workplace	■ 68.81 (a) The owner or operator shall investigate each incident that resulted in, or could reasonably have resulted in, a catastrophic release of a regulated substance

The Process Hazard Analysis Process hazard analysis methodologies are basic components of these systems. Even if a business or process is not regulated by these requirements, hazard analysis methodologies are useful tools when applied to pollution prevention assessment (Table 4.12).

In selecting an appropriate process hazard analysis methodology, consider such factors as:

- Capabilities and skills of individuals participating in the analysis
- Time frame budgeted for conducting the analysis
- Availability of suitable process hazard information
- Documentation capabilities

■ **TABLE 4.10 EPA Risk Management Program Eligibility Criteria**

Program 1	Program 2	Program 3
No offsite accident history (five years prior to submitting RMPP)	—	Process is subject to OSHA PSM
No public receptors in worst-case circle	The process is not eligible for Program 1 or 3	Process is in SIC Code 2611, 2812, 2819, 2821, 2865, 2869, 2873, 2879, or 2911
Emergency response coordinated with local responders	—	—

SIC Code	Industry
2611	pulp mills
2812	chlor-alkali production
2819	production of industrial inorganics
2821	production of plastics and resins
2865	production of cyclic crudes
2869	production of industrial organics
2873	production of nitrogen fertilizers
2879	production of agricultural chemicals
2911	petroleum refineries

Eligibility criteria are process-specific.

The deadline for initial submission of Risk Management Prevention Plans under this rule is August 20, 1999.

- Whether the process uses hazardous materials
- Whether the process generates hazardous wastes

There are standard engineering software packages that can be purchased and used to support the hazard analysis process. However, with the growing availability of inexpensive word processing software with sophisticated, but easy to use, matrix and table capabilities, documentation of results is easier without a significant financial investment in software or learning curve for software implementation.

Documented conclusions from a process hazard analysis can be ranked by priority with regard to their impact on the organization. This ranking provides a basis for preparing alternative preventive actions to cost and evaluate for full implementation.

Preventive actions are steps that are taken to either stop a pollution event from ever occurring or eliminate or minimize its potential impact. After all, if there is no pollution event, then there are no costs for remediation.

■ TABLE 4.11 Comparison of Program Requirements: EPA Risk Management Program Rule

Program 1	Program 2	Program 3
Hazard Assessment		
■ Worst-case analysis	■ Worst-case analysis	■ Worst-case analysis
■ —	■ Alternative releases	■ Alternative releases
■ 5-year accident history	■ 5-year accident history	■ 5-year accident history
Management Program		
—	■ Document management system	■ Document management system
	Prevention Program	
■ Certify that no additional steps needed	■ Safety information	■ Process safety information
	■ Hazard review	■ Process hazard analysis
	■ Operating procedures	■ Operating procedures
	■ Training	■ Training
	■ Maintenance	■ Mechanical integrity
	■ Incident investigation	■ incident investigation
	■ Compliance audit	■ Compliance audit
		■ Management of change
		■ Pre-startup review
		■ Contractors
		■ Employe participation
		■ Hot work permits
	Emergency Response Program	
■ Coordinate with local responders	■ Develop plan and program	■ Develop plan and program
	Risk Management Plan Contracts	
■ Executive summary	■ Executive summary	■ Executive summary
■ Registration	■ Registration	■ Registration
■ Worst-case data	■ Worst-case data	■ Worst-case data
■ 5-year accident history	■ 5-year accident history	■ 5-year accident history
	■ Prevention program data	■ Prevention program data
	■ Emergency response data	■ Emergency response data

■ TABLE 4.12 Hazard Analysis Techniques for Process Evaluation

Technique	Methodology	Application to Pollution Prevention Assessment
What If? Analysis	Examines possible hazard scenarios and consequences by posing a series of questions. There is no prescribed set of "What If. . ." questions.	■ Root out possible mechanical sources of pollution with regard to a piece of production equipment by asking questions that relate to catastrophic failure, lack of maintenance, or incorrect maintenance. ■ Root out human causes of pollution by asking questions that relate to skill level of equipment operator, workplace environment (e.g., lighting, ambient temperature, etc.)
HAZOP (Hazard Operability) Study	A multidisciplinary team uses brainstorming to evaluate selected impacts on the mechanical components of a process operation. Guide words (e.g., no, more, less, reverse) are systematically applied to process criteria, such as flow or pressure, in order to identify consequences of deviations from design intent.	
Failure Modes, Effects, and Criticality Analysis	A system or piece of equipment is evaluated to identify how it could fail (failure mode). Failure modes are ranked by criticality. This results in a decision on the modes most likely to cause a serious accident.	
Fault Tree Analysis	This method uses deductive reasoning to reconstruct the causes of an accident. Starting at the accident, the investigation moves backward in time to graphically reconstruct the combination of equipment failures and operator errors.	Using a case study or an accident scenario in a brainstorming session with a multidisciplinary team, new potential causes of a pollution event can be identified.

Source: U.S. Environmental Protection Agency, *Managing Chemicals Safely*, EPA-510-K-92-001, April 1992.

SUMMARY

This chapter detailed the regulatory framework for hazardous waste management within the United States. It discussed the process of environmental auditing and the components of an effective auditing program. Finally, it introduced the concept of developing a comprehensive pollution prevention assessment driven by the needs of your organization. Developing and ranking such an assessment for your organizational activities, products, or services provides a foundation for defining and costing specific pollution prevention tasks or possible solutions.

Exercise

Objective: Gain experience in planning and implementing an audit through an objective evaluation of activity, product, or service waste streams.

1. Select an activity, product, or service subject to a hazardous waste management compliance audit.
2. Develop a scope for the audit, to include:
   - **a.** Time frame for the audit (e.g., waste streams generated from January 1, 1995, through June 30, 1996)
   - **b.** Regulatory structure (e.g., federal or state-specific).
3. Develop an audit plan. A typical plan looks at a sampling of paperwork and verifies such items as:
   - **a.** Retention of waste analysis test records (three years)
   - **b.** Retention of signed copies of manifests, land disposal restriction notices (three years)
   - **c.** Designated TSDF is authorized to receive hazardous waste, and applicable section of manifest completed, signed, dated
   - **d.** Manifest copies returned include signature of all transporters
   - **e.** Waste stream properly classified per EPA and state
   - **f.** Waste stream lists contain correct DOT shipping description
   - **g.** Discrepancies between waste streams shipped off-site and waste stream received at TSDF are documented and explained
   - **h.** Biennial report sent by March 1 of even-numbered years
   - **i.** Biennial report copy kept for three years

4. Conduct an audit of hazardous waste management activities per your audit plan.
   a. Assemble hazardous waste manifests, biennial report submittals, and other paper evidence of both hazardous and nonhazardous waste disposal activities.
   b. Interview process personnel to determine if all potential sources of waste activity have been accounted for. Questions can elicit information on such typical problems areas as:

      - Spilled materials
      - Rejected product
      - Sample management
      - Damaged raw material freight
      - Customer returns
      - Maintenance residues

CHAPTER 5

MEASURING P^2 PERFORMANCE AND CHANGE

Performance measurement and accountability are basic management tools. This chapter focuses on the application of measurement in order to enhance:

- Performance assessment
- Control of costs
- Assignment of unbiased accountability

Effective implementation of cost-controlled P^2 requires the understanding and use, as appropriate, of quantitative methods and techniques that promote the collection of objective evidence, the bias-free indicator of performance. Included are discussions on:

- Statistical process control
- Environmental cost accounting

APPLIED STATISTICAL PROCESS CONTROL

What is statistical process control and how can it help, rather than hinder, work performance?

Statistical process control (SPC) relates to the methods through which statistics (quantitative data) are collected and analyzed with the objective of controlling process parameters. In other words, it is a way to collect numeric data in an effort to improve work process performance.

An everyday application of simple SPC appears on many monthly utility bills. Data on fuel or water consumption are presented (and sometimes graphed), providing a comparison to the usage from an earlier time period (usually the previous year). Based on the data given, the consumer then makes informed decisions on future utility usage.

While SPC is most commonly applied to manufacturing processes, it is also applicable to service functions. To that end, process monitoring data and statistical techniques can be used to identify trends and point out variables or deviations in such things as delivery time, volumes shipped, or numbers of flaws.

Eldec Corporation, an electronics manufacturer in Martha Lake and Bothell, Washington, while working to get a better handle on environmental costs, uncovered an ad hoc rule on the schedule and procedures for the service function of preventive maintenance and cleaning of equipment. The result was a high generation (and cost) of hazardous waste. The company applied statistical process control to analyze and redesign the preventive maintenance system. It adopted a modified schedule and cleaning techniques that reduced the use of hazardous solvents. Costs were cut concurrent with maintaining product quality and reliability.[1]

When to Use SPC

Like any tool, SPC it is used most cost-effectively when it is applied judiciously. There is no need to collect data just to have data, or because the organization has software that enables the data to be collected.

For example, one company got a new phone system with elaborate job tracking capabilities. After the system was installed, managers and supervisors began to receive unsolicited monthly reports detailing all types of telephone usage from every phone line within the managers' geographic areas of responsibility. Several managers, after reviewing these a few times, saw that they provided no useful data and implied a level of distrust and control over

1. Ditz, Daryl, Janet Ranganathan, and R. Darryl Banks, "Accounting for Environmental Costs," Pollution Prevention Northwest, Pacific Northwest Pollution Prevention Research Center, July/August 1995, p. 7.

line employees that was unnecessary and hurt morale. After several months of complaint, the managers eliminated the use of that statistical control methodology.

Primary reasons for using SPC include:

- To ensure quality of product (or, in the case of pollution prevention, the output from a process stream)
- To monitor that regulatory parameters for emissions are not exceeded
- To identify operating areas where the process can be improved
- To satisfy customer demands

Table 5.1 illustrates some typical objectives and process conditions that can be supported using SPC methods.

What Is a Control Chart?

A chart is a graphic or visual display of numeric data. A control chart is a graph displayed so that changes in process parameters are easily read and

■ TABLE 5.1 Applications of SPC for Pollution Prevention

Objective	Environmental Element	Typical SPC Methodology
Monitoring and control of a suitable work environment	■ Temperature ■ Pressure	■ Local: thermometer, pressure gauge ■ Remote: resistance temperature device (RTD), thermocouple pressure transducer
Compliance with regulatory parameters	■ Wastewater discharge	■ Grab sample and lab analysis ■ Continuous sample and lab analysis ■ On-line analyzer
Monitoring and control of process parameters	■ Turbidity ■ Identification of uncontrolled/ undesirable reactions	■ Manual action: cause identification, correction ■ Auto action: automatic shutdown
Maintenance affecting process	■ Replacement of O-rings and gaskets ■ Cleanup of production line from production residue	■ Replacement correlated with hours of service ■ Scheduled maintenance

understood. Figure 5.1 exemplifies a typical use of a control chart. The purpose of this chart is to identify when the process stream is "out of control" with regard to pH. The upper limits and lower limits selected were based on U.S. EPA requirements for designating a waste stream as hazardous waste.

In this example, the plot of Sample g indicated that the process stream was heading for an out-of-control situation. The chart identified the trend and the organization was able to take corrective action. Sample p, however, went out of control. Because the out-of-control condition was identified in a timely manner, though, actions were taken to get the process stream back into control.

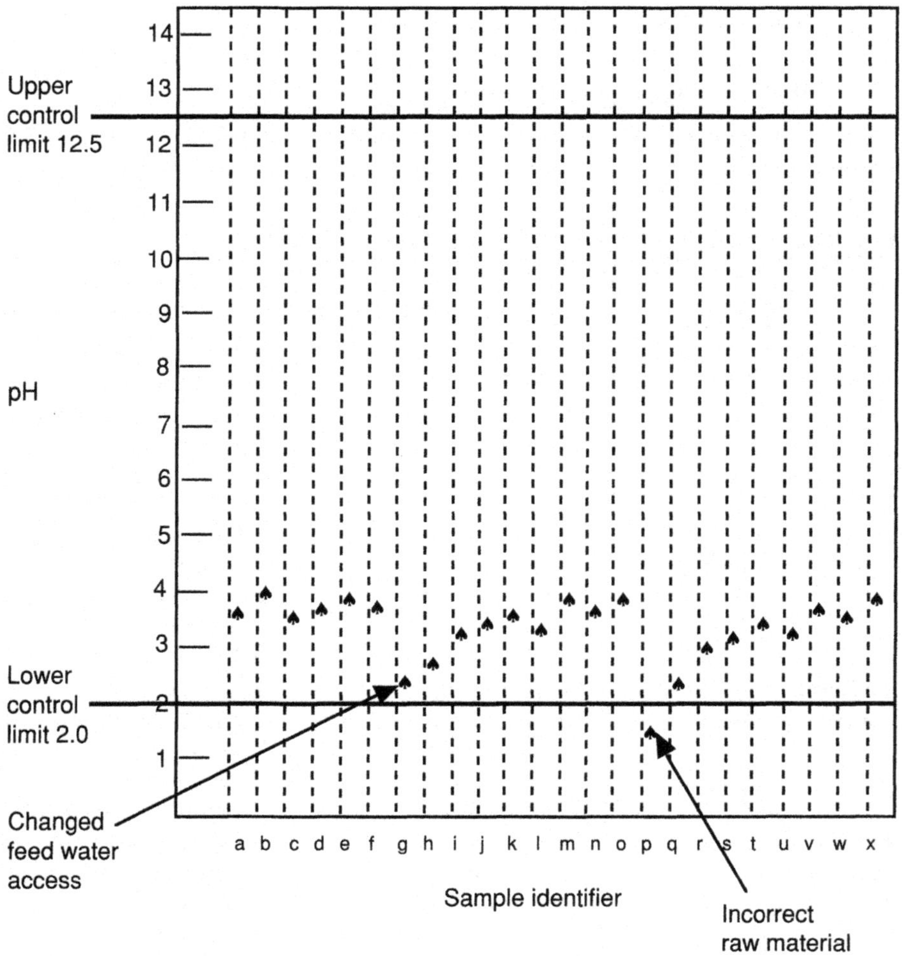

■ **FIGURE 5.1 Sample Control Chart pH of Aqueous Process Stream**

There are various methods for charting (graphing) statistical information (Figures 5.2 through 5.5). Plotting can be as simple as a few marks on graph paper; more complex displays can be produced using computer software. It is not uncommon for many Macintosh or PC-based spreadsheet packages to bundle in charting capabilities. Additionally, some automated process monitoring includes charting capabilities.

Cost-effective use of control charts takes into account budget (time and money) and the business need for the chart. Don't waste a lot of time on fancy charting formats just because the software capability exists. Also, be cautious, because some charts may not be an accurate or useful rendition of the data (Figure 5.6).

If you know what you want to track and why, collect and organize statistical data in a manner that facilitates its conversion to a chart. For example, if data are collected manually, design the layout of the collection log so that it also serves as an easy-to-use data entry document. If in doubt, speak to the people who may be performing the data entry and solicit their ideas on collection format.

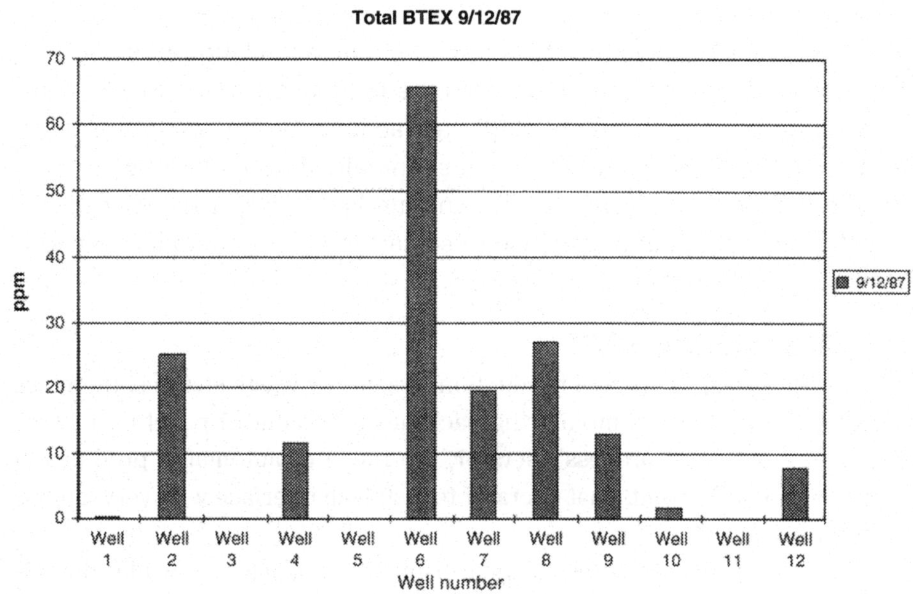

■ **FIGURE 5.2 Effective Charting: Several Data Points at One Point in Time**

Type: Bar Chart

Comment: Easy to develop using graph paper. Very easy for most people to understand.

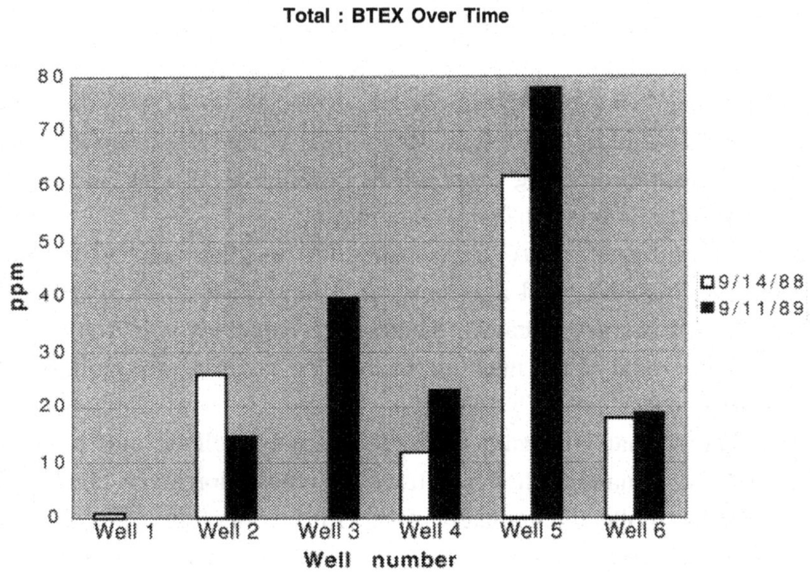

■ FIGURE 5.3 Effective Charting: Several Data Points at Several Points in Time

Type: Bar Chart

Comment: This shows trends at any one data acquisition point (Wells). At the same time, it enables the reader to make a quick evaluation of overall activity at all data acquisition points. In this example, Well 3 is of interest because initially it generated no readings (data). One year later it generated a high reading relative to most other data points. Because this chart integrates data it is more difficult to understand than Figure 5.2.

Automating SPC

With regard to process production streams or waste streams, there are many standard process monitoring packages with control panel user interfaces. For the small process operator, the range of automated products is even greater and permits that operator to install an interface with very limited data input.

In evaluating the various standard software packages, consider the following as they relate to your organizational pollution prevention objectives and resources:

- Is the package Macintosh (Apple), Windows or DOS-based for use on a PC, or is it UNIX, which requires a special workstation? If UNIX, does this give you better and required data security?

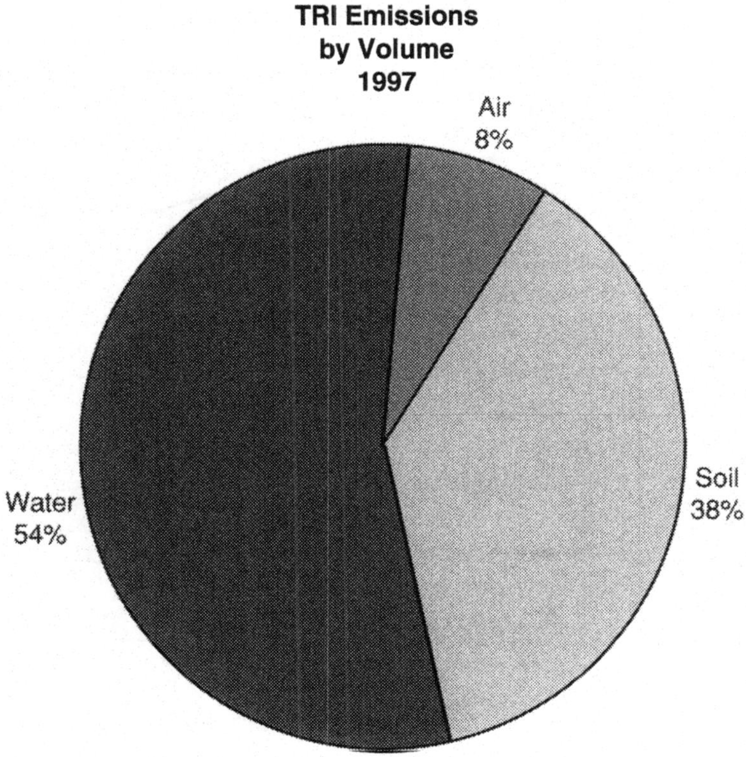

■ FIGURE 5.4 Effective Charting: Percentages
Type: Pie Chart
Application: Compares different data sets (environmental receptors) at one point in time.
Comment: These data can be used to set priorities for future P^2 activities by environmental receptor. Pie charts are easy to understand. They cannot be used to integrate data points over time (e.g. emissions over time).

- Compare the recommended memory to the required memory. The required memory may work, but may be inadequate for a real-time operation.
- Does the data output permit you to move it to the spreadsheet of another software package to facilitate assembly of trend data over time?
- How long are data stored before they are wiped out or replaced? This is significant when troubleshooting to find the root cause of a problem.

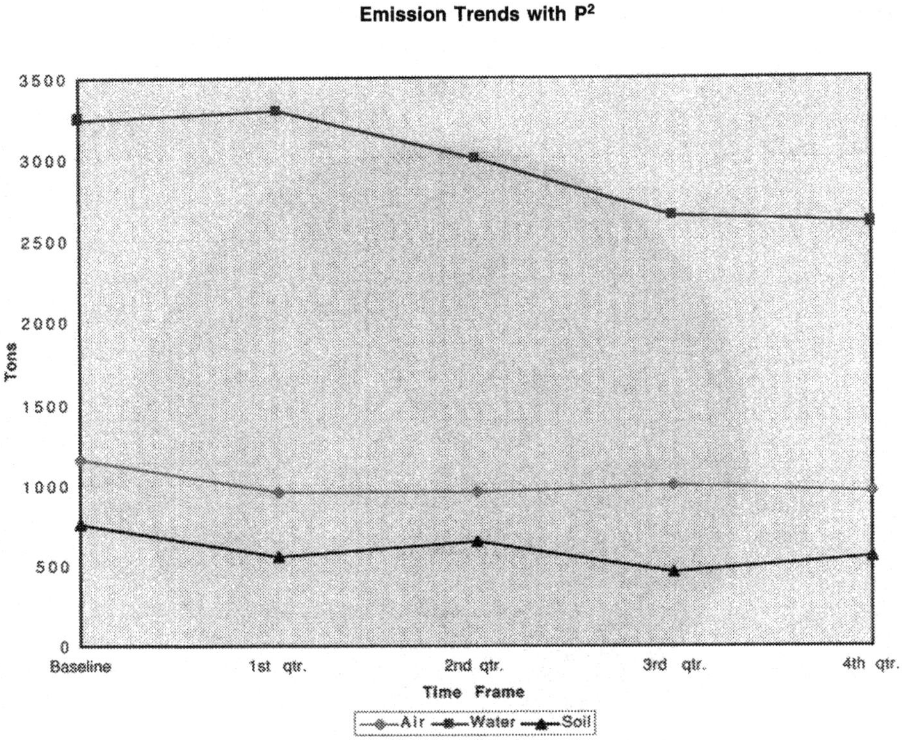

■ **FIGURE 5.5 Effective Charting: Trend Data**

Type: Line

Application: Permits comparison of several types of data over time.

Comment: In this layout, the data are laid out to make comparison between media volumes easy. It does not illustrate total volume for all media (cumulative data). To show that, the Y axis would have to be larger.

Don't become trapped in addictive, compulsive behavior that demands the latest, hottest, fastest version of software. Separate out your software/hardware "need/must have" from your "want/desire/it would be nice to have" lists.

Software is most cost-effective when used to make it easier to collect data you would otherwise collect and array by hand. Make time to learn how to use its basic capabilities before trying to use more sophisticated characteristics.

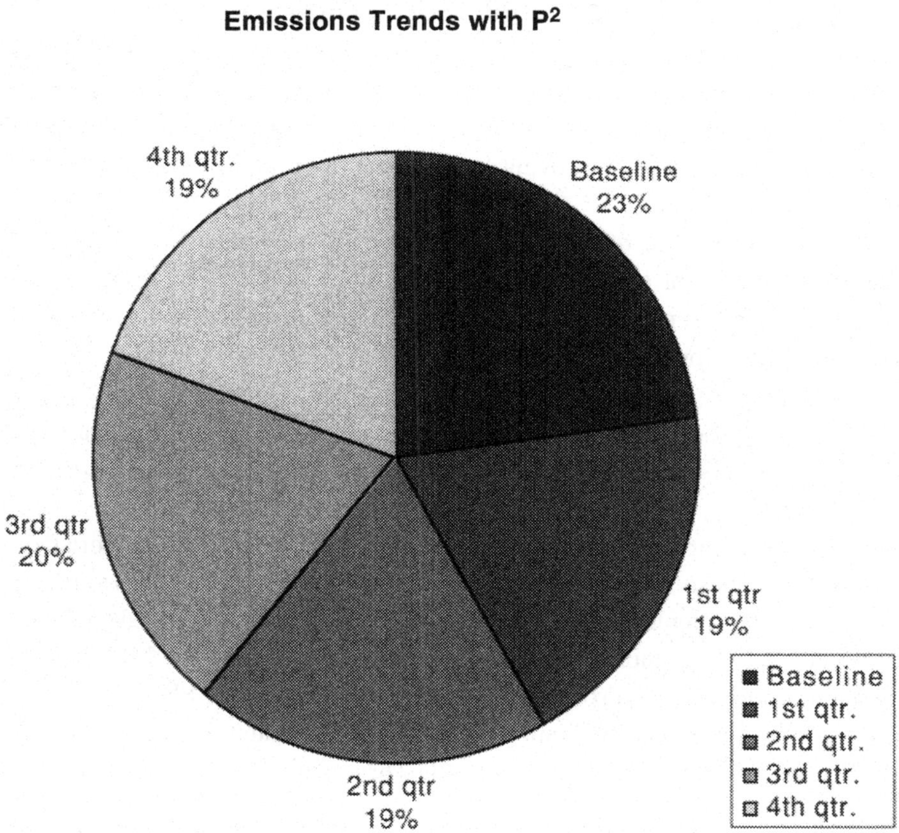

■ **FIGURE 5.6 Chart Abuse**

Type: Pie Chart

Application: An example of how information can be displayed improperly with the potential to lead to false conclusions and errors.

Comment: Displaying the same data as appears in Figure 5.5, this chart looks very attractive and is easy to read, but impossible to interpret. It doesn't say anything of substance.

Finally, designate and communicate responsibilities for maintaining the integrity of data. Without assignment of data "owners" who are responsible for ensuring that data are collected accurately and maintained according to a prescribed schedule, an expenditure in software and hardware becomes a waste of money and could result in a legal liability. Monitor the work of these data owners to ensure that integrity is kept and information is maintained in a timely fashion.

Application to QS-9000 Requirements

One of the major sources of nonpoint pollution in the United States is the automobile. In response to this, the Federal government and many states have passed regulations affecting that industry with a distinct pollution prevention intent, such as improving fuel consumption and reducing air emissions.

A review of the requirements prepared by automakers in their joint quality management specification, QS-9000, illustrates several areas where SPC is required or encouraged to support the pollution prevention needs of the Big Three suppliers' network.

> Government Safety and Environmental Regulations: A supplier shall have a process to ensure compliance with all applicable government safety and environmental regulations, including those concerning handling, recycling, eliminating or disposing of hazardous materials.[2]

QS-9000 further presents the concept of "special characteristics" (Table 5.2). These are process characteristics that require process control and additional attention because "excessive variation in them might affect a product's safety, compliance with government regulations . . ."[3] This additional attention can be provided through the application of SPC.

SPC is a mandatory component of preventive maintenance for the QS-9000 supplier. These suppliers to the Big Three must:

- Identify the equipment or machine used to fulfill an automaker's purchasing requirement
- Develop a preventive maintenance schedule for that equipment or machine
- Ensure that the schedule correlates SPC to actual preventive maintenance activities

In order to accomplish this, the SPC plans must be designed to collect information that is useful to establishing and monitoring equipment/machine preventive maintenance. For example, parameters on the quality of lubricating oils can be monitored to make the decision about when lubricating oil should be changed or replaced.

SPC is encouraged in process monitoring. However, the implementation of a new process requires a preliminary process capability study for desig-

2. Chrysler Corporation, Ford Motor Company, General Motors Corporation, *Quality System Requirements QS-9000*, Second Edition, February 1995, p. 27.

3. ibid.

■ TABLE 5.2 QS-9000 Implementation Using Special Characteristics to Address Safety or Legal Considerations

	Key Characteristic (With Safety or Legal Consideration)
General Motors	Nomenclature: Safety/compliance Is a product characteristic for which reasonably anticipated variation could significantly affect the product's safety or its compliance with government regulations (such as: flammability, occupant protection, steering control, braking, etc.), emissions, noise, radio frequency interference, etc.
Ford	Nomenclature: Critical Characteristic Are those product requirements (dimensions, specifications, tests) or process parameters which can affect compliance with government regulations or safe vehicle/product function and which require specific producer, assembly, shipping or monitoring actions and inclusion on the control plan.
Chrysler	Nomenclature: Shield Safety characteristics are defined as engineering-designated specifications or product requirements applicable to component material, assembly operation(s) which require special manufacturing control to assure compliance with governmental vehicle safety, emissions, noise, or theft prevention requirements.

Source: Industry requirements are taken from Chrysler Corporation, Ford Motor Company, General Motors Corporation, *Quality System Requirements QS-9000*, Second Edition, February 1995, p. 82.

nated special characteristics. If they are not designated, ". . . a Ppk value $\geq$ 1.67 should be achieved for preliminary results (less than 30 production days) and for chronically unstable processes."[4]

Application to Aerospace Industry Requirements

In January 1997, a new aerospace document on quality was made effective by the Boeing Corporation, the largest commercial airplane producer in the world and one of the principal U.S. exporters. Referred to as D1-9000, Rev. A, it is a detailed description of an Advanced Quality System, as it relates to subcontractors providing products to Boeing Corporation. In addition to its Basic Quality System, which uses ISO 9002 as a framework, the document includes specific Boeing Advanced Quality System Requirements on statistical process control. While not designed specifically for environmental issues, it includes parameters for waste and defect reduction.

4. ibid., p. 29.

Specifically, these suppliers to Boeing are required to have established internal measures of defects or waste. Measurements are then tracked to illustrate trends from some designated baseline.

In the simplest terms, if a manufacturing process produces fewer defective products and waste, then some additional level of pollution prevention has been achieved.

The goal over time is for application of Basic Quality and Advanced Quality Systems, including SPC, in order to achieve a 90 percent reduction in defects and waste from the original baseline.[5]

SPC in Life-Cycle Analysis

As the methods for conducting life-cycle analysis (chapter 2) become more defined and mature, greater reliance is placed on the application of statistical techniques in order to remove bias from the analysis process. During the impact assessment phase of an analysis, impacts are classified, characterized, and given a valuation. "The aim of the valuation is to determine the significance of the results of the impact assessment and may involve interpretation, weighting and ranking of the data produced from the inventory analysis or characterization step."[6]

Can SPC Help Cut Costs?

A ubiquitous concern in any organization is, "Can the effort save me money?" Expect a certain amount of additional up-front effort and expense for planning, collecting, and analyzing statistical data. But success after success show that the application of simple statistical techniques and controls does save money, when realistically implemented.

To make SPC successful:

- Decide what the problem or problem area is that needs (or may need) improvement.
- Determine what measurement(s) can be simply made to evaluate performance (e.g., pounds reduced, fewer breakdowns, and so on).
- Collect that measurement as baseline (control) performance data for a defined period of time (e.g., one week).
- Analyze the data and decide what type of improvement (e.g., corrective action) it is appropriate to introduce.

5. The Boeing Company, *Advanced Quality System® D1-9000®*, July 1996, p. 2-Q-20.
6. Secretariat AFNOR, *Life Cycle Assessment* ISO/TC 207/SC 5 N 44, draft 3/15/95, p. 13.

- Introduce the improvement.
- Measure to quantify how great the improvement is.
- Put a price tag or dollar value on the improvement (e.g., reduced scrap by 2,500 pounds, or 15 percent, at $3/pound, for a savings of $7,500).
- Broadcast the results within the organization.

You can find that the simple act of targeting a problem or issue, can be a very inexpensive way to bring about improvement, because it ensures that people know more clearly what is important to the organization.

ENVIRONMENTAL COST ACCOUNTING SIMPLIFIED

Accounting is as much art as science. Although it is hoped that, eventually, all the numbers balance and reconcile, the intellectual process that goes into developing financial statements is influenced by the vagaries and interpretations of tax codes, generally accepted accounting principles (GAAP), the policy desires of the organization, and possibly even accounts payable clerks' subjective designations of account codes to invoices being processed.

Accounting Essentials

The act of accounting has long been understood as a necessary business function. Keeping accurate track of financial assets and liabilities are basic to business management. Accounting permits the organization to know where it stands with regard to profit and loss throughout its business cycles.

Traditional accounting systems have been formatted, in part, to address these general internal management needs, as well as external requirements for:

- Tax reporting
- Assembly of financial statements for lenders
- Reporting to shareholders

However, the emphasis on meeting external needs has often resulted in accounting systems with limited capability to address, or even acknowledge, changing internal business management requirements. Thus, the business professional, concentrating on resolving environmental problems, experiences financial data tracking limitations due to the artificial constraints of an accounting system.

Managing the Cost Impact of Delayed Reports

A business organization retained an environmental engineering firm to conduct and report on performance audits for 35 to 50 hazardous waste management facilities (TSDFs) located throughout the United States. Because the information was to be used to determine to which facilities the business organization would send its hazardous waste streams, it was viewed as essential that the reports be received in a timely fashion so outstanding issues could be identified and resolved quickly.

The project manager for the business organization had specified a requirement that a draft report be received within one week of the on-site audit. The final report would be due two weeks later. After a few false starts, the project manager introduced the simple, inexpensive management technique of maintaining a log showing when reports were due, compared to when reports were received. In providing objective evidence to the engineering firm, a reduction in final costs was negotiated due to consistent lack of timeliness on the part of the engineering firm.

The impetus for introducing environmental cost accounting methods comes from several realizations:

- There are benefits that can accrue to the bottom line once the true costs associated with environmental concerns are fully understood.
- There are opportunities for offsetting known environmental costs by generating income (value) from previously unidentified environmental revenue streams.
- Once clearly identified and understood, some environmental costs can be reduced or eliminated.

Historically, environmental costs have been buried in the accounting black hole identified on charts of accounts as "general administration." Thus, limited attention and priority have been placed on modifying business practices that influence and improve environmental concerns.

Environmental costs have not always been understood as being controllable. The response to them has been to complain about the government and get out the checkbook. When such things as remediation costs have been funded by insurance companies, there has been even less incentive on the

Tracking P^2 Costs

During the 1980s, the growing popularity of solar energy and lower prices for natural gas resulted in many homes and businesses converting heating systems from home heating oil as a fuel. Since many of these customers were on an automatic "keep filled" program, unless they remembered to cancel oil delivery and/or to completely remove the fill pipe from their structures, the oil truck made a delivery. Occasionally this resulted in basements filled with oil.

At the same time, one heating oil distributor began a concerted effort to increase its capability to efficiently respond to these and other types of petroleum releases occurring during oil deliveries. Instead of relying on old rags and luck, delivery trucks were equipped for response with oliophilic, hydrophobic spill pads, pillows, and socks.

With a fleet of more than 30 trucks, the initial investment for this consumable equipment was high. The cost of these cleanup products was far greater than that for the traditional rags. In order to remove resistance to the spill response products, the first purchase was conducted at a corporate level. When the invoice arrived, the loss control manager coded the invoice approval as a safety expense. The accounting department questioned the expense. After all, what was the purpose of buying pillows and socks, as the invoice specified? When the legitimacy of the expense was explained, the concept of tracking the cost as a safety expense was denied, and the dollars were buried in financial statements as a general operating cost.

part of an organization to manage the business side of projects. Cost-effective should not mean cheap or low bid. It should mean appropriate costs for the goods and services being provided.

How Do You Do It?

Environmental cost accounting (Table 5.3) requires cross-functional input. Accountants, environmental managers and production/operations managers must coordinate their ability to collect data with their requirement to supply valid data that add value and facilitate organizational decision making.

Understand the type of accounting software package used by the organization, including the definitions applied to the chart of accounts. Does this

■ **TABLE 5.3 Using Net Present Value for Economic Analysis**

Objective: Calculate the net present value of a current process and a pollution prevention alternative.

Instructions:

For each of the five years:

Row 1	Enter expected depreciable investment. This is an expense, or negative number. If your total equipment expense greatly exceeds $17,500, then select a depreciation schedule most appropriate for your business (usually the fastest possible) and enter the annual depreciation in Row 7. Consult your tax accountant for the current maximum allowable equipment expense deduction, and for selection of an appropriate depreciation schedule.
Row 2	Enter the expected revenue. This is income, or a positive number. If revenues will not be affected by either alternative, leave this row blank.
Row 3	Write in the expected costs. Note that this category includes equipment lease payments. These are expenses, or negative numbers.
Row 4	Enter the expected compliance costs. Compliance costs are those that are necessary to comply with environmental regulations. If a facility can be penalized for not spending the money, it is a compliance cost.
Row 5	Fill in expected oversight costs. These are all extra costs incurred because a facility uses hazardous substances or generates wastes. Examples are regulatory research, permit applications, and inspections.
Row 6	Subtract operating expenses (rows 3, 4, 5) from income (row 2) to calculate annual operating income.
Row 7	Enter annual depreciation of this income. Because the IRS changes this deduction, check with your accountant for the current amount allowed under Section 179. If the total equipment expense for a facility is less that the allowed yearly deduction, add the investment to year operating expenses and enter "0" for depreciation in row 7.
Row 8	Subtract depreciation (row 7) from operating income (row 6) to get taxable income.
Row 9	Calculate and enter the amount paid in business taxes on taxable income.
Row 10	Enter the depreciation amount again (row 7). Depreciation is not a real cash expense, but is used for tax accounting purposes.
Row 11	Add depreciation (row 10), subtract depreciable investments (row 1) and taxes (row 9) from the taxable income (row 8) to get actual annual cash flow from the process.
Row 12	Choose a discount rate to decide the value of future cash flows today. The rate usually reflects the expected inflation rate plus the opportunity cost of capital (that is, the rate potentially earned by investing the money elsewhere). Using the present value factor (PVF) information included in Table 1, enter the discount factors for the chosen discount rate. For example, you may want to discount future cash flows at 10 percent. Enter the factors in years one through five for the 10% row in the PVF table into row 12 of the cash flow table.
Row 13	Multiply the annual cash flow by the discount factor to find the present value today of the cash flow in each year.

(Continued)

■ **TABLE 5.3** (*Continued*)

Row 14 Sum all of the annual discounted cash flows to learn the net present value of the process. This is how much the process is worth to you over the next five years, in today's dollars.

		Year 0	Year 1	Year 2	Year 3	Year 4	Year 5
1	Subtract investments						
2	Add revenues						
3	Subtract usual costs						
4	Subtract compliance costs						
5	Subtract oversight costs						
6	Total operating income						
7	Subtract depreciation						
8	Total taxable income						
9	Subtract taxes						
10	Add back depreciation						
11	Total cash flow						
12	Present value factor (PVF)						
13	Multiply cash flow X PVF						
14	Sum row 13	= Net Present Value					

Source: Washington Department of Ecology, *Cost Analysis for Pollution Prevention*, Publication 95-400, revised August 1995.

system already contain suitable data? Are the data collected at a level of use for desired environmental cost accounting applications (i.e., by designated activity, product, or service), or are they only aggregated at a corporate level? Does the system have the flexibility to aggregate the data at a more useful level, such as a specific activity, product, or service?

Keep in mind that many of the data limitations you appear to face may cease to exist after a thorough discussion and planning session with the appropriate accounting authority in the organization.

For example, one company set up a new profit center to track its sales performance in a newly reorganized hazardous waste transportation and storage business it provided. Although the accounting system was old and somewhat inflexible, a cross-functional team studied the issue. Within three months of starting the renovated business venture, a functional account tracking system was fully implemented. Although it did not have every bell and whistle desired, and it was awkward to use, it did offer some simple baseline information on the types of business transactions performed (e.g., for incineration, recycling, and the like) and the costs associated with those transactions. Without this effort, it would have been several years before a new, more flexible accounting software system was adopted.

Implementation Issues

Table 5.4 lists some of the types of environmental costs that can be considered for performance tracking purposes. If the numbers are captured in an accounting system, they serve as important devices for measuring performance. But the reality is that few organizations can collect or would benefit from collecting all the data.

Successful implementation of environmental cost accounting addresses what is feasible and realistic for the organization. It begins by understanding what the purpose is for collecting certain data in an accounting format. Then it identifies the data, collects it, and uses it to make better, more informed management decisions.

The primary purpose is measurement in order to improve business performance. Look at the list of regulatory costs in Table 5.1. Of these, taxes and fees are among the most common found in a traditional accounting setup.

Because of the societal pressures of deficit reduction, governments at all levels have become more creative and prolific about designating and collecting taxes and user fees. Within your organization, how easily are the environmentally oriented taxes and fees recognized so the true environmental costs of a business activity are better understood?

SPECIAL ENVIRONMENTAL ISSUES DURING MERGERS AND ACQUISITIONS

Surprises are not cost-effective. Contingency costs associated with the management of environmental surprises are the bane of corporate officers and shareholders throughout the United States.

How can you keep surprises to a minimum? Performing "due diligence" (ASTM Practice E 1527, *Phase I Environmental Site Assessment*) has become the standard for assessing potential environmental liabilities and pollution events associated with property transfers, including those associated with business mergers and acquisitions.

In practice, the data supplied through these due diligence reports have been used by both lenders and potential buyers. Lenders use them as a way to limit Superfund liability and as basis for loan decisions. Potential buyers use them for negotiating purposes to bring down purchase prices based on environmental risks.

While due diligence data are useful, from a pollution prevention planning standpoint they require augmentation with other detail. By collecting

■ TABLE 5.4 Types of Environmental Costs

Potentially Hidden Costs

Regulatory	Up-front	Voluntary (Beyond Compliance)
■ Notification	■ Site studies	■ Community relations/outreach
■ Reporting	■ Site preparation	■ Monitoring/testing
■ Monitoring/testing	■ Permitting	■ Training
■ Studies/modeling	■ R&D	■ Audits
■ Remediation	■ Engineering and procurement	■ Qualifying suppliers
■ Recordkeeping	■ Installation	■ Reports (e.g., annual environmental reports)
■ Plans		■ Insurance
■ Training	Conventional costs	■ Planning
■ Inspections		■ Feasibility studies
■ Manifesting	Capital equipment	■ Remediation
■ Labeling	Materials	■ Recycling
■ Preparedness	Labor	■ Environmental studies
■ Protective equipment	Supplies	■ R&D
■ Medical surveillance	Utilities	■ Habitat and wetland protection
■ Taxes/fees	Structures	■ Landscaping
■ Financial assurance	Salvage value	■ Other environmental projects
■ Pollution control		■ Financial support to environmental groups and/or researchers

(Continued)

■ **TABLE 5.4 (*Continued*)**

Regulatory	Back-End	Voluntary (Beyond Compliance)
Potentially Hidden Costs		
■ Spill response		
■ Stormwater management	■ Closure/decommissioning	
■ Waste management	■ Disposal of inventory	
■ Taxes/fees	■ Post-closure care	
■ Environmental insurance	■ Site survey	
Contingent Costs		
■ Future compliance costs	■ Remediation	■ Legal expenses
■ Penalties/fines	■ Property damage	■ Natural resource damages
■ Responses to future releases	■ Personal Injury damage	■ Economic loss damages
Image and Relationship Costs		
■ Corporate image	■ Relationship with professional staff	■ Relationship with lenders
■ Relationship with customers	■ Relationship with workers	■ Relationship with host communities
■ Relationships with Investors	■ Relationships with suppliers	■ Relationships with regulators
■ Relationships with insurers		

Source: U.S. Environmental Protection Agency, *An Introduction to Environmental Accounting As A Business Management Tool: Key Concepts and Terms*, EPA 742-R-95-001, June 1995.

The Model Toxics Control Act in Washington State

This was established to:

- Clean up state hazardous waste sites
- Plan and manage solid and hazardous waste
- Manage water and environmental health protection

Administered by the state Department of Ecology, it is self-funded through a hazardous substance tax on established businesses. The tax is on three categories of hazardous substances:

- Petroleum products
- Chemicals listed by U.S. EPA in the federal Comprehensive Environmental Response, Compensation, and Liability Act (CERCLA)
- Pesticides registered under the Federal Insecticide, Fungicide, and Rodenticide Act (FIFRA)

The tax is paid by the first person who possesses the hazardous substance in Washington State. These include:

- Businesses that purchase them from out-of-state sources
- Persons who take consignment for sale either in their own name or the name of an out-of-state consignor
- Persons who manufacture hazardous substances within the State
- Out-of-state sellers who keep a stock of hazardous substances within Washington

Where purchases are controllable and substitute materials can replace hazardous substances, the direct and indirect tax bills are reducible. The direct tax bill is the fee. The indirect tax bill is the cost of establishing and maintaining an method of collecting data and calculating the tax fee.

and evaluating specific types of information, a potential purchaser can make a more informed decision about the true environmental costs associated with a new purchase. The P^2 manager can then develop better plans for reducing pollution risks or transferring the costs of those risks to the original business or property owner.

The objective of this approach is to construct a more comprehensive history, with costs (estimated or actual). Data that the potential seller can make available for review can include:

- TRI (Toxic Release Inventory) report submittals
- Tier 2 inventory submittals
- Hazardous waste manifests
- OSHA 200 log (employee injury and illness)
- Contingency plans

Table 3.3 provides a reference list of additional suitable reports, all of which are public record. So if the seller won't provide them, get them directly through the appropriate regulatory agency.

Interviews with facility neighbors and the local fire department can also be helpful. If pollution events are identified, then reviewing old issues of local newspapers may provide other details about the facility operation.

Following an analysis of these data, additional areas of pollution concern can be identified. Costs for compliance and pollution prevention efforts can then be forecast with greater accuracy.

Additionally, these data form a baseline for measuring performance of future P^2 improvements.

SUMMARY

There are many factors that drive the total costs associated with pollution prevention. This chapter illustrated the application of process performance measurements and simple environmental cost accounting techniques to:

- Assess environmental risks
- Estimate costs of managing environmental risks
- Identify trends in environmental performance based on both environmental process characteristics and costs

Techniques such as statistical process control and environmental cost accounting, and the software and hardware that support them, are tools. They are a means to an end, not an end in themselves. As such, they are cost-effective when applied thoughtfully. They save the most money when a specific opportunity for environmental improvement is either considered highly desirable by the organization or demanded by a regulatory enforcement authority.

Exercise

Objective: Establish targets for improvement as they relate to environmental aspects of a process.

1. Select an activity, product, or service.
2. Using data sources such as waste reports, original equipment manufacturer (OEM) recommendations, or utility use records, identify at least two pollution prevention performance parameters (characteristics) for the activity, product, or service.
3. Compile baseline costs (direct and indirect) associated with the activity, product, or service.
4. Establish quantifiable targets for performance improvement.

CHAPTER 6

IMPLEMENTING P^2 PURCHASING DECISIONS

"Measure twice, cut once." Whether speaking of construction or ice cream production, preparation and planning play a vital role in successful cost control. The purchasing process is a significant component of that effort. Furthermore, as organizations choose to outsource more activities beyond their basic business competencies, the need for control of purchasing expands.

Ad hoc, informal purchasing processes are ripe for errors and fraud by suppliers and employees. These processes can also create regulatory problems such as:

- Failure to collect and remit sales taxes in an accurate and timely manner
- Inadequate or incompetent service delivery, resulting in the creation or exacerbation of a pollution event

This chapter considers implementation of purchasing decisions associated with pollution prevention. It addresses such features as:

- Traditional purchasing controls
- Deciding when to outsource

- Effective use of small suppliers
- Contract administration
- Managing contractors
- Invoice auditing

PURCHASING CONTROL

From one standpoint, purchasing is similar to budgeting. The P² professional has little influence over the administrative process of purchasing. The organization has a purchasing system, and all activities are moved through that system, irrespective of who has made the decision to make the purchase.

In a well managed organization, the typical purchasing process is filled with control features:

- Purchasing specification sheets
- Central purchasing bureaucracies
- Expenditure limitations (with levels of authority)
- Authorized suppliers
- Sequentially numbered purchase orders
- Automated accounts payables processing
- Competitive bidding

Specification sheets detail what is being purchased. But it is the controls of a purchasing authority that bring structure and a level of efficiency to the purchasing process.

Designated expenditure limitations enable an individual employee to make certain types of purchases when needed, without creating an unproductive delay in the work process. For example, transient technical work crews that make service calls to repair and maintain equipment may be authorized to stop at a hardware store, if necessary, to cost-effectively buy small amounts of ancillary supplies, such as screws or washers.

Expenditures for capital projects, on the other hand, typically require approval at higher levels, depending on the size of the expenditure. If an expenditure is less than $1,000, it may be possible for a local operations manager to approve it. Capital expenses greater than that, however, may require officer authorization. The limits and levels are up to the organization. However, capital lenders, such as banks, can require a very structured process.

The use of authorized suppliers is growing, as organizations work to control costs by reducing the number of suppliers they purchase from. This makes sense for many types of purchasing transactions. Costs are reduced through minimizing administrative expenses associated with purchasing. An authorized supplier may send a consolidated bill, reducing the number of checks that must be cut. Better pricing can be negotiated because of the quantities purchased. In some cases, it may be possible to automate invoice receipt and payment, further lowering internal processing costs for a purchase.

As a traditional control mechanism, competitive bidding becomes a theoretically objective process. Many organizations, however, have learned that the lowest bid is not necessary the most cost-effective, as explained later in this chapter.

Recognizing that the P^2 project must be managed within the existing organizational framework for purchasing, this chapter presents practical ideas for improving cost control in making and implementing purchasing decisions. It focuses on the purchase of goods and services required to design and implement a capital investment in P^2. The three types of suppliers these ideas are most helpful with include:

- Consulting services
- Original equipment manufacturers (OEMs)
- Construction contractors

WHEN IS OUTSOURCING A GOOD CHOICE?

From a general management perspective, outsourcing is done when:

- No internal capability exists
- The cost of developing and staffing the internal capability exceeds the anticipated need for or benefit from that internal capability
- On a strategic basis, the organization does not want to expand its core competencies
- The internal capability cannot meet the schedule required
- The external source can provide the product or service more quickly, more innovatively, and at a lower cost
- There is a strategic need for using a third-party provider because of regulatory requirements and/or security reasons

In any or all of these cases, a purchase of an external product or service becomes necessary. Thus, the organization finds itself in the position of needing to manage a purchasing process to meet a functional need.

If the organization has no internal pollution prevention capability, should it make the business decision to outsource? In order to be effective, pollution prevention must be integrated into the basic management structure of the organization. Some initial, internal designation for overall responsibility and accountability for P^2 must be made. The organization looks to the following, in cases where there appears to be no one with the management capacity to tackle P^2 as a business issue.

Type of Organization	**P^2 Responsibility/Accountability**
Publicly traded	Board of directors Officers
Privately held	Board of directors Officers
Nonprofit	Executive board Advisory board Executive management
Government agency	Elected executive with responsibility Appointed directors Citizens' advisory board

These management groups make the business decisions about the amount of resources allocated for P^2. In doing that, they provide general direction on whether the activities, from planning to implementation, are fully handled internally or whether and what type of outsourcing may occur.

For example, if the P^2 need is seen as being one-time and short-term (e.g., substitute a nonhazardous raw material for a hazardous material), then the entire project may be outsourced, except for management of the outsourced contract.

Similarly, if the organization does not see pollution prevention as a core competency it can sell to others, outsourcing of a project such as recycling of office waste may be optimal.

When the need for P^2 grows from a crisis, such as a regulatory violation, or linkage to a capital resource access requirement (e.g., insurance), outsourcing may be necessary to :

- Meet an aggressive schedule
- Show intent to make rapid progress
- Achieve an innovative solution

- Address an external requirement for a third party to perform the function because of lack of regulatory agency trust and/or security reasons

The decision to outsource is dependent on factors unique to the organization and its P^2 targets and objectives. When an activity, product or service is outsourced, however, achieving the pollution prevention opportunity is made more complex due to contract issues and a relative reduction in control.

THE CASE FOR THE SMALL SUPPLIER

An organization determines it has a need for a good or service. It then finds a supplier and makes a purchase. Despite management controls, however, purchasing in the free-market economy is not always that simple and remains a highly subjective and competitive process.

A certain organizational inertia results from having previously been supplied a good or service by a particular vendor. If the service or product was satisfactory, and the price relatively competitive, it is not always worth the administrative cost of locating and evaluating other vendors.

It may be that the person authorizing the purchase may not have the skill in a selected area of need and may rely on a supplier to give credible recommendations for materials or services (e.g., pumps, advertising), or even fill out the scope of a request for proposal. In other cases, a potential supplier

Can Outsourcing Fail?

ValuJet, Inc., a discount air carrier, was the darling of stock analysts, shareholders, and the air traveling public until a crash in the Florida Everglades killed 110 people in May 1996. The Federal Aviation Administration grounded the airline in June 1996 and only reissued its operating license in September 1996, after major changes in management and operations.

Extensive outsourcing of airplane maintenance, while cost-effective in the short term, had proved a fatal weakness. To get its operating license back, it became essential for ValuJet to revamp. Maintenance was pulled in-house. Quality assurance procedures were made uniform. An internal program for systematically analyzing possible operating hazards was established.

may be the golfing buddy of the company owner, so the purchasing organization has little choice over where to shop.

Small suppliers, from welding operations to engineering firms, offer a number of advantages over large national organizations, with the common result of more competitive pricing for work of equal value or better.

Typically, the businesses operated by these small suppliers:

- Identify more creative solutions
- Are more conscious of your scheduling requirements and are flexible in accommodating them
- Have substantially lower overhead costs
- Can easily access a team of outside contract talent with the experience and skills to address short-term, custom projects
- Make better use of technology to meet your needs

Unfortunately, a typical abuse of this small supplier comes from the organization that issues a thinly written request for proposal. It is thinly written because the organization doesn't know precisely what it wants and plans to pick the brains of its potential small suppliers, at no direct cost. It receives a well-thought-out proposal, filled with materials specifications and detailed work schedules. This proposal is then taken and "shopped" to other potential, more favored, less creative suppliers for a lower bid.

This doesn't mean that the other potential suppliers/vendors have the capability to actually perform to the proposal, but they have won the bid and can come back later with change orders to ensure their profitability.

CONTRACT REVIEW

A contract (verbal or written) serves as a formal definition of a customer/supplier relationship with regard to a specific piece of business. With the increasing visibility and significance of such management systems as Total Quality Management and ISO 9000, the value of contracts has moved beyond being a piece of paper.

In ISO 9001, a basic contract review process is required to ensure the quality of product (or service) delivered from a supplier to a customer. As it relates to a pollution prevention capital project, consider the types of items that can be included in the purchaser's definition of a quality product or service:

- Depth of detail in the project proposal/plan
- Ability to meet schedule
- Ability to stay within budget
- Ability to supply specified equipment
- Forethought to plan for routine contingencies and reflect this in the final price of the contract

Under the ISO 9001 standard, contract review consists of the following steps:

- Contract review and coordination
- Adequate definition and documentation of order requirements
- Agreement to order requirements before their acceptance
- Resolution of discrepancies between contract and tender (or request for proposal)
- Confirmation that the supplier has the capability to meet the contract[1]

While the steps appear simple, their implementation is not. The cautious manager works to ensure that the definition of product or scope is clear and in sufficient detail to avoid miscommunication between the purchaser and the supplier.

Unfortunately, to certain types of organizations, such as government agencies and some major corporations, the emphasis is on detail over clarity, sometimes to the detriment of the cost-effectiveness of the project.

While small and medium-sized business owners can work with their own legal counsel to develop certain standard boilerplate protective language, the pollution prevention professional wants to ensure that there is clarity in describing the product and/or scope of work.

This is especially important because many contractors make their profit not on the original bid, but on the "change order." That is, whenever an effort is required outside of the original scope and project definition, a change order can be filed by the contractor for the purchaser's approval. While the nature of construction, particularly a retrofit, make many change orders valid, others present an easy way for a contractor to pad a project. Changes create stress in project management. To not approve the change may delay other activities in a critical path to completion.

1. American National Standards Institute, *Quality Systems—Model for Quality Assurance in Design, Development, Production, Installation, and Servicing*, ANSI/ASQC Q9001–1994, p. 3.

Don't Forget to Follow Through

One company operating the storage component of a TSDF had spent a year working with the regulatory community to get a permit to upgrade the structure used as the hazardous waste storage facility. The new facility was to become a real pollution prevention showcase, with single-pour concrete floors and berms segregating incompatible waste streams. The permit process was slow because, in addition to negotiations with regulatory agencies, there were mandatory public comment periods that had to be observed. The permit was finally granted and money allocated by the organization to pay for the work. The only trouble was that no one ever had ever told the local management to get construction bids and start the work. Eventually, the permit was invalidated because construction was not started by the previously negotiated date.

WORKING WITH CONTRACTORS

Check for Complaints

Most states have licensing requirements for the construction industry. If a construction contractor is licensed and bonded, there is some additional assurance that the work can be performed, or risks/costs can be transferred if nonperformance becomes an issue. Additionally, the purchaser can check with the state licensing authority to determine if there have been any complaints against the contractor that should be considered in making a purchasing decision.

Insurance

Caveat emptor, let the buyer beware. In the litigious American society, the wise purchaser obtains verification that the supplier has current, adequate insurance for the good or service being supplied. By asking that the organization be named an additional insured, claims management proceeds more smoothly in the event of a loss.

A discussion with your insurance agent or broker can help to refine your insurance requirements. To help in that discussion, the following are typical types of insurance coverage suppliers can be asked to have and prove through a certificate of insurance.

Supplier Type: Consultants and Engineers

- Commercial general liability
- Errors and omissions
- Workers' compensation

Supplier Type: General Contractors

- Commercial general liability (to include products and completed operations)
- Environmental impairment liability
- Property (to include all risk, goods of others in contractor's care)
- Workers' compensation

Supplier Type: Construction Subcontractors

- Commercial general liability (to include Products and completed operations)
- Environmental impairment liability
- Property (to include all risk, Goods of others in contractor's care)
- Workers' compensation

Supplier Type: Public Warehouses

- Warehouse legal liability (adequate to cover value of bills of lading)
- Environmental impairment liability
- Workers' compensation

Supplier Type: Waste Disposal Facilities

- Commercial general liability
- Environmental impairment liability
- Workers' compensation

Supplier Type: Hazardous Waste/Hazardous Materials Carriers With the deregulation of trucking, there is more competitive pricing of transportation services. From a pollution prevention standpoint, the lowest price does not necessarily ensure appropriate service. The selection of a transporter for hazardous materials or hazardous waste shipments should consider various safety and financial fitness criteria (Table 6.1) . The MCS 90 Form (Appendix D.2) is an endorsement to an insurance policy. It stipulates the amount of coverage a carrier has with regard to public liability. In this case, public liability refers to liability for cleaning up releases of hazardous materials or hazardous waste occurring during transportation.

Performance Bonds

It is not uncommon on large capital projects, especially public projects, that the contractor be asked to post a "performance bond." The theory behind the performance bond is that if the contractor fails to meet contract obligations, the bond is "called" and the purchaser receives remuneration from the bonding company to complete the project.

■ TABLE 6.1 Evaluate the Hazardous Materials/Waste Transporter

1. Carrier Identification:

a) Carrier Name: ______________________________

b) Carrier Address: ______________________________

c) Carrier Phone: ____________________

d) Carrier HazMat Registration Number (49 CFR 107.608) : ______________

e) Registration Current to 6/30/xx: __ Yes __ No

f) Carrier EPA Identification Number: ____________________

2. Evidence of Financial Responsibility:

a) Certificate of Insurance for General Liability and Workers' Compensation

b) Current MCS 90 Form

3. Driver Qualification:

a) Annual Calendar Year Summary Information for Driver Controlled Substance and Alcohol Testing (49 CFR 382.403), or "EZ" report

b) Do drivers assigned to hazardous materials/waste transportation activities have a CDL with an "H" (Hazardous Materials) endorsement? __ Yes __ No

c) Are drivers assigned to hazardous materials/waste transportation activities current to training requirements per 49 CFR 172.704? __ Yes __ No

d) Does the Carrier use Contract Carriers? __ Yes __ No

e) If yes, how are Contract Carriers qualified? ______________________

4. Safety Fitness:

a) Carrier's DOT Rating?

__ Satisfactory __ Conditional __ Unsatisfactory — Unrated

b) In the event of a spill or release of a material shipped by the company, briefly describe Carrier's process for:

1. conducting emergency response

2. notifying the company

In application, collecting on called bonds is not that simple. Bonding companies are profit-making businesses and take whatever steps they can to delay payment on a called bond. Through the process of delay, they often persuade the organization that will receive the proceeds to settle for a lower amount in the interests of expediency and lower legal fees.

Contractor Liens

When dealing with construction projects, an organization may actually have the work performed through a general contractor. In these cases, the general contractor is expected to subcontract out most of the actual labor. Where this presents a problem is when the general contractor fails to pay the subcontractors that perform the work. The subcontractors can then file a "mechanic's lien" against the owner of the property worked on. The actual filing process and applicability vary from state to state.

There are several ways to reduce the chances of a mechanic's lien being filed. These include:

- Acting as your own general contractor, by soliciting bids for different phases of the project (electrical, concrete, and so on), selecting subcontractors, and paying them directly as any invoice would be paid
- Paying the general contractor's subcontractors directly, rather than passing the money through the general contractor
- Obtaining a periodic (monthly) accounting of expenses and payables from the general contractor before issuing further payments

Bids

The requirements you establish for your bidding process can be used to promote more realistic bids, with fewer opportunities for change orders. For example, you may ask for the bid to be priced "not to exceed" a predetermined amount without prior approval. By then having that "not to exceed" schedule include unit pricing, you have a better idea of any additional charges incurred through an increase in units. This leaves less room for the contractor to maneuver higher margins.

Other final construction bids specify an "allowance" for certain types of construction activity or equipment. Recognize that if the allowance is exceeded, you pay for the "extra." If the allowance is not exceeded, or perhaps not even taken (for example, replacement of a certain water line is not necessary), the purchaser may get a reduced final price, depending on how the contract price and payment terms are structured.

Pre-Startup Environmental Review

Bringing in contractors, especially for retrofit construction, is a disruptive process. But it becomes more so if incidents and accidents occur. When working with contractors, be especially conscious of safety and environmental concerns. Avoiding injuries and pollution events helps to control costs. Table 6.2 is a tool used by a general contractor involved in cleaning tanks containing hazardous sludge. Its objective is to reduce various environmental risks at the job site. Thus, costs are controlled more effectively.

Negotiate Areas of Potential Hidden Costs

Some areas are more likely than others to have contractors transfer a cost of doing business to you, without your knowledge. Keep the following in mind when writing the contract and auditing work progress:

- Utilities (sewer, water, electrical, gas)
- Retention of consumables remaining at project completion
- Hazardous waste disposal fees
- Retention of reusable equipment purchased to support the project

INVOICE AUDITS

Tedious as they may seem, invoice audits are essential for cost control. Duplicate billings, fake vendors, arithmetic errors, overcharges, and charges for materials or services never received are all business management issues that hit the bottom line.

One worker group dramatically affected by the trend to corporate downsizing has been the accounts payable clerk. By replacing the hands-on reviewer of invoices with computers, invoice processing costs have gone down, but a major cost control feature has been lost. There is more opportunity for fraud and missing small details, such as discounts for quick payment. Estimates from the University of Southern California, Los Angeles, point to a 30 percent annual growth in these types of losses since 1994: "With staff reductions, we can't always check paperwork face-to-face with the department head ordering the goods or services . . . . A computer only spots an error if it's programmed specifically to catch the exact mistake."[2]

Some of the features of invoice control take place before an invoice is even written. It is during the contract stage that agreement should be reached

2. "Many Firms Cut Staff in Accounts Payable and Pay a Steep Price," *Wall Street Journal*, September 5, 1996, p. 1.

■ **TABLE 6.2 Organization: PDQ Electroplating Pre-Startup Environmental Review Checklist**

Job Site: 345 Fillmore, Anywhere USA Contractor: Ready Quick Welding

Date of Conference: 3/25/96 Contractor Supervisor: Jan Smith

Checklist completed by: Signature on file ____________
signature print name

Organization to provide:

× **1.** List of Organization's hazardous chemicals that may affect Contractor work area.
potassium cyanide

× **2.** Copies of MSDS(s) for appropriate hazardous chemicals supplied to Contractor.

× **3.** Identification of precautionary measures Contractor must take to protect its employees during organization's normal operation.
Be aware of chemical bath locations; raw materials storage.
Know how to evacuate.
Avoid welding while batch is being transferred; see shift supervisor to coordinate.

× **4.** Identification of precautionary measures Contractor must take to protect its property during organization's normal operation.
Use mobilization site in south end of parking lot for storage of all equipment when not in use.
Keep welding gases out of plating room unless shift is over.

× **5.** Identification of precautionary measures Contractor must take in the event of an organizational emergency.
× alarm signals × escape routes × assembly points
__ emergency shutdown (describe) __ other (describe)

× **6.** Decision on whose Confined Space Entry Permit system:
__ Organization __ Contractor × Not applicable

× **7.** Decision on whose Hot Work Permit system:
__ Organization × Contractor __ Not applicable

× **8.** Decision on whose Lock-Out/Tag-Out Permit system:
__ Organization × Contractor __ Not applicable

× **9.** Description of Organization's hazardous chemical labeling system at the site:
manufacturer's labels left on drums; NFPA signal system used at plating baths

Contractor to provide:

× **1.** List of hazardous chemicals that Contractor may bring on-site.
oxygen, acetylene

× **2.** Copies of MSDS(s) for Contractor hazardous chemicals.

× **3.** Identification of precautionary measures Organization must take to protect its employees during Contractor's normal operation.
Keep welding area clear of combustible materials.
No pedestrian pass-through at welding area during project.

(*continued*)

×4. Identification of precautionary measures Organization must take to protect against accidental environmental releases during Contractor's normal operation.

Drain and purge all tanks.
Remove cyanide storage from proximity to welding.

×5. Identification of precautionary measures Organization must take in the event of a Contractor emergency.

× alarm signals __ escape routes __ assembly points
× emergency shutdown (describe) __ other (describe)
Follow procedure 3-16 for emergency shutdown.

×6. Description of Contractor's hazardous chemical labeling system at the site:
× rely on mfg/distributor labels __ NFPA signals __ HMIS __ Other (describe)

____

____

×7. Verification of appropriate Contractor employee training certifications.

Certification Description	Identification of Contractor Employees
A. OSHA HAZWOPER (29 CFR 1910.120)	Mike Jaffe
Effective Through: 12/31/96	Cheryl Allen
__ 40 Hour Level	Tom Fox
__ Emergency Response Level	
× Awareness	
__ Operations	
__ HazMat Technician	
__ On-Scene Commander	
__ Manager/Supervisor of Operations	
__ Instructor/Trainer	
B. OSHA Process Safety (29 CFR 1910.119)	Mike Jaffe
Effective Through: 12/31/96	Cheryl Allen
__ Process Hazard Analysis	Tom Fox
× Contractor Responsiblities	
C. OSHA Confined Space Entry	Mike Jaffe
(29 CFR 1910.146)	Cheryl Allen
Effective Through: 2/15/97	Tom Fox
× Authorized Entrant	
× Attendant	
× Entry Supervisor	
__ Rescue and Emergency Service	
D. OSHA Marine Safety (29 CFR 1915.7)	
Effective Through: __/__/__	
__ Competent Person (29 CFR 1915.7)	
E. DOT (49 CFR 172.704)	
Effective Through: __/__/__	
__ General Awareness	
__ Safety	
__ Commerical Driver's License	
__ Function Specific: __	
F. Other (describe)	
Certified Welder	Cheryl Allen
	Tom Fox

with vendors on what constitutes documentation of labor, travel, and expenses.

Get control of these things by requiring the details before the invoice can be reviewed for approval. The details include:

- Time sheets or other verification of time expended (especially important for a time and materials contract)
- Copies of signed expense reports
- Original receipts for plane tickets, rental cars, hotels, and meals
- Original receipts of equipment suppliers, rental agencies, etc.

Similar to subcontractors, general contractors also purchase and rent equipment in order to perform the work. If your organization has accounts with these ancillary suppliers, direct billing to the organization can be desirable. If it doesn't have accounts, it may want to consider setting them up. Direct billing:

- Ensures that these ancillary vendors get paid
- Eliminates the contractor markup
- Reduces the chance of invoice mistakes or fraud
- Builds a relationship between the ancillary vendor and the organization that ultimately uses the product or service

This relationship can be very important because it may enable a lower up-front price to be negotiated based on the potential for additional long-term business. Even warehouse stores, such as Price Costco, have introduced the concept of an annual rebate to businesses using their Price Costco credit card. Why should the rebate go to the contractor, when the organization has paid the bill?

On the other hand, some contractors enjoy excellent pricing and preferential service because of the amount of business they already bring to the ancillary vendor. Ask and compare to get the best value for your organization.

A major disadvantage of direct billing may be inherent in the organization itself. That is, how much does it cost to process an invoice? If the fixed cost is high, then it can be more cost-effective to get a detailed, consolidated bill from the contractor. This consolidated billing is one of the marketing tools used in setting up certain types of accounts.

When an invoice comes in for approval and payment, the items that should be reviewed include:

- Pricing (Is it per quote or contract?)
- Accuracy of calculations, including totals and extensions
- Terms (Is it per quote or contract?)
- Sales tax (Is it charged? Should it be? Is the charge correct for your tax jurisdiction?)
- Identifier of person who authorized purchase (and/or purchase order number, as appropriate)

If discrepancies are noticed, they should be reported, both verbally and in writing, to the contractor in a timely fashion. It is advisable to withhold final payment until discrepancies are resolved. This is the best leverage to encourage the contractor to satisfactorily respond to a discrepancy. If you have protracted problems, however, you may want to report the contractor to the appropriate licensing authority.

Types of Contracts

From a financial administration viewpoint, there are several types of contracts:

- Fixed price (all-inclusive)
- Fixed price for labor, materials and expenses at cost or cost plus
- Fixed "not to exceed" price
- Time and materials (at cost or "cost plus")

The contract ripest for abuse is the time and materials contract with a "cost plus" feature. "Cost plus" is the practice of paying the real cost and an additional handling fee for the contractor because of additional costs associated with financing and managing invoices. Typically, this feature is some percentage of the costs (excluding labor). This is a disincentive for the contractor to keep costs competitive and controlled.

Cost plus is reasonable from the standpoint that there are expenses associated with processing invoices and keeping the cash flow moving. Fifteen percent is not an unrealistic number. But the actual percentage is negotiable. This author remembers once receiving an invoice for several consumables associated with a project that contained some interesting pricing. Among the most interesting was 25 feet of ordinary garden hose that was invoiced at $75. That "up charge" was a gouge.

SUMMARY

Purchasing decisions and the administrative activities associated with purchasing have a substantial impact on cost control. In addition to traditional purchasing controls, other factors help reduce total costs associated with pollution prevention implementation. When outsourcing is an appropriate choice, a logical qualification and selection of vendors is needed. This process, coupled with the implementation of strong contract administration and management, enables a project to move forward in a timely manner. When the investigative features of invoice auditing are added, the least-cost approach has been achieved.

Exercise

Objective: Practice using objective methods for making a vendor determination.

1. Select a project that may require the use of an outside vendor or contractor.
2. Based on the characteristics of the project, develop an initial list of vendor evaluation criteria.
3. Review the initial list and determine where it may be appropriate to include numeric performance criteria (e.g., amount of insurance, etc.).
4. Organize the criteria into a survey format suitable for:
   a. Attaching to a request for proposal
   b. Easy compilation of results for evaluation

CHAPTER 7

A Pollution Prevention Catalog

The purpose behind pollution prevention is to solve a real or perceived problem. There are two categories for pollution prevention solutions or alternatives in any organization: structural and nonstructural.

Structural, or engineering, solutions address equipment, design, instrumentation, or related items that result in the construction of something tangible. Their implementation is task-focused and results are easily measured.

Nonstructural, or behavioral, solutions focus on human actions (routine and nonroutine). While these alternatives are less tangible and generally more difficult to measure or evaluate, they have a profound impact on any work process.

An important implementation interface between structural and nonstructural pollution prevention alternatives is the contract. Contracts translate a structural want into a structural solution through the activities (behaviors) of human beings. With cost control a ubiquitous concern in most organizations, improved management of the contract at the structural/nonstructural interface can create bottom-line savings for the organization.

This chapter lays the foundation for pollution prevention implementation. It details selected key topics that have a significant bearing on total costs. These include:

- Contracts
- Operator error
- Training
- Emergency response planning
- Construction, architecture, and landscape design

The chapter concludes with a simple-to-use catalog. Divided into operational tasks common to many organizations, the catalog presents both structural and nonstructural alternatives for the tasks listed.

Remodel That Landfill

With more than 400 unlined landfills in the United States that don't meet current federal requirements, does closure represent the only pollution prevention alternative? Two commercially viable techniques are worth consideration.

In many older landfills developed before current recycling laws, the fills can be mined, with recyclables separated out, recovered, and sold. The remaining waste, stream is then reduced in size through compaction; the excavated portion of the landfill is lined, refilled with the remaining waste, and capped. One mining project in Edinberg, New York, saved an estimated $300,000 to $400,000 in closure costs (including post-closure monitoring) by reducing the size of its landfill by 60 percent.

While landfill remodeling is similar to mining, it focuses even greater attention on reopening additional landfill capacity and extending the life of the landfill. This has the potential to be particularly cost-effective in areas of the country where landfill space is in tight supply. One project, in Fairhaven, Massachusetts, expects to regain 70 percent of its landfill capacity, raising further revenues while adding 10 years to the life of the facility, thereby deferring closure costs.[1]

Reducing total waste volume, recovering recyclables, and extending the useful life of fixed facilities are noteworthy P^2 benefits.

1. "Land Fill Remodeling May Work Wonders on that Old Dump," *Wall Street Journal*, September 9, 1996.

DEFINING REQUIREMENTS THROUGH CONTRACTS

Every day, we make, break, or otherwise participate in the implementation of contracts. Contracts can be written, verbal, or implied.

Within the business culture of the United States, written contracts are among those most commonly recognized and understood. They are the written rules, specific to the scope of the situation being managed through the contract. Acceptance of and adherence to contracts is a bedrock for free-market activity in the United States.

Since all wording in contracts is inevitably subject to interpretation, contract disputes are sometimes settled by arbitration or through litigation in the courts. Over time, these court rulings have created legal precedents in interpretation. These precedents are the reason that many contracts are so full of "legalese." The language is there, in theory, to protect parties to the contracts with the ready-made interpretation of legal precedents. However, this language is subject to local, state, and/or federal interpretations based on earlier judicial interpretations or legal precedents. Therefore, it is prudent to work with an attorney with a working knowledge of local, state, and federal interpretations in the jurisdiction where the contract is implemented.

The Uniform Commercial Code is one federal regulatory device used to systematize the interpretations of certain rights and responsibilities as they relate to contracts. It clarifies such business concepts as F.O.B. (freight on board), which are important in establishing ownership issues.

For example, Party A purchases a hazardous substance from Party B, F.O.B. the Party B loading dock. While in transit, the hazardous substance is released by accident. Who "owns" the release? Who has responsibilities relative to the release?

In this example, the release is "owned" by Party A, because the purchase was F.O.B. Party B's dock. Had it been F.O.B. delivered, then Party B would have "owned" the release. In this example, some responsibility also rests with the transport company, based on both the Uniform Commercial Code and Department of Transportation regulations. But these responsibilities are also defined through contracts and regulations.

In making a delivery of oil from a vessel into a storage tank farm, a hose is run from the vessel to the tank farm. If the hose breaks during delivery, who "owns" the release? It depends on who owns the hose. If it is the tank farm's hose, the tank farm owns the release. If it is the vessel's hose, the vessel owns the release, although the tank farm still has a release "problem" to face at its dock.

Contracts are only as good as the desire and capacity to abide by them. In one case, an environmental auditor for a chemical storage warehouse asked to see evidence that the sprinkler system was maintained and tested at least annually. In this case, the responsibility was the landlord's and additional rent was paid on the lease to have the landlord conduct the tests. In checking with the landlord, it was learned that the system had not been tested or maintained in three years, in violation of the lease. As a result, the system was tested, and the risk of a pollution event from fire decreased. The tenant recovered a rebate for the landlord's failure to meet contractual responsibilities.

Uncertainty is more inherent in the nature of verbal and implied contracts. While the cost control opportunities are large, it is the uncertainty that causes organizations to ignore them. However, this uncertainty is a flexible feature that can facilitate inexpensive improvements to cost control efforts.

It is the verbal or implied contract that results in an organization not putting contract work or purchases out for competitive bids. While it is not universally cost-effective to put everything out for these bids, occasionally testing these implied relationships through a competitive bid process can be healthy for the bottom line.

Faced with the jargon of standard contracts and the desire to take an action or step, pollution prevention professionals can view contracts as a hindrance to workplace improvement. In fact, they offer great flexibility and opportunity for cost savings.

THE OPERATOR ERROR CONUNDRUM

The objective analysis and critique of accident and incidents can lead to conclusions of "operator error." People are not machines programmed to respond in a uniform manner to a given situation or set of indicators. Human innovation and creative response are both their strength and, occasionally, their weakness.

Over the last two decades, changes in educational format and content, coupled with enormous social change, have served to create even greater public sensitivity to the disintegration of performance capabilities of individuals completing their education at the elementary, secondary, and college level.

While some of the changes are very favorable, such as increased emphasis on the use and operation of computer technology, others are less so. Decreased skills in reading comprehension, mathematics, analytical reasoning, and verbal communication have become a challenge to the workplace.

Can the Employee Read?

The lack of reading and comprehension skills may be the last social stigma still hiding in the workplace closet. According to estimates on American adults from the U.S. Office of Education:

- 20 percent are functionally illiterate
- 34 percent are marginally illiterate
- 50 percent of all industrial workers read below the ninth-grade level[2]

These stark numbers have been confirmed further by READ RIGHT Systems, a Washington State–based educational services provider. Sixty percent to 80 percent of employees at its industrial clientele require some type of reading enhancement.[3] These are not all functional illiterates, but they are people who have not been properly taught how to read.

The people without these skills reside at all levels in an organization, including senior management. Many of them adapted to the manual workplace and functioned at a high level of performance by developing exceptional memory skills and critical thinking skills.

Similarly, they are very creative in the manner by which they hide their lack of reading. They "read" newspapers during break. They use others to read and write for them.

An employee at a Weyerhauser sawmill had risen to the level of team leader during his thirty-one years with the company despite poor reading comprehension skills. He compensated by listening and having his wife help him with difficult materials. But the demands of new technology made it more and more difficult for him to perform his job. "With all the automation we've installed," he says, "if you can't understand how the equipment works, it really slows you down. If you can't read today you probably won't make it in a sawmill."[4]

But as more automation appears in the workplace, these skills of memory, critical thinking, and subterfuge are not sufficient. In fact, in many cases,

2. READ RIGHT Systems, *Information About Read Right®*, 1993, p. 3.

3. READ RIGHT Systems, interview with Leo Frank, September 4, 1996.

4. *Industry Week*, "Raising the Reading Curve," Penton Publishing, Inc., November 20, 1995.

these automated systems have been designed to capture the operational memory of those individuals and overlay automated control over specific aspects of the work process.

An organization's investment in adult literacy training is less expensive than automation. Although measuring the payback is difficult, it has enormous payback potential from:

- Improved individual worker productivity
- Reduced rates of pollution event accident/incidents
- The capacity to defer or eliminate selected expenditures for additional, complex control automation

"The extent to which the workplace literacy training saved the company money was examined by tracking monthly scrap and rework costs for a one-year period before, during, and after the training occurred . . . . As of December 1991, Magnavox had saved $262,000 on scrap costs over 1990, a 35 percent decline, and had saved $74,000 on rework costs, a 25% decline . . . ."[5]

What Knowledge Base Must the Worker Have?

The challenge to employers is to have an employee perform a work activity properly. Every work activity or work process requires the individual responsible for that activity to have and apply a certain knowledge base. Preparation of a job hazard analysis is one way to identify the knowledge base applicable to a work process.

This analysis can be as simple as a job evaluation through use of a matrix. As a tool, the matrix is designed to take into account such things as specific process procedures, engineering data, or chemicals in use in the process (Table 7.1).

In a more detailed process, each step of a job activity is recorded, in an effort to:

- Identify existing or potential hazards
- Decide the best way to perform the activity in order to reduce or eliminate the pollution event risks.[6]

The job hazard analysis is similar in scope to the process hazard analysis (Chapter 4), but puts its emphasis on hazards to personnel rather than hazards of equipment.

5. Jack J. Phillips, Editor, *Measuring Return on Investment*, "Excerpts from Case 6: Three Rs in the Workplace," Donald J. Fox, Magnavox Electronic Systems Company, 1994, American Society for Training and Development.

6. *Job Hazard Analysis*, OSHA Publication 3071, 1981.

■ **TABLE 7.1 Job Hazard Evaluation Matrix**
For Selection of Personal Protective Equipment (OSHA 29 CFR 1910, Subpart I, Effective 7/4/94)

Survey Date:___/___/___ Evaluator:________________
Job Category/Title/Description:________________________

	Hazard Categories	Impact	Penetration	Compression (rollover)	Chemical	Harmful Dust	Light (optical radiation)
SOURCES							
Motion (machinery, collisions, etc.)							
High Temperature							
Chemical Exposure							
Harmful Dust							
Light Radiation							
Falling/ Dropping Objects							
Sharp Objects							

(continued)

■ TABLE 7.1 (*Continued*)

	Hazard Categories	Impact	Penetration	Compression (rollover)	Chemical	Harmful Dust	Light (optical radiation)
SOURCES							
Rolling/ pinching objects (crush feet)							
Workplace Layout							
Electrical Hazards							

Personal Protective Equipment Selection

Body Part Requiring Protection	Equipment Recommendation
Eye	
Face	
Head	
Foot	
Respiratory System	
Hand	

Numerous regulatory requirements have a training component. Appendix C.5 is a checklist suitable for use in determining whether the organization is subject to some type of mandatory training because of a federal regulation. The citations provided are limited to federal regulatory requirements. Depending on what state the activity is being conducted in, there may be more stringent requirements. For example, in the state of Washington, the OSHA HAZWOPER 40-hour certification is insufficient. Employees working in hazardous waste activities in Washington State must have 80 hours of initial training.

TRAINING PROGRAM MANAGEMENT

No matter what the subject matter, training is essential to contemporary organizations. Operating the training function as any other business function gives better return on the expenses incurred. Poor training is a waste of money. Good training adds value.

While not traditionally viewed as being within the purview of pollution prevention professionals, costs of P^2 implementation are controlled effectively when a certain level of process worker competency exists and is maintained.

Make the Training Useful

An objective analysis of pollution events can reveal that, even though training had been conducted, the adequacy of the training was questionable.

Among the contributing factors to inadequate training are:

- Failure to identify minimum knowledge base for activity or process
- Lack of recognition of depth of knowledge needed to perform activity or process consistently
- Failure to recognize or maintain currency of knowledge base as activity or process evolved
- Inattention to curriculum content
- No follow-up evaluation of training effectiveness
- Unmotivated students

Unfortunately, in some cases, management addresses training as a necessary overhead evil designed to satisfy a regulatory requirement. It is not considered a component of a successful management system. Thus, it assigns training instruction responsibilities to individuals who have not worked out well in other parts of the organization. They may:

- Know nothing about the subject
- Discover the curriculum is nonexistent or poorly prepared
- Have no background in the idiosyncrasies of teaching adult learners
- Be boring

Shoving a video into a VCR is not training. The model management system for organizational training is comprehensive, and addresses:

- Identification of training needs
- Specification of training goals and objectives
- Development of learning activities (curriculum)
- Training implementation
- Evaluation (measurement) of effectiveness
- Curriculum improvement

Whether the source of training is internal or external, in order to get value (cost-effectiveness) from a commitment of time to training, these same comprehensive criteria apply.

Internal training is accomplished under the direct control of the organization. This classification applies even though the training may be designed and conducted by a third-party contractor. A high degree of customization is expected from internal training, although it may use generic materials (videos, books, tests) obtained from an outside source, such as a library or training supply company.

Under this definition of internal training, use of an externally purchased training course, such as a mail-order self-study curriculum, is internal training if the organization is actively involved in coaching or assessing participation.

The advantages of internal training include:

- Scheduling flexibility
- High degree of customization
- Easy access to workplace venues for hands-on applications
- Interaction of employees (team building, exchange of common problems)

The following disadvantages of internal training are organizational problems, and do not exist in a well-managed program:

- Inbreeding of course content/delivery
- Failure to update course content
- Failure to implement training

- Political disincentives for employees to express opinions or identify problems candidly

External training is less directly controlled by the organization. It includes, but is not limited to, participation in appropriate courses at colleges and universities. Depending on the industry group of the organization, it also includes seminars, lectures, and workshops sponsored by trade associations, training supply companies, vendors, or customers.

Its principal advantages include:

- Exposure of employees to individuals with ideas, thoughts, problems, and solutions from outside the organization
- Sequestering of employee participants from workplace distractions

Disadvantages of external training include:

- The prerogative of the training provider to cancel the course
- Lack of direct applicability of the material to the organization's needs

Also among the disadvantages of external training are items that reflect a lack of serious commitment to training by the purchasing organization. These include:

- Failure to define for the participants desired goals and objectives of the organization
- Selection of external training as a perk, rather than because of the expected results
- Failure to track employee participation

Organizational Training Objectives

Perhaps one of the poorest methods for meeting a real or perceived training need is to purchase a book and give it to someone with a comment such as, "Read it, you'll learn something." But, every day, hundreds of books get bought and disseminated and are never read, thought of, or discussed again.

No matter what training technique is used, it is most effective when the organization and its participants understand what the goals and objectives of the training are. This forces an organization to use training as an essential component of the management process. Through articulation of course-specific training goals and objectives, it is clearly understood what the organization wants the employee:

- To do
- To do better
- To stop doing

To the extent possible, these objectives should be stated in a concise, measurable manner. Of the following objectives, which is the most useful to the developer of course curriculum and course participants?

Course Application: Satisfy training per 29 CFR 1910.120 (q)(6)(i) for Hazardous Waste Operations and Emergency Response, First Responder: Operations Level.

- Sample Objective 1: Train an emergency response team.
- Sample Objective 2: Present basic information and a practice exercise that enables students to respond at the First Responder Operations Level to a release or potential release of a hazardous substance.

Objective 2 is better, because it clearly identifies what participants must do to demonstrate what they have learned (exercise to show response capability at First Responder Operations Level).

Set Priorities for Training

Cost-effective training occurs not because a certain training program is available or inexpensive to participate in, but because it is useful to the individuals being trained.

A model of the U.S. Occupational Safety and Health Administration, while devised to identify employee populations at high levels of occupational risk, is also helpful in identifying individuals at risk of creating or being exposed to pollution events (Table 7.2).

■ TABLE 7.2 Employees "At Risk" of a Pollution Event

Variable	"At Risk Characteristics"
Age of employee	Younger
Length of time in work activity	Newer
Size of firm	Medium-sized
Type of work	Incidence and severity per SIC Code
Use of hazardous substances in work activity	Use increases risk

Based on U.S. Occupational Safety and Health Administration.

In using this model, it is important to recognize the hazardous environmental conditions under which a work activity is performed. When the work environment involves hazardous conditions and materials or processes that could create a pollution event, a pollution prevention opportunity exists. Hazardous conditions include:

- Noise
- Heat or cold
- Shift work (particularly rotating shifts)
- Inadequate illumination
- Confined spaces
- Proximity to energy sources
- Proximity to or use of hazardous materials or substances

The objective is to give priority training to "at-risk" employees in an effort to promote the safe performance of work within a hazardous work environment.

In addition to these hazardous conditions, other variables correlate to a high rate of worker illness and injury. To the extent that personnel meeting the criteria of these variables are in activities that have the potential to create a pollution event, they also have a greater need for training. Thus, the return on the training dollar investment is greater for these at-risk workers.

Special Needs

Communication is a two-way process involving both the sending and receiving of messages. Risk communication training targets improved understanding of risks associated with various types of activities, many of which have the potential to create a pollution event.

Can risk communication play a role in controlling costs associated with pollution prevention? It can and does. That is the intent behind such functions as the employee involvement requirement and the provision of material safety data sheets in the OSHA Process Safety Standard.

Employees who are trained and knowledgeable in the risks associated with their activities make rational choices with regard to how they conduct an activity day in and day out. Every day, these personnel make hundreds of decisions that determine whether or not a pollution event is created or controlled. By openly communicating the risks associated with a specific work environment, they gain an awareness of their responsibility for themselves, other employees, the community at large, and the natural environment.

During organizational mergers, acquisitions, and divestitures, uncertainty and anxiety increase the risk of pollution events, no matter what the training or experience level of employees. The employee stress level is high because of:

- Lack of knowledge about or the uncertainty of pending changes
- Inability to control or influence pending changes
- Introduction of new managers or supervisors
- Introduction of new or modified equipment
- Introduction of new or modified work activities
- Introduction of new or modified work procedures

Recognizing these conditions, the pollution prevention professional responds by reorganizing training priorities to address concerns in a timely manner. The return on investment in these cases can be substantial.

One chemical company, being acquired by a major competitor, counseled its employees in a special communication session. They were asked to draw a picture of their personal dragon (i.e., what bothered them most about the situation). By acknowledging the existence of these legitimate concerns, a great deal of anxiety was released and the transition to a new owner was incident-free.

Flexibility in Recurrent Training

Many times, employees are subject to recurrent training requirements because of an outdated company policy, or a mandatory regulatory requirement. Recurrent training is not always cost-effective. As it relates to pollution prevention, all recurrent training should be periodically reevaluated to determine:

- If it is still needed
- What format is the most appropriate for delivery
- Whether an alternative approach can be substituted

This type of systematic review enables an organization to document and use on-the-job evaluations, rather than classroom time, to satisfy recurrent training. Further, ". . . when the hazard and assigned duty remain the same, the use of a properly designed, annual written test may suffice to meet annual training requirements. A written test by itself may not be sufficient, however, for training that requires demonstration of competency (e.g., donning and doffing of personal protective equipment that is infrequently used)."[7]

7. OSHA Interpretive Letter, May 2, 1994.

Judgment needs to be applied in determining whether and what alternative to use. For example, recurrent training for newly merged departments can be modified to facilitate team building or address identified gaps in knowledge. These gaps are often overlooked if the decision makers understand, and think everyone else does as well.

Curriculum Design

The curriculum design component results in the production of the details of training. It describes learning activities, including lecture notes, visual aids, and testing. When information tools such as a job hazard analysis are used, the curriculum can be customized to be even more effective.

People learn in different ways and this should be considered when designing curricula. Depending on the individual, a person processes some types of communications messages more effectively than others. The principal senses that control this individual preference are:

- Hearing
- Seeing
- Touching/feeling.

Throughout history, the oral tradition has been the predominant method of communication. Some people would tell stories, others would listen, and knowledge was transferred.

With the advent of modern communications systems though, the social culture of the United States has evolved. In contemporary Western society, seeing or visualization has become the dominant form of communication. In application, this means most people are less likely to retain a fact that is heard (for example, a lecture) than one that is seen (for example, through the use of visual aids or reading).

When visual communication is coupled with the performance of an activity (touching/feeling), retention of the concept is even higher. Thus, the construct of interactive education presents a structure for practicing a communication message in order to improve both the comprehension and retention of the message.

Participant learning motivation is enhanced by applying the following practices to both curriculum design and delivery:

- Explain the goals and objectives of instruction.
- Relate instruction to the interests, skills, and experience of the participants.

- Outline the main points to be presented during the training session(s).
- Communicate the desired benefits of the training.

Another element of curriculum design is the specification of how participants demonstrate that they have acquired the desired skills or knowledge. There are several purposes for this demonstration or testing:

- Repetition and reinforcement of information presented
- Identification of subject areas that require further clarification
- Compliance with a regulatory requirement

Good test design does not focus on causing people to fail (trick questions, etc.). Effective tests serve as objective tools for evaluating the competency of individuals and serve as an evaluation of the curriculum itself.

In selecting an appropriate type of testing, the test designer needs to consider learning limitations of its participants. The multiple choice test enjoys several advantages:

- It repeats and thus reinforces the correct answer.
- It is inexpensive to prepare and administer.
- It is easily given orally to address needs of nonreaders.
- It provides participants with a written reference of correct answers.

Correcting tests out loud within the classroom structure, provides the following benefits:

- Additional reinforcement of correct answers
- Verbal repetition of correct answers for nonreaders who rely on memory
- Encouragement of discussion in areas where errors occurred

In its training curriculum guidance (Appendix F.1), the U.S. Occupational Safety and Health Administration recommends a base of at least 50 questions in order to cover a topic comprehensively.

Tracking Training Implementation and Effectiveness

An administratively well-run training program is documented to provide the records necessary to verify compliance with regulatory requirements and function as a performance measurement. Such documentation is student-specific, including:

- Student name
- Course title

- Description, location of training materials used (curriculum tracking)
- Course date
- Date recurrent training needed
- Length of course (in hours, for regulatory compliance purposes)
- Statement that the student has successfully completed the course (certification of training and testing)
- Name and address of the training provider
- Individual identification number, if certificate issued

With the advent of inexpensive relational databases and the availability of Mac/PC-based training tracking software, these data, once collected, can be sorted to produce valuable reports (measurements) for management control (i.e., cost control) purposes.

For example, driven by the HAZWOPER training recordkeeping requirements, one company with multiple locations implemented an automated tracking system, which became the official legal record of training activities. By consolidating training data into a single source, environmental auditors were able to obtain records before traveling to a facility for a compliance audit. With this objective evidence, auditors could work in cooperation with local managers to flag potential problems created by employees deficient in certain prescribed requirements.

Evaluation

Among the techniques for training evaluation are:

- Analysis of test scores
- Participant questionnaires
- Supervisors' observations of performance
- Identification of workplace improvements

Training evaluation is conducted to determine whether the desired value was obtained from a specific program of training. As an element of cost control, it assesses whether the program gave the organization its money's worth and sets a direction for future training expenditures.

The evaluation process (Table 7.3) validates what the participants learned or fail to learn. In order to direct future program improvements and ensure that the organization got value for its training dollar, its objective is to assemble information that identifies training components that are superfluous, confusing or distracting or totally lacking.

■ TABLE 7.3 Training Evaluation Checklist

1.	Are the program objectives clearly stated?
2.	Is the program accomplishing its objectives?
3.	Are appropriate facilities and staff available?
4.	Is there an appropriate mix of classroom, demonstration, and hands-on training?
5.	Is the program providing quality employee health and safety training that fully meets the intent of regulatory requirements?
6.	What are the program's main strengths?
7.	What are the program's main weaknesses?
8.	What is recommended to improve the program?
9.	Are instructors instructing according to their training outlines?
10.	Is the evaluation tool current and appropriate for the program content?
11.	Is the course material current and relevant to the target group?

Source: U.S. Occupational Safety and Health Administration.

Smart organizations develop a survey form for evaluating all organization-sponsored training. Such a form is used irrespective of whether the training is internal or external to the organization. A simple approach is to ask a short series of questions (no more than 10, to promote response) and have employees rank their response to each individual question on a common scale (e.g., 1 to 5, 1 = poor and 5 = excellent).

The actual questions can be tailored to the organization, and address such items as:

- Applicability to the person's job (current and future functions)
- Applicability of skills used to job
- Appropriateness of curriculum
- Suitability of training tools
- Instructor preparation
- Instructor feedback on performance during training
- Most useful aspects
- Least useful aspects

EMERGENCY RESPONSE PLANNING

Fifteen Minutes for Cost Control

When incidents that have the potential to create a Pollution Event occur, effective emergency response is critical to the cost control effort. The capacity of an organization to provide competent internal response within the first 15 minutes of an incident determines what the bottom-line costs of an event are going to be.

It's in the first 15 minutes that you can put on a patch, turn a container, shut off a pump, sound an evacuation alarm, or use a fire extinguisher in order to contain the incident itself.

At one home heating oil distributor, oil delivery drivers had a plan and were drilled to "keep the oil from all waterways, even storm drains." So, when a hose ruptured in the street of a posh residential neighborhood, the driver applied common sense and used his rubber raincoat to protect the storm drain. Did the company have a cleanup expense? Of course. But it was significantly smaller, since the oil was able to pool onto the raincoat protecting the storm drain.

Integrated Contingency Planning

Unfortunately, in many organizations, the emergency response planning required to promote this kind of trained 15-minute reaction is conducted as a halfhearted effort. Such planning concentrates its limited resources on pulling a plan together to keep a regulator happy, rather than developing something that really makes a difference when it is needed.

Unfortunately, the regulatory community bears some responsibility for the paper monster that has been created because of various requirements from different agencies and at different levels of government (federal, state, and local). Plans have become too thick and unwieldy to work in a crisis.

In order to improve plan development, review, and utility, four federal agencies adopted a guidance to improve emergency response planning. Dubbed "One Plan," this 1996 Integrated Contingency Plan Guidance Document offers a more cost-effective system for coordinating emergency response planning requirements for the:

- U.S. Environmental Protection Agency
- U.S. Department of Transportation (U.S. Coast Guard and Research and Special Programs Administration)
- U.S. Department of Interior Mineral Management Service
- U.S. Department of Labor Occupational Safety and Health Administration

Its objectives are to minimize duplication of planning efforts at single facilities and improve economic efficiency in planning and plan use.

As a guidance document, One Plan does not solve all duplication issues or address all emergency responses a company may want to consider. Depending on where facilities are located, they can still be subject to additional state and local emergency response planning requirements. However, One Plan is a step in the right direction.

The integrated plan has three sections. Section I is administrative in orientation, consisting of:

- Purpose and scope of plan coverage
- Table of contents
- Current revision date
- General facility identification and information

Section II presents the essential steps needed to respond. Therefore, it must be concise and easy to read. "The rule of thumb is that the core plan should fit in the glovebox of a response vehicle."[8]

It includes procedures for:

- Discovery of the emergency situation
- Initial response (first 15 minutes)
- Sustained response actions
- Termination and follow-up actions

Section III is the largest section because of the level of detail required to support the core plan. In structure it is administrative, rather than action-oriented. Its annexes address:

- Geographic information about the facility and community
- Notification procedures, internal and external
- Response management structure
- Incident documentation
- Records of training, exercises, and drills
- Response critiques; plan reviews and modifications
- Prevention
- Exercising the plan

If only regulatory requirements are considered in the planning, the organization can expect some gaps with regard to what its emergency response needs may be. If the following situations are applicable to your organization, address them in your plan:

- Does your facility ship or receive hazardous materials from off-site using common or contract carriers?

8. 61 Fed. Reg. 28644, #109. June 5, 1996.

- Does your facility store hazardous materials in a public warehouse, including bulk storage facilities?
- Does your facility use freight forwarders for import or export of hazardous materials?
- Does your facility, or other facilities in your organization, supply products that contain hazardous materials to customers?
- Would your organization expect your facility to provide emergency response backup services to other facilities in the organization?
- Does your facility contain any type of radioactive materials, including small-dosage machines used for scanning equipment?
- Does your facility work with or produce any biological pathogens?

A Practical Guide to Making the Emergency Response Plan User-Friendly

Emergency response plan management has been simplified by the inexpensive nature of and software choices for personal computers. If using a personal computer, the following document control issues can improve plan maintenance:

- Identify a management individual with the responsibility for "ownership" of the computer disks and paper files containing the plan and plan support information.
- Address unique computer files in a rational way that simplifies locating the master file for updating; develop a description of the logic that is used and ensure that several people know how to retrieve it.
- Create each section of the plan (I, II, III) as separate master documents in unique computer files.
- Back up the master file.
- Adopt a common header/footer to include on each page:

1. Document name
2. Section number
3. Date plan version is effective
4. Date of plan that is superseded, as in "supersedes 6/1/96"
5. Page number, as in "Page 1 of 15"

The following collection of ideas has been assembled from years of experience managing responses at fixed facilities, on customer premises, during transportation emergencies, earthquakes, tornadoes, hurricanes, and volcanic eruptions. It applies most appropriately to the core plan elements discussed in a preceding section:

- Issue plan in a rugged three-ring binder. A yellow binder is easier to find in a real emergency.
- Ensure that the binder has at least one pocket and contains a notepad, pen or pencil, and two quarters taped to the inside, in case someone needs to use a pay phone.
- Inside each binder pocket include a sturdy (20#), laminated standard letter-sized card sign for placement in vehicle windows. Customize the language on the sign but include the following type of identifier: "Emergency Response Personnel for XXX Company."
- Issue two copies to each plan holder, one for home/car, one for work.
- Advise plan holders to carry business cards in the plan for identification and ease in crossing police lines.
- Include a copy of the current U.S. DOT Emergency Response Guidebook with each copy of the plan.
- Periodically audit plan holders to ensure that they still have the information they need.

Exercise the Plan

Plan exercises do not have to be complex or expensive to work. They are simply another version of the type of employee training described earlier in this chapter.

Exercises are conducted to:

- Provide training and experience in using the plan
- Internally test the plan
- Fulfill a regulatory requirement
- Participate in an external (e.g., local government) test of a community plan

If the plan has never been tested, a tabletop exercise is an excellent first test choice. Its advantages include:

- Ease in scheduling
- Defined time limit for conducting
- No special equipment requirements
- Inexpensive to conduct

Tabletop Exercise The tabletop is prepared in advance and includes:

- Statement of exercise objectives
- Several short scenarios
- Several questions (and potential responses) for each scenario
- Critique questions

Its personnel requirements consist of:

Typical Personnel Assignment	Responsibility
Exercise facilitator	Runs the exercise May assume several "acting" roles during the exercise
Emergency response team ■ Incident commander ■ First responders: awareness level ■ First responders: operations level ■ First responder: hazardous materials technician	Individuals assume the roles they would normally have during an emergency, unless they are being introduced to other skills Depending on the size of the organization, they can be rotated as observers
Media coordinator	Handles internal communications with employees and external communications with the media, public
Various "external" roles, depending on scenarios chosen, or availability of "extras" ■ On-scene Incident commander (fire department) ■ News media ■ General public	Per the scenario
Observer	Observe and make notes of response to assist in the critique These may be emergency response team members who rotate in

Participants bring their individual copies of the emergency response plan. Other materials, if needed, are provided during the exercise by the facilitator. Material safety data sheets or emergency response guides are typically the only additional materials required.

After the participants have been assembled, the facilitator informs them of the exercise objectives and describes the format that is used. Roles are assigned at this time. At the option of the facilitator, participants may receive a written copy of the first scenario. They are never given the questions, but respond to them verbally as they are presented, sequentially. When the questions are completed, the second scenario is presented and the process is repeated.

The exercise is critiqued following the completion of all scenarios feasible within the time limit for conducting the exercise. If the critique results in corrective action assignments, the observer makes a record to include as formal documentation of the event.

Appendix G.1 presents a sample tabletop exercise that can be adapted to your organization.

CONSTRUCTION, ARCHITECTURE, AND LANDSCAPE DESIGN

The easiest, most cost-effective time to integrate structural pollution prevention alternatives is during the facility design phase. Although retrofitting is a possibility, modifying the physical aspects of an organization's infrastructure becomes more costly and faces more limitations and contingencies.

Local building codes, such as the Uniform Building Code, are often viewed as the single foundation for architectural and landscape design. Depending on a facility's location, these can include, for instance, requirements for low water usage plumbing appliances and energy-efficient lighting. However, in order to get the best pollution prevention benefit, the concept of pollution prevention must also be a design driver. Whether the issue is facility layout, materials of construction, heating and lighting systems, or the inclusion of sidewalks, various types of pollution prevention goals can be fostered.

For example, Kodak Park, the home of Eastman Kodak, is an eight-mile-long industrial facility located in Rochester, New York. Operating since the 19th century, this industrial area uses a dual piping system providing both potable and nonpotable water. The nonpotable, untreated water is plumbed to toilets, hand-washing sinks and other areas of the facilities that do not require potability or other special treatment, resulting in the use of water treatment chemicals only when needed for sanitation.

The Fiesta Mall commercial area in the city of Mesa, Arizona, illustrates several types of design appropriate to pollution prevention. This area is at a hub of several major thoroughfares and just a short distance from a freeway entrance. Automobiles are a virtual way of life. However, several design components have been implemented to promote a reduction in car trips.

Why Make Access to Public Transportation Difficult?

The United States has been required to subsidize public transportation systems heavily to keep them operating. The public policy motivation for this has been reducing air pollution and congestion, as well as providing an affordable system of transportation to those who are low on the economic scale.

Unfortunately, the private sector makes decisions that directly contradict this public policy approach, inhibiting the pollution prevention benefit. A Seattle company, for instance, relocated from a downtown office to a plush suburban office campus. In the one day it took to move, it went from having 85 percent of its employees using public transit to get to work, to having less than 5 percent being able to realistically use public transit. But it did agree to pay for the parking permits required at the new facility.

Then there is the tragic case associated with the Galleria Mall in Cheektowaga, New York, a suburb of Buffalo. The Galleria is the largest mall in that metropolitan area, attracting droves of price-conscious shoppers from Canada. Like most malls, it is congested. But unlike most malls, it did not permit the public transportation system to have a bus stop on the mall's private property.

The walk to the mall buildings from the perimeter bus stop, after crossing a major highway, was long and dangerous in good weather. But in this notoriously snowy climate, it became treacherous. In early 1996, a young store clerk, proceeding to a bus stop after work at night, was killed by a motor vehicle while trying to cross the busy roadway outside the mall property. After considerable controversy, Galleria Mall began to permit buses to stop within the mall premises.

How convenient or safe is it for employees at your facility to get access to public transit services? Can you work with the public transit authority to improve this situation?

For public transit officials, are park-and-ride lots designed to promote safe walking and/or biking and riding?

There is handy public bus access across from the hotels in the area. The commercial development is also laid out with sidewalks designed to encourage walking between establishments. So, if you are a business traveler using one of the several hotels located a block off the main streets, you are not trapped. You have a pleasantly landscaped walk to several types of restaurants and shops without having to walk on the busy roads. Commercial district employees of every economic class can be observed riding their bicycles to work.

The landscaping relies heavily on xeriscaping. Through this approach, hardy, native plants or other drought-tolerant plant species are used to create a comfortable sidewalk micro-climate. These plants survive and thrive in the desert environment (110° F) with minimal irrigation requirements. Vegetative shade is a useful, inexpensive component that creates the comfort in the sidewalk micro-climate.

The savings from these types of design approaches are long-term and incremental. In small ways, on a day-to-day basis, less energy is used; pollution-free activities, such as walking or bicycling, are encouraged and made safe; water is used judiciously; fewer fertilizers or pesticides are needed to maintain landscape integrity.

Whether the setting is an industrial park, a commercial district, or a governmental campus, pollution prevention design factors make sense. Dur-

Design: Foundation for Cost-Effective Pollution Prevention

Exterior facility design:

- Landscaping uses native plants when possible
- Landscaping is laid out to promote energy efficiency in structures (e.g., trees are placed to shade windows and doors during hot part of the day in the summer, shrubs are placed as a windbreak around foundations in cold, winter climates, etc.)
- Irrigation systems are water-efficient (bubbler in lieu of spray type: in spray systems, 90 percent of the water is evaporated without reaching a plant)
- Sidewalk construction and layout encourage walking, bike riding, and access to public transportation
- Exterior lighting is energy efficient and strategically placed to illuminate sidewalks, especially during the dark months of the peak of winter

Interior facility design:

- Makes use of natural daylight from regular windows and skylights (if executives need those windowed spaces to appease their egos, "relights" can be added to bring daylight into building interiors)
- Includes toilets, sinks, and showers that are water-efficient and low-flow
- Does not overlight and uses interior lighting ballasts that are energy-efficient
- Takes into consideration hazardous operations (e.g., chemical handling) and designs the layout of the work space to reduce the risks of leaks (e.g., does not require a processor in a plating shop to drip a basket of newly plated materials across a long area of floor in order to get to the drying area)

Lighting is used to ensure that people can see what they are doing. Some simple tips on improving visibility without increasing energy use include:

- Placing light ballasts where they do the most good and don't cast shadows on work spaces
- Keeping light bulbs clean
- Painting facilities light colors to reflect light from walls and ceilings
- Replacing inefficient ballasts with lighting ballasts that give required lumens for visibility, but use substantially less energy

Many office buildings have been over-designed for lighting. Some light fixtures can be shut off without creating a work environment with insufficient lumens to operate safely.

ing the early 1980s, a chemical storage facility located within a few yards of San Francisco Bay started a retrofit of an old tank farm and yard storage area. The chief engineer on the project was a skilled construction engineer, but had come of age during the Great Depression. His design bias was to meet established code requirements, but not spend any more money than necessary.

One day, during construction, an issue arose with regard to yard paving. The quandary facing this chief engineer was: Should the asphalt in the yard be three inches thick (minimum code) or five inches thick? The vice-president responsible for paying for the construction made the call with pollution prevention in mind. Five inches was not a hard decision to make, since he was a corporate officer and "designated jailee." Besides, in the earthquake-prone

San Francisco Bay area, additional structural strength can pay itself back very quickly in reduced maintenance.

The moral of this story is to work with facility engineers, designers, and architects to ensure that pollution prevention considerations are a consistent component of construction design.

Retrofit for Efficiency

Such plumbing features as low-flow toilets, flow restrictors in shower heads, and sinks with automatic shutoffs significantly reduce water usage

Disincentives for Pollution Prevention

Metropolitan Phoenix, Arizona, is a growing, dynamic area reaping the economic benefits of refugees from California looking for a mild climate and less congestion. But that growth also breeds air pollution. Its major source is nonpoint, coming from auto emissions. In 1995 the air pollution risk in Arizona air was upgraded from "moderate" to "serious."

U.S. EPA, in an effort to promote improvement, proposed withholding highway construction funds to Maricopa County (Phoenix). This would serve as an incentive for the community to make changes and reduce pollution levels. But money and politics go hand in hand. In September 1996, EPA announced it had abandoned this plan and replaced it with one calling for two highway signs directing motor vehicles to perimeter highways around, rather than through, the metropolitan area, and greater enforcement action against smog-producing vehicles. Instead of simple warning letters to the drivers of smog-producing vehicles, the letters would direct them to fix their vehicles and go through an emissions test.

Oddly, though, a politician is encouraging a return to enforcement tactics over behavioral changes to achieve air quality improvement. Representative Jim Kolbe expressed his satisfaction that EPA is working to clean up dirty air through ". . . tougher enforcement, not sanctions like the loss of highway dollars."[9]

If an organization uses motor vehicles in its operation, regular maintenance ensures that its pollution prevention devices operate effectively.

9. "EPA Scuttles Stringent Plan to Clean Air," *The Arizona Republic*, September 27, 1996, p. 12.

without decreasing the quality of life for the plumbing user. If you operate an older plumbing network, some devices can be inexpensively installed without any special building permits or licenses. Conduct a water usage audit of the facility and see which of the following simple improvements are usable:

- Repair leaking water pipes or faucets
- Replace existing old-style hardware with low-flow faucets and shower heads, or add flow restrictors
- If toilets have a tank system, put a plastic container (1-2 quarts, depending on tank configuration) filled with water inside the tank to displace some of the flow when flushed
- Protect outside faucets and lines from damage due to freezing (drain lines and protect faucets with insulating caps available at home and garden stores)

Energy audits are also effective ways to save money and resources.

- Have heating/air conditioning systems maintained at least annually.
- Change air filters frequently, or clean electrostatic air filters at least monthly.
- Add insulation to attics and ceilings (insulation settles over time).
- Replace single-glaze windows with double- or triple-glaze.
- Select plastic or wood replacement windows, which are more energy-efficient than metal windows.
- Close off heating and air conditioning systems in spaces where they are not needed (e.g., seldom-used offices, storage areas, etc.).
- If the facility has zoned heating and cooling, set the heating/cooling requirements for each zone as required.
- Install paddle fans to circulate air, if appropriate.
- Install attic fans to pull hot air out.
- Use a dehumidifier to remove moisture from the air, and empty it regularly.
- Encourage personnel to dress according to the season and their personal body requirements.

Public utilities recognize that it is less expensive to reduce energy use through conservation than to build new power plants. Therefore, many offer information and tips on increasing energy efficiency. The more progressive send special technicians to an industrial facility to conduct detailed, site-specific energy audits. Some even have low-cost financing to help with structural implementation.

Determine who your energy suppliers are (heating, cooling, electrical) and contact them for their ideas on how to improve your bottom line.

Storing and Transferring Bulk Liquids

Controlled movement and storage of liquids take place through either piping systems or hoses leading into and out of tanks, tank cars, and tank trucks. Common failure points of such systems include:

- Permanent fixed connection points, such as welds
- Stress points, such as piping bends or hose kinks
- Flanged connections
- End points
- Buried piping and tanks

The risks associated with these stress points are increased when coupled with other conditions that promote structural deterioration, such as:

- Weather extremes (e.g, wind, differential heating and cooling, ice buildup)
- Earthquake activity or other types of earth subsidence
- Wear caused by walking on piping systems
- Direct contact with the earth or standing water
- Degradation of insulation

What, then, are some of the techniques that can be applied to reduce the pollution risk associated with these stress points?

- Design piping layouts to minimize the number of pipe bends.
- Use quick disconnects (air pressure or spring loaded) at especially vulnerable locations where hoses or piping systems are routinely changed.
- Mount tanks on platforms, such as concrete pads or saddles, to prevent direct contact with the soil or standing water.
- Install pressure relief valves on tanks.
- Run piping systems on saddles to prevent direct contact with soil or standing water.
- Build crosswalks at logical locations over aboveground piping systems to reduce the opportunity for stepping on the piping.
- Pressure-test hoses at regularly scheduled intervals (e.g., once a year) to assess hose deterioration and the need for replacement.
- Replace gaskets on hoses on a regularly scheduled basis.

- Use gravity feed, rather than air pressure to transfer fluids.
- Use a suction pump rather than gravity feed to transfer fluids.
- Conduct regularly scheduled, nondestructive pressure tests of tanks and fixed piping.

Underground Systems

Despite a massive removal of old underground storage tanks (USTs) during the 1980s, there will always be situations in which an underground tank is more appropriate than an aboveground system. Where space is tight, or the commodity stored highly volatile, underground storage is strategically safer, even if there is a risk of leakage. Additionally, in parts of the country where heating oil is a major fuel source, the widespread elimination of underground tanks and piping is unrealistic.

To assist in managing these systems with pollution prevention in mind, the U.S. Environmental Protection Agency has produced several useful tools (Table 7.4), which range from methods for reconciling inventory to alternatives for leak detection.

IDENTIFY RECYCLING OPPORTUNITIES

Recycling and the use of products made from recycled materials can make a significant contribution to pollution prevention. Prime contenders for the gold medal in the recycling Olympics are the city of Seattle and King County in Washington State (Figures 7.1 and 7.2).

Past practices provided the initial incentive for recycling of glass, aluminum, paper, and even certain plastics. It seems that, in the mid-1980s, the residents of Seattle found themselves the not-so-proud owners of two Superfund sites, which had previously been landfills. The closure costs for these landfills was $100 million and was borne by local residents, who found themselves facing the highest garbage disposal rates in the country.

Drawing on the frugal nature of its community, as well as the strong environmental ethic of its populace, the city was able to develop and successfully implement an alternative waste management effort of curbside recycling.

While the actual rules change periodically (depending on access to recycling markets), residents and businesses are able to recycle at their curbs at scheduled times. There are two methods for collecting recyclables at the curb: either sorted and weekly, or less sorted and monthly. The method used is based on where the home or business is located.

■ **TABLE 7.4 UST Management Resources**

Location	Publication Name	Summary
Appendix E.6	A Summary of Federal Regulations for Underground Storage Tank Systems	Federal requirements on leak detection; corrective action; closure and financial responsibility are presented for the UST owner/operator.
Appendix E.7	Introduction to Statistical Inventory Reconciliation for Underground Tanks	Statistical Inventory Reconciliation (SIR) uses computer software to consistently analyze tank inventory, delivery, and dispensing data. This booklet explains the SIR process, including factors that affect data quality. Its concepts are also applicable to aboveground tank inventory management.
Appendix E.8	Leak Detection Methods for Petroleum Underground Storage Tanks and Piping	This publication explains several methods for UST leak detection, including automatic and manual tank gauging, vapor and groundwater monitoring.
Appendix E.9	Financial Responsibility Requirements for Underground Storage Tanks	Five categories of UST owners/operators face specific financial responsibility requirements. Insurance and other financial options are presented to assist organizations in selecting the most effective way to meet financial tests.

There is curbside pickup of yard wastes (called Clean Green) which are then shredded, composted and sold. Residents can even get worm bins and special green plastic containers for on-site management of food wastes that would otherwise go down the garbage disposal.

While convenience helps promote the program, the biggest incentive is cost. Each individual facility is given the ability to control its waste disposal costs. All households must pay for solid waste service, even if they don't use it. However, they can select a level of service based on household needs. There are mini-cans, regular cans, and special tags for bulky or heavy items. Collection service is privately contracted, but administered by the city.

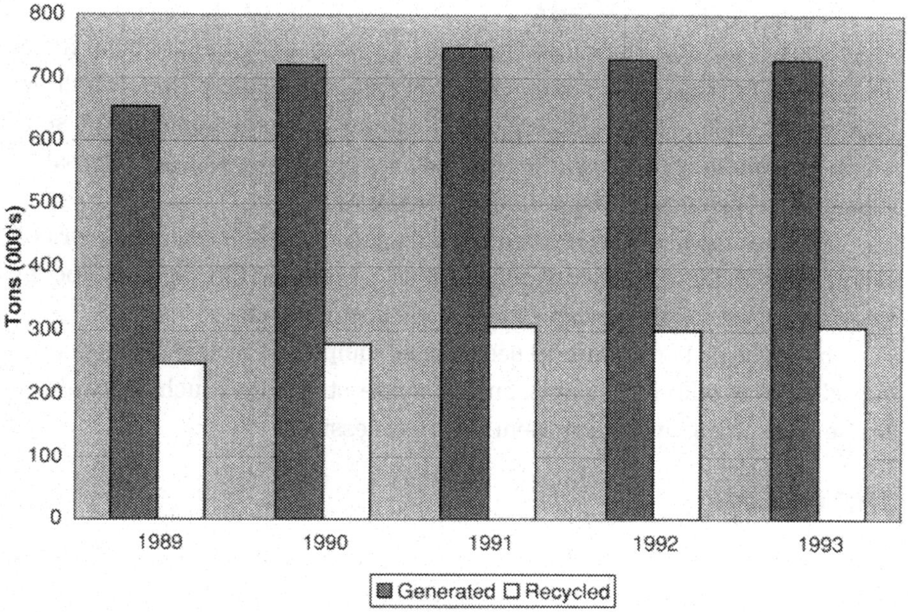

■ **FIGURE 7.1 Waste Management: Seattle, WA**

Source: Seattle Solid Waste Utility, Seattle, WA

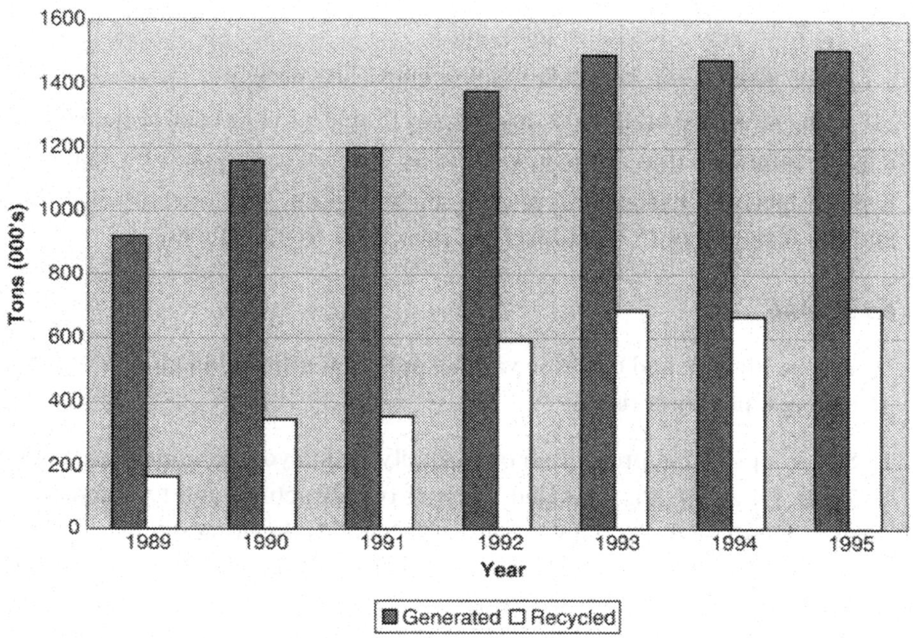

■ **FIGURE 7.2 Waste Management: King County, Washington**

Source: King County, Washington, Solid Waste Division

What Can Be Done?

The act of recycling is not limited to a reprocessing requirement. It can be also mean reuse. Many state and local solid waste authorities can direct industry to waste exchanges, organizations that serve as conduits for solid waste information. They work to link businesses with unwanted, but usable, materials to businesses that can use the material.

Office supply companies, for instance, have developed services for collecting and refilling toner cartridges from computer, fax, and photocopy systems. Other companies specialize in electronic scrap.

In Texas, it is common to see special equipment at highway construction sites that take old cement and concrete structures, such as outdated bridge abutments, and crush them for reprocessing.

THE P^2 LOG

As an implementation foundation, this chapter reviewed both structural and behavioral alternatives for pollution prevention:

- Contracts
- Operator error
- Training
- Emergency response planning
- Construction, architecture, and landscape design

The table that concludes this chapter (Table 7.5) has been assembled as a quick reference that can help you create your own strategies for improvement. It has been designed to get you thinking about your own organization, and the diversity of P^2 solutions that exist for it to implement.

Exercise

Objective: Identify and describe various pollution prevention alternatives for managing a pollution risk.

1. Select a pollution prevention opportunity within your organization. Using Table 7.4 as an aid, if needed, identify two structural and two nonstructural alternatives for eliminating, reducing, or mitigating the risk.
2. For each item identified in #1, define whether the alternatives selected are corrective or preventive in nature.
3. Prepare a scope statement of no more than two sentences, which describes the proposed project.

TABLE 7.5 Catalog of P² Alternatives

Description of Process	P² Alternatives
Function: Shipping and Receiving	Structural (engineering): **1.** Berming to protect storm drains in vicinity **2.** Sloping of loading bay to capture leakage **3.** Energy-efficent lighting system
Areas of Pollution Risk: ■ Damaged hazardous materials freight ■ Fire ■ Energy use (forklifts, lighting)	Nonstructural (behavioral): **1.** Train/certify employees per 49 CFR 172 ■ General awareness ■ Safety ■ Function-specific: emergency response **2.** Emergency response training (29 CFR 1910.120) **3.** Emergency response equipment
Function: Raw Materials Storage: Tanks	Structural (engineering): **1.** Tanks constructed above ground **2.** Tanks placed on pads or saddles **3.** Berm construction to contain size of largest tank + 10 percent **4.** Alternative secondary containment **5.** Air emissions collection/treatment system **6.** Cathodic protection of tanks/piping systems **7.** Overflow alarm system **8.** Berming and impervious flooring at tank truck or rail tank car delivery area **9.** Secondary containment below surface of tank truck or rail tank car delivery area **10.** Fencing **11.** Catchment basins at all permanent connection points **12.** Mechanical pumping with emergency shutoff

(Continued)

TABLE 7.5 (*Continued*)

Description of Process	P^2 Alternatives
Areas of Pollution Risk: ■ Tank failures (rupture or leak) ■ Hose or piping system failure (rupture or leak) ■ Air emissions ■ Leaks at connection points ■ Sludge from tank cleaning ■ Vandalism	Nonstructural (behavioral): **1.** Preventive maintenance to include: ■ Nondestructive testing of tanks, hoses, and piping systems ■ Valve and gasket replacement ■ Exterior/interior coating applications **2.** Emergency response training (29 CFR 1910.120) **3.** Emergency response equipment **4.** Routine equipment inspection
Function: Raw Materials Storage: Packaged Goods (hazardous materials)	Structural (engineering): **1.** Automatic fire alarm **2.** Sprinkler system **3.** Storage over solid flooring
Areas of Pollution Risk: ■ Releases from punctured or fatigued containers ■ Container failure in fire ■ Aged inventory	Nonstructural (behavioral): **1.** Storage on pallets or pallet racks **2.** Emergency response training (29 CFR 1910.120) **3.** Emergency response equipment **4.** Routine equipment inspection
Function: Research and Development Laboratory	Structural (engineering): **1.** Automatic fire alarm **2.** Sprinkler system **3.** Emissions control device for laboratory hoods **4.** Containment for shelf storage **5.** Energy efficient lighting **6.** Sinks drain into containment basin

Areas of Pollution Risk: ■ Releases from punctured or fatigued containers ■ Container failure in fire ■ Air emissions from lab hoods ■ Water use ■ Energy use ■ Compressed gas cylinders used in processes ■ Biological hazards ■ Aged inventory	Nonstructural (behavioral): **1.** Emergency response training (29 CFR 1910.120) **2.** Emergency response equipment **3.** Routine equipment inspection
Function: Product Processing Areas of Pollution Risk: ■ Releases from punctured or fatigued containers ■ Container failure in fire ■ Air emissions ■ Water use ■ Energy use ■ Discharge to sewer treatment system ■ Discharge to waterways ■ Waste production	Structural (engineering): **1.** Wastewater treatment **2.** Air stack scrubbers Nonstructural (behavioral): **1.** Energy audit **2.** Water use audit

(*continued*)

■ **TABLE 7.5** ***(Continued)***

Description of Process	P^2 Alternatives
Function: Packaging Areas of Pollution Risk: ■ Waste production ■ Releases from punctured or fatigued containers ■ Container failure in fire ■ Overpackaging	Structural (engineering): **1.** Energy-efficient lighting Nonstructural (behavioral): **1.** Train/certify employees per 49 CFR 172 ■ General awareness ■ Safety ■ Function-specific: packaging **2.** Emergency response training (29 CFR 1910.120) **3.** Emergency response equipment
Function: Facility Maintenance Areas of Pollution Risk: ■ Fire/explosion ■ Chemical reaction ■ Equipment leaks and failures	Structural (engineering): **1.** Primary and secondary containment systems Nonstructural (behavioral): **1.** Preventive maintenance schedule **2.** Environmental and general liability coverage from contractor **3.** Contractor environmental review work plan **4.** Daily pre-startup, post-activity environmental inspection
Function: Storage of Product in Process (hazardous materials)	Structural (engineering): **1.** Automatic fire alarm **2.** Sprinkler system **3.** Storage over solid flooring

Areas of Pollution Risk: ■ Releases from punctured or fatigued containers ■ Container failure in fire	Nonstructural (behavioral): **1.** Storage on pallets or pallet racks **2.** Emergency response training (29 CFR 1910.120) **3.** Emergency response equipment **4.** Routine equipment inspection
Function: Storage of Finished Product (hazardous materials)	Structural (engineering): **1.** Automatic fire alarm **2.** Sprinkler system **3.** Storage over solid flooring
Areas of Pollution Risk: ■ Releases from punctured or fatigued containers ■ Container failure in fire	Nonstructural (behavioral): **1.** Storage on pallets or pallet racks **2.** Emergency response training (29 CFR 1910.120) **3.** Emergency response equipment **4.** Routine equipment inspection
Function: Emergency Response	Structural (engineering): **1.** Fixed containment devices
Areas of Pollution Risk: ■ Emissions into all media ■ Water, energy use for cleanup	Nonstructural (behavioral): **1.** Emergency response training (29 CFR 1910.120) **2.** Emergency response equipment

(*Continued*)

■ **TABLE 7.5 (*Continued*)**

Description of Process	P^2 Alternatives
Function: Shipping Product (hazardous materials) Areas of Pollution Risk: ■ Damaged containers ■ Fire ■ Energy use (forklifts, lighting)	Structural (engineering): **1.** Berming to protect storm drains in vicinity **2.** Sloping of loading bay to capture leakage **3.** Energy-efficent lighting system Nonstructural (behavioral): **1.** Train/certify employees per 49 CFR 172 ■ General awareness ■ Safety ■ Function-specific: emergency response **2.** Emergency response training (29 CFR 1910.120) **3.** Emergency response equipment
Function: Shipping Waste Areas of Pollution Risk: ■ Damaged containers ■ Fire ■ Energy use (forklifts, lighting)	Structural (engineering): **1.** Alarm systems **2.** Containment in staging and shipping areas Nonstructural (behavioral): **1.** train/certify employees per 49 CFR 172 ■ General awareness ■ Safety ■ Function-specific: emergency response **2.** Emergency response training (29 CFR 1910.120) **3.** Emergency response equipment
Function: Acquisitions	Structural (engineering): **1.** Change all locks **2.** Assess and upgrade fire protection and alarm systems as needed **3.** Construct temporary berms, catchment basins until capital plan prepared and implemented

Areas of Pollution Risk: ■ Equipment maintenance ■ Fire ■ Waste accumulation ■ Aged inventory ■ Inefficient processing equipment ■ Poor lighting ■ Inefficient use of water ■ Unknowns in inventory	Nonstructural (behavioral): **1.** Upgrade work practices in processing area **2.** Upgrade waste management **3.** Implement maintenance evaluation of all equipment **4.** Implement preventive maintenance schedule **5.** Determine training gaps and/or upgrade training to minimum compliance standards **6.** Install emergency response equipment **7.** Energy audit **8.** Water use audit
Function: Miscellaneous (office, retail space, etc.) Areas of Pollution Risk: ■ Energy use ■ Water management ■ Solid waste management ■ Mis-shipments of hazardous materials ■ Hazardous materials use in maintenance of office equipment, facilities	Structural (engineering): **1.** Low-flow resistrictors for all plumbing fixtures **2.** Energy-efficient lighting **3.** Automatic timers for selected lighting systems **4.** Structural insulation **5.** Trash segregation (for recyling) **6.** Trash compaction (to reduce volume) Nonstructural (behavioral): **1.** Energy audit **2.** Water use audit

(*Continued*)

■ **TABLE 7.5** *(Continued)*

Description of Process	P^2 Alternatives
Function: Outside Contractor Activity	Structural (engineering):
	1. Shoring and trenching
Areas of Pollution Risk:	Nonstructural (behavioral):
■ Fire hazards ■ Hazardous materials (including compressed gases) on-site ■ Generation of hazardous waste	**1.** Environmental and general liability coverage from contractor **2.** Contractor environmental review work plan **3.** Daily pre-startup, post-activity environmental inspection
Function: Environmental Remediation	Structural (engineering):
	1. Site-specific per the work plan
Areas of Pollution Risk:	Nonstructural (behavioral):
■ All media ■ Shipment or on-site treatment of contaminated materials	**1.** Train/certify employees per 49 CFR 172 ■ General awareness ■ Safety ■ Function specific: emergency response **2.** Emergency response training (29 CFR 1910.120) **3.** Emergency response equipment
Function: Audits and Inspections (internal and external)	Structural (engineering):
	1. Emergency shutoff switches
Areas of Pollution Risk:	Nonstructural (behavioral):
■ All media	**1.** Lock-out/tag-out procedure

CHAPTER

8

ACHIEVING P^2 IMPLEMENTATION EXCELLENCE AND COST SAVINGS

When does pollution prevention fail? It fails when it does not or cannot be implemented. Common sense tells us that preventing problems is better than correcting them once they occur. But common sense isn't always the determining factor in a decision to take an action.

Human nature creates significant social, political, and legal barriers. Coupled with the expected challenges of physical constraints and nature, getting a pollution prevention idea into complete and continuous implementation is not always easy.

This chapter presents a discussion on recognizing and managing barriers to change. Its orientation is toward the behavioral aspects of implementation. Thus, it asks the pollution prevention practitioner to recognize and respond to these aspects.

BARRIERS TO CHANGE

Public Perception of Business

A limiting factor facing business in the United States is the perception that business cannot be trusted. As commonly portrayed in the media, business is considered big, dishonest, rich in capital, and intent on polluting the environment for additional profit.

These concepts and ideas, whether true or not, are instilled in the public psyche through repetition. We see them in movies and on TV. They are represented in children's cartoon shows. We read about them in papers and magazines. Bad news catches people's attention. Bad news sells papers. Tragic or catastrophic news sells even more. So both the media outlets and their advertisers are happy.

But this pervasive perception begins to haunt us in unexpected ways. Consider the reaction of the owner of a privately held chemical distribution company. He painfully described how his six-year-old daughter, in a conversation about an upcoming Earth Day event, asked her father if he polluted for a living.

The attitude of mistrust is also found within the regulatory enforcement community. By definition they are faced with evidence of mismanagement, error, and worse. They have seen the sloppy operators, experienced the missed deadlines for improvement, and received public criticism for being too easy on business. Concurrently, many regulators are career public employees, with limited firsthand knowledge of the fundamentals of running a business activity that adds value to the economy.

Can an organization overcome this anti-business bias? Can the regulating community become a stakeholder in effective problem solving?

To be successful in crossing this barrier, organizations benefit from admitting that this type of thinking exists. Once acknowledged, strategies can be designed and adopted to counteract its effects. They can be applied to build a higher level of trust between business organizations and the communities in which they are located or conduct business.

The components of success become open and honest communication coupled with systematic, consistent action toward improvement. This isn't glamorous, and it probably won't make the evening news. But it can result in lower pollution prevention costs, as regulatory authorities receive community feedback and are therefore encouraged to promote greater flexibility in problem solving.

Environmental Racism

Environmental racism is an intellectual argument that concludes that poor and minority members are more likely to live in or near areas of high pollution risk. Whether this is true or not, in fact, is still subject to intense debate. However, the basic communications psychology in the United States suggests that perception is reality. Therefore, if environmental racism is perceived, it is real to those potentially affected by it.

Why should this be of interest to the pollution prevention practitioner? Certainly, there should be concern about the environmental impacts of plant sabotage, whether overt or covert in nature. The damage can be just as serious from a focused, single community "enemy" as from the uncontrolled force of a rioting mob.

Recognize, too, that there are so-called social activists who channel their intellectual energies, and sometimes their entire lives, at closing down a business activity of one type or another. They can orchestrate a strategic bombardment of negative publicity.

The free-market system, over time, results in rational economic behavior. It balances costs and benefits with competition and demand to achieve the right price for a good or service. That is the price the market will bear.

It has been suggested: "The only way to achieve an equitable distribution of environmental costs and benefits is to charge those benefitting from the land use while financially compensating those who host the undesirable facility."[1]

Two Decades of Public Pressure

Beginning in the mid-1970s, Portland General Electric (Portland, Oregon) was publicly hounded by one single-issue activist, who focused on a nuclear power plant operated by the company. The plant, known as Trojan, was closed in 1992, several years ahead of schedule. The cost of maintaining operating equipment, coupled with low competitive prices from natural gas, made continued facility operation uneconomical.

Closures and permanent shutdowns bring their own special pollution management requirements. Closures under public pressure are not generally cost-effective. Even planned actions, such as RCRA closures or plant demolitions, can result in unanticipated costs for managing complex environmental remediations.

1. Boerner, Christopher and Thomas Lambert, "Environmental Justice Can Be Achieved through Negotiated Compensation," in *Environmental Justice*, Greenhaven Press, Inc., 1995, p. 85.

The thought of compensation raises red flags of high costs and out-of-control litigation. But the free-market system is not inherently litigious. The type of compensation can include such creative ideas as:

- Purchasing residences in the surrounding community and subsidizing below-market rents for the occupants
- Promoting employment opportunities for the community at the facility
- Offering adult literacy training and other types of enrichment education so the surrounding community can be more employable at the facility
- Contributing financial and in-kind donations to local neighborhood associations or community churches
- Making selected areas of the facility available for meetings at below-market rates to community activities, such as the Girl or Boy Scouts
- Providing off-street, secure parking on evenings and weekends

Organizational Sabotage

In an earlier chapter, the concept of actively working to achieve organizational buy-off was introduced. It spoke of group participation in order to accomplish consensus decision making. Its theory is that when consensus is reached, implementation proceeds without problems.

The reality, however, is that people change their minds, even after reaching so-called consensus. So, although individuals involved in a consensus decision may appear to agree, they can then work to cause a project or program to fail by undermining its implementation.

Clearly, it is more cost-effective to introduce a pollution prevention project or program and have it work than to have it fail. Thus, implementation planning needs to address both the conscious and unconscious organizational sabotage that may occur which can result in project or program failure.

This type of implementation planning is an honest assessment of organizational strengths and weaknesses conducted in the interests of achieving complete and continuous project/program implementation success. Questions to ask in the assessment include:

- Where did the idea for the project or program orginate?
- Are there any internal political barriers created by this origin?
- Is management committed to supporting the project or program and all the changes required?

- What specific actions can management take to reinforce its commitment through personal action?
- Who may want the project or program to fail?
- Why?
- What actions can weaken the success of project or program implementation?
- What can be done to prevent these actions from occurring?
- If the actions occur, what can be done to minimize their impact?

People react negatively or resist change because they are afraid, or because their own sense of security is threatened. They may feel that they are powerless and have lost control. They may actually be embarrassed and work to cause problems to distract attention from this embarrassment.

But perhaps the most important reason change is resisted is because people have inadequate information on which to assess the full positive implications of the change. If you detect resistance, determine what you can do to improve the true flow of useful information throughout the organization.

The objective is to overcome the resistance by creating a win-win situation. That is, it becomes a win for the individual as well as the organization.

It can help to distill the essense of the project or change to one or two simple sentences. Then, begin to repeat that message whenever the opportunity presents itself within the organization. Repetition is the hallmark of every successful marketer.

MAKING P^2 WORK

Communicate with the Community

The community is diverse, and perhaps larger than you think. Understand that each community member is a stakeholder in pollution prevention success or failure. So recognize the existence of all these stakeholders, including:

- Employees
- Retired employees
- Former employees
- Residents of the community
- Political leaders in the community
- Suppliers

- Critics
- Regulatory agency personnel

Think of the way a small business, such as a local real estate or insurance office, "works" the community. Its sales and management personnel join business service organizations, such as Kiwanis or Rotary. They organize and volunteer for local neighborhood cleanup campaigns or public service fundraisers. While they do this, they are not selling an insurance policy or a listing, but they are selling themselves and learning more about their community in the process.

Many obstacles to change and improvement are made because of inconsistent, incoherent, or misunderstood communication. Where it concerns the ability to implement a desired pollution prevention project or program, this means continually marketing the item to all the audience groups or stakeholders. Just because you believe in something doesn't mean that other potential stakeholders automatically believe or have the information needed to make a rational choice.

Build community trust, one action at a time. Community events, such as Earth Day (Table 8.1) or Take Your Daughters to Work Day, can be used to deliver a positive message about the organization's contributions to the community and the environment. They don't have to be expensive, but to be successful, they should address several factors:

- Who is the target audience for the message?
- What is a one-sentence statement of the message?
- What is the time frame for delivering the initial message?
- Are follow-up messages required or desirable?

These types of activities lay a groundwork in the community that expands the options that can be considered when working with regulatory enforcement personnel to specify a pollution prevention solution.

People Respond to Incentives

There are inherent weaknesses in traditional command-and-control management systems. These have become glaringly apparent as global competition in production and services intensifies. While useful in times of crisis or in noncompetitive markets, certain practices represent enormous burdens in responding to emerging worldwide economic pressures.

These burdens include:

■ TABLE 8.1 Ideas for Working with Community Stakeholders

Target Audience	Description of Communication Event	Message	Time Frame	Follow-up
Elementary school students and teachers	Conduct “environmental” workshop for 2nd/3rd graders	Products can be made/used in an environmentally sensitive way	Earth Day project	Annually, or with different schools
Fire department: local first responders	Facility tour with an emphasis on areas of high environmental risk	Working together, we can protect the community	When SARA Tier 2 report filed	Annually
Local community leaders	Speech with question and answer session	Here is what we do to apply pollution prevention in our activities	Periodic; at least once per calendar year is recommended	With different types of community groups, from senior citizen social clubs to social service clubs

- Legislatively prescribed solutions to technical problems
- Leverage for change based solely on power and force
- Inflexibility and bias

The best incentives for change and improvement are market-driven. The free, unencumbered market offers the best opportunities for progress. In practice, this does not mean chaos and a breakdown of order. Rather, it means eliminating barriers created by prescriptive, self-limiting responses concurrent with the promotion of problem resolution.

The prescriptive and punitive nature of many environmental regulations limits the solutions that are initially applied to some environmental issues. Others create a disincentive to identifying where environmental problems (and, therefore, pollution prevention opportunities) might exist.

Be creative with your solutions. The point is to resolve and/or minimize a problem. If you have a better solution, but it doesn't seem to fit the language of a prescribed regulation, sit down with the regulators and learn what flexibility exists that you may not understand.

Involve regulators at all levels as necessary (local, state, and federal). In some cases, involve elected officials. They know who pays the taxes and who votes. They have been elected to represent us.

Much of our environmental regulatory structure permits negotiation and the granting of discretionary exceptions. While the process may seem long and complex, it can be made more flexible, without having to involve expensive legal counsel.

How can you create leverage without force? Think about the things that marketers do to persuade you to buy a certain product. They can give a direct incentive or create an indirect incentive.

Direct incentives are easiest to understand and involve something tangible to motivate a purchase or "buy-off." Safety award programs, environmental incentive bonuses, even a free pizza party can all be effective. A combination of both long-term (annual bonus) and short-term (pizza party for a spill-free month) actions is effective, since it brings gratification quickly.

Indirect incentives are less tangible in nature. They can't be held in the hand and looked at, but they are extremely powerful. They are also more personal in nature—what is an intangible incentive for one person may not be for another.

For example, some people may become more environmentally responsible because they like to hunt and fish; others, because they are aggravated by traffic congestion, and still others, because of the feeling of accomplishment associated with being finally able to read the instructions on any consumer product purchased.

Open the Tool Box

Regulatory inspections may seem to create, rather than solve, problems in an organization. However, a systematic review of such inspections can be used to measure performance of existing programs concurrent with providing insights into additional opportunities for improvement. Application of standard evaluation tools (Appendix C.7) organizes the elements of these third-party inspection activities into data from which future management decisions may be derived.

Likewise, when problems occur, a thoughtful, ordered investigation opens the door to better solutions (Appendix C.8). By using an objective tool for fully assessing an incident, blame is avoided, factual evidence is developed, and all issues are considered. The resultant conclusions lead the organization to development of corrective and preventive actions that are more cost-effective than those reached by biased, reactive means.

Sometimes, companies change, but their administrative systems fail to follow suit. In addition to the operational audits presented earlier in this book, systematic audits of such things as documented procedures (Appendix C.9), or even forms, can identify gaps, inconsistencies, conflicts, or organizational inefficiencies not previously realized.

One chemical company, when such an audit revealed immense administrative confusion and duplication, appointed a "Forms Goddess," who functioned to keep track of what internal forms the company invented. Additionally, she ensured that each form template had a designated owner and control number. This owner approved all changes or could have the form eliminated from use. The owner also specified retention requirements, such as where, and for how long, records developed from the forms were kept.

Keep P^2 Projects Simple

Many projects condemn themselves to failure by either being, or appearing to be, too complex. In communicating a project or program, keep it simple.

- Break complex projects into easily understood components, each of which can be explained by itself.
- Keep the end in sight, and relate the project component to the goal.

- Speak in the vocabulary of the listener.
- Reinforce the benefits (incentives) for change.

Keep in mind that one of the most challenging groups to communicate with can be your senior management. You need them on your side to ensure that funding is available and their intangible support of the project or program is shown consistently.

Senior management understands the language of money. Keep them informed about progress and costs during project implementation.

After a project is in place, communicate periodically about its effects. Speak in terms of money saved (less waste), money earned (lower insurance costs), or new business obtained.

SELECT THE P² DIRECTION FOR CHANGE, THEN MAKE IT HAPPEN

Pollution prevention costs can be controlled while achieving established pollution prevention targets. Using the U.S. regulatory structure as a basis for action, this book has considered the integration of financial information into the pollution prevention process.

Programs such as Responsible Care® and ISO 14001 provide alternative institutionalized approaches for identifying, quantifying, and costing environmental exposures.

The total cost-effectiveness of proposed corrective and preventive actions is more objectively evaluated through the application of such tools as environmental cost accounting, environmental auditing, and statistical process control. These same tools can later provide measurements for evaluating the true effectiveness of a pollution prevention activity.

Change is inevitable. But controlled change is not automatic. It takes a concerted effort to move along a path and create the future.

Exercise

Objective: Begin to improve the structure and value of administrative systems that support pollution prevention.

1. Assemble a list and sample copy of internally developed forms used to facilitate recordkeeping as it relates to a specific activity, product, or service that represents an environmental exposure for the organization.

2. Evaluate each form. For each, specifically determine if the form:
   a. Has a designated owner
   b. Is current, as indicated by a visible preparation date or revision number
   c. Is current, as it relates to existing environmental regulations and/or documented internal requirements or procedures
   d. Has a defined retention time
   e. Is retained in accordance with the defined retention time
   f. Is still suitable and appropriate for the organization
3. From the assembled list, select one form and describe what the impact would be to the organization if the form were completed incorrectly.

Responsible Care®
A Public Commitment

GUIDING PRINCIPLES

As a member of the Chemical Manufacturers Association, this company is committed to support a continuing effort to improve the industry's responsible management of chemicals. We pledge to manage our business according to these principles:

- To recognize and respond to community concerns about chemicals and our operations.
- To develop and produce chemicals that can be manufactured, transported, used and disposed of safely.
- To make health, safety and environmental considerations a priority in our planning for all existing and new products and processes.
- To report promptly to officials, employees, customers and the public, information on chemical-related health or environmental hazards and to recommend protective measures.
- To counsel customers on the safe use, transportation and disposal of chemical products.
- To operate our plants and facilities in a manner that protects the environment and the health and safety of our employees and the public.
- To extend knowledge by conducting or supporting research on the health, safety and environmental effects of our products, processes and waste materials.
- To work with others to resolve problems created by past handling and disposal of hazardous substances.
- To participate with government and others in creating responsible laws, regulations and standards to safeguard the community, workplace and environment.
- To promote the principles and practices of Responsible Care® by sharing experiences and offering assistance to others who produce, handle, use, transport or dispose of chemicals.

Appendix A is reprinted with the permission of the Chemical Manufacturers Association.

■ APPENDIX A.2 Pollution Prevention Code of Management Practices

Responsible Care®
A Public Commitment

POLLUTION PREVENTION
CODE OF MANAGEMENT PRACTICES

Purpose

This Code is designed to achieve ongoing reductions in the amount of all contaminants and pollutants released to the air, water, and land from member company facilities. These reductions are intended to respond to public concerns with the existence of such releases, and to further increase the margin of safety for public health and the environment.

The Code is also designed to achieve ongoing reductions in the amount of wastes generated at facilities. These reductions are intended to help relieve the burden on industry and society of managing such wastes in future years.

In implementing the Code, each company should strive for annual reductions, recognizing that production rates, new operations, and other factors may result in increases. Despite these fluctuations, however, the goal is to establish a long-term, substantial downward trend in the amount of wastes generated and contaminants and pollutants released. Quantitative reduction goals will be established for giving priority to those pollutants, contaminants and wastes of highest health and environmental concern.

This code also includes practices that address the broader waste management issues beyond source reduction and other waste and release reduction efforts. Each member company must manage remaining wastes and releases in a manner that protects the environment and the health and safety of employees and the public.

This Code complements, and should be implemented in conjunction with current and future Codes of Management Practices. Key terms are defined in the Glossary, which should be consulted for assistance in interpreting the provisions of this Code.

Relationship to Guiding Principles

Implementation of this Code helps achieve the following Guiding Principles:

- To recognize and respond to community concerns about chemicals and our operations.

- To develop and produce chemicals that can be manufactured, transported, used and disposed of safely.

- To make health, safety, and environment considerations a priority in our planning for all existing and new products and processes.

- To report promptly to officials, employees, customers and the public, information on chemical-related health or environmental hazards and to recommend protective measures.

- To operate our plants and facilities in a manner that protects the environment and the health and safety of our employees and the public.

- To extend knowledge by conducting or supporting research on the health, safety, and environmental effects of our products, processes, and waste materials.

- To promote the principles and practices of Responsible Care® by sharing experiences and offering assistance to others who produce, handle, use, transport, or dispose of chemicals.

- To work with others to resolve problems created by past handling and disposal of hazardous substances.

- To participate with government and others in creating responsible laws, regulations and standards to safeguard the community, workplace and environment.

- To promote the principles and practices of Responsible Care® by sharing experiences and offering assistance to others who produce, handle, use, transport or dispose of chemicals.

Management Practices

Each member company shall have a pollution prevention program which shall include:

1. A clear commitment by senior management through policy, communications, and resources, to ongoing reductions at each of the company's facilities, in releases to the air, water, and land and in the generation of wastes.

2. A quantitative inventory at each facility of wastes generated and releases to the air, water, and land, measured or estimated at the point of generation or release.

3. Evaluation, sufficient to assist in establishing reduction priorities, of the potential impact of releases on the environment and the health and safety of employees and the public.

4. Education of, and dialogue with, employees and members of the public about the inventory, impact evaluation, and risks to the community.

5. Establishment of priorities, goals and plans for waste and release reduction, taking into account both community concerns and the potential health, safety, and environmental impacts as determined under Practices 3 and 4.

6. Ongoing reduction of wastes and releases, giving preference first to source reduction, second to recycle/reuse, and third to treatment. These techniques may be used separately or in combination with one another.

7. Measurement of progress at each facility in reducing the generation of wastes and in reducing releases to the air, water, and land, by updating the quantitative inventory at least annually.

8. Ongoing dialogue with employees and members of the public regarding waste and release information, progress in achieving reductions, and future plans. This dialogue should be at a personal, face-to-face level, where possible, and should emphasize listening to others and discussing their concerns and ideas.

9. Inclusion of waste and release prevention objectives in research and in design of new or modified facilities, processes, and products.

10. An ongoing program for promotion and support of waste and release reduction by others, which may, for example, include:

    a. Sharing of technical information and experience with customers and suppliers;

    b. Support of efforts to develop improved waste and release reduction techniques;

    c. Assisting in establishment of regional air monitoring networks;

    d. Participation in efforts to develop consensus approaches to the evaluation of environmental, health, and safety impacts of releases;

    e. Providing educational workshops and training materials;

    f. Assisting local governments and others in establishment of waste reduction programs benefiting the general public.

11. Periodic evaluation of waste management practices associated with operations and equipment at each member company facility, taking into account community concerns and health, safety, and environmental impacts and implementation of ongoing improvements.

12. Implementation of a process for selecting, retaining, and reviewing contractors and toll manufacturers taking into account sound waste management practices that protect the environment and the health and safety of employees and the public.

13. Implementation of engineering and operating controls at each member company facility to improve prevention of and early detection of releases that may contaminate groundwater.

14. Implementation of an ongoing program for addressing past operating and waste management practices and for working with others to resolve identified problems at each active or inactive facility owned by a member company taking into account community concerns and health, safety, and environmental impacts.

Industry Trend Data

To develop and maintain statistical industry trends, CMA will collect currently available data. Each company shall report annually to CMA, or its designated agent, for each facility:

- Releases of substances as reported under SARA Section 313; and
- Wastes generated, as defined and reported in CMA's annual waste survey.

Member Self-Evaluation

Each member company shall report annually to CMA, or its designated agent, the stage of implementation of each management practice in this Code. The reports shall be on the member self-evaluation form attached as Attachment A.

Glossary of Terms

As used in this Code, key terms are defined as set forth below. Note that these definitions may be broader than regulatory definitions, and that adherence to this Code does not relieve a company of the obligation to meet Federal, state and local regulatory requirements.

Facility - A site used for chemical manufacturing, processing, refining, packaging, R&D, distribution or related commercial activity.

Recycle - A practice which regenerates or processes a material from a process to recover a useable product or material for reuse.

Release - Any emission, effluent, spill, discharge or disposal to the air, land, or water, of any pollutant or contaminant, whether routine or accidental, at or from a facility. The term does not include shipment or distribution of chemical product, nor release to the environment as part of normal and intended use of a product by the consumer.

Reuse - A practice that reemploys a material from a process either as an ingredient in a process to make a product, or as an effective substitute for a commercial product in a particular function or application.

Source Reduction - A practice that reduces the amount of any release or waste generated at the source, including closed loop recycle and reuse before exit from a process. The term includes, among other practices, equipment and technology modifications, process and procedures modifications, reformulation and redesign of products, substitution of raw materials, and improvements in housekeeping, maintenance, training and inventory control.

Treatment - A practice, other than recycle or reuse, that alters the physical, chemical, or biological characteristics or the volume of a waste through a process or activity separate from the production of a commercial product or the provision of a service.

Waste - Any gas, liquid, or solid residual material at a facility, whether hazardous or non hazardous, that is not used further in the production of a commercial product or provision of a service and which itself is not a commercial product.

■ APPENDIX B.1 Standards-Writing Organizations

American National Standards Institute
11 West 42nd St.
New York, NY 10036

Data Interchange Standards Associations, Inc.
Suite 355
1800 Diagonal Rd.
Alexandria, VA 22314-2852

National Fire Protection Association
1 Batterymarch Park
Quincy, MA 02269

Chemical Specialties Manufacturers Association
1913 Eye St. NW
Washington, DC 20006
phone (202) 872-8110
fax (202) 872-8114

The American Society for Nondestructive Testing, Inc.
Columbus, OH 43271-0901

American Welding Society
P.O. Box 351040
Miami, FL 33135
(305) 443-9353

NSF International
P.O. Box 130140
Ann Arbor, MI 48113-0140
1-800-NSF-MARK

■ APPENDIX B.2 Industry Standards Applicable to Hazardous Waste Activities Incorporated by Reference into 40 CFR 260

"ASTM Standard Test Methods for Flash Point of Liquids by Setaflash Closed Tester," ASTM Standard D-3278-78, available from American Society for Testing and Materials, 1916 Race Street, Philadelphia, PA 19103.

"ASTM Standard Test Methods for Flash Point by Pensky-Martens Closed Tester," ASTM Standard D-93-79 or D-93-80. D-93-80 is available from American Society for Testing and Materials, 1916 Race Street, Philadelphia, PA 19103.

"ASTM Standard Method for Analysis of Reformed Gas by Gas Chromatography," ASTM Standard D 1946-82, available from American Society for Testing and Materials, 1916 Race Street, Philadelphia, PA 19103.

"ASTM Standard Test Method for Heat of Combustion of Hydrocarbon Fuels by Bomb Calorimeter (High-Precision Method)," ASTM Standard D 2382-83, available from American Society for Testing and Materials, 1916 Race Street, Philadelphia, PA 19103.

"ASTM Standard Practices for General Techniques of Ultraviolet-Visible Quantitative Analysis," ASTM Standard E 169-87, available from American Society for Testing and Materials, 1916 Race Street, Philadelphia, PA 19103.

"ASTM Standard Practices for General Techniques of Infrared Quantitative Analysis," ASTM Standard E 168-88, available from American Society for Testing and Materials, 1916 Race Street, Philadelphia, PA 19103.

"ASTM Standard Practice for Packed Column Gas Chromatography," ASTM Standard E 260-85, available from American Society for Testing and Materials, 1916 Race Street, Philadelphia, PA 19103.

"ASTM Standard Test Method for Aromatics in Light Naphthas and Aviation Gasolines by Gas Chromatography," ASTM Standard D 2267-88, available from American Society for Testing and Materials, 1916 Race Street, Philadelphia, PA 19103.

"ASTM Standard Test Method for Vapor Pressure-Temperature Relationship and Initial Decomposition Temperature of Liquids by Isoteriscope," ASTM Standard D 2879-86, available from American Society for Testing and Materials, 1916 Race Street, Philadelphia, PA 19103.

"APTI Course 415: Control of Gaseous Emissions," EPA Publication EPA-450/2-81-005, December 1981, available from National Technical Information Service, 5285 Port Royal Road, Springfield, VA 22161.

"Flammable and Combustible Liquids Code" (1977 or 1981), available from the National Fire Protection Association, 1 Batterymarch Park, Quincy MA 02269

"Screening Procedures for Estimating the Air Quality Impact of Stationary Sources, Revised", October 1992, EPA Publication No. EPA-450/R-92-019, Environmental Protection Agency, Research Triangle Park, NC.

"ASTM Standard Test Methods for Preparing Refuse-Derived Fuel (RDF) Samples for Analyses of Metals," ASTM Standard E926-88, Test Method C Bomb, Acid Digestion Method, available from American Society for Testing Materials, 1916 Race Street, Philadelphia, PA 19103.

API Publication 2517, Third Edition, February 1989, "Evaporative Loss from External Floating-Roof Tanks," available from the American Petroleum Institute, 1220 L Street, Northwest, Washington, DC 20005.

"ASTM Standard Test Method for Vapor Pressure-Temperature Relationship and Initial Decomposition Temperature of Liquids by Isoteniscope," ASTM Standard D 2879-92, available from American Society for Testing and Materials (ASTM), 1916 Race Street, Philadelphia, Pennsylvania 19103.

"Test Methods for Evaluating Solid Waste, Physical/Chemical Methods," EPA Publication SW-846 [Third Edition (November 1986), as amended by Updates I (July 1992), II (September 1994), IIA (August 1993), and IIB (January, 1995)]. The Third Edition of SW-846 and Updates I, II, IIA, and IIB (document number 955-001-00000-1) are available from the Superintendent of Documents, U.S. Government Printing Office, Washington, DC 20402, (202) 512-1800. Copies may be inspected at the Library, U.S. Environmental Protection Agency, 401 M Street, SW, Washington, DC 20460.

The following 47 analytical testing methods are contained in the Third Edition of "Test Methods for Evaluating Solid Waste, Physical/Chemical Methods" EPA Publication SW-846 (November 1986) and its Revision I (December 1987), which are available for the cost of $110.00 from the Government Printing Office, Superintendent of Documents, Washington, DC 20402, (202) 783-3238 (document number 955-001-00000-1):

0010 Modified Method 5 Sampling Train
0020 Source Assessment Sampling System (SASS)
0030 Volatile Organic Sampling Train
1320 Multiple Extraction Procedure
1330 Extraction Procedure for Oily Wastes
3611 Alumina Column Cleanup and Separation of Petroleum Wastes
5040 Protocol for Analysis of Sorbent Cartridges from Volatile Organic Sampling Train
6010 Inductively Coupled Plasma Atomic Emission Spectroscopy
7090 Beryllium (AA, Direct Aspiration)
7091 Beryllium (AA, Furnace Technique)
7198 Chromium, Hexavalent (Differential Pulse Polarography)
7210 Copper (AA, Direct Aspiration)
7211 Copper (AA, Furnace Technique)
7380 Iron (AA, Direct Aspiration)
7381 Iron (AA, Furnace Technique)
7460 Manganese (AA, Direct Aspiration)
7461 Manganese (AA, Furnace Technique)
7550 Osmium (AA, Direct Aspiration)
7770 Sodium (AA, Direct Aspiration)
7840 Thallium (AA, Direct Aspiration)
7841 Thallium (AA, Furnace Technique)
7910 Vanadium (AA, Direct Aspiration)
7911 Vanadium (AA, Furnace Technique)
7950 Zinc (AA, Direct Aspiration)
7951 Zinc (AA, Furnace Technique)

9022 Total Organic Halides (TOX) by Neutron Activation Analysis
9035 Sulfate (Colorimetric, Automated, Chloranilate)
9036 Sulfate (Colorimetric, Automated, Methylthymol Blue, AA II)
9038 Sulfate (Turbidimetric)
9060 Total Organic Carbon
9065 Phenolics (Spectrophotometric, Manual 4-AAP with Distillation)
9066* Phenolics (Colorimetric, Automated 4-AAP with Distillation)
9067 Phenolics (Spectrophotometric, MBTH with Distillation)
9070 Total Recoverable Oil and Grease (Gravimetric, Separatory Funnel Extraction)
9071 Oil and Grease Extraction Method for Sludge Samples
9080 Cation-Exchange Capacity of Soils (Ammonium Acetate)
9081 Cation-Exchange Capacity of Soils (Sodium Acetate)
9100 Saturated Hydraulic Conductivity, Saturated Leachate Conductivity, and Intrinsic Permeability
9131 Total Coliform: Multiple Tube Fermentation Technique
9132 Total Coliform: Membrane Filter Technique
9200 Nitrate
9250 Chloride (Colorimetric, Automated Ferricyanide AAI)
9251 Chloride (Colorimetric, Automated Ferricyanide AAII)
9252 Chloride (Titrimetric, Mercuric Nitrate)
9310 Gross Alpha and Gross Beta
9315 Alpha-Emitting Radium Isotopes
9320 Radium-228

[46 Fed. Reg. 35247, July 7, 1981, as amended at 50 Fed. Reg. 18374, Apr. 30, 1985; 52 Fed. Reg. 8073, Mar. 16, 1987; 52 Fed. Reg. 41295, Oct. 27, 1987; 54 Fed. Reg. 40266, Sept. 29, 1989; 55 Fed. Reg. 8949, Mar. 9, 1990; 55 Fed. Reg. 25493, June 21, 1990; 56 Fed. Reg. 7206, Feb. 21, 1991; 58 Fed. Reg. 38816, July 20, 1993; 58 Fed. Reg. 46040, Aug. 31, 1993; 59 Fed. Reg. 62896, Dec. 6, 1994; 60 Fed. Reg. 3089, Jan. 13, 1995; 60 Fed. Reg. 17001, April 4, 1995.]

■ APPENDIX C.1 Estimating Financial Impacts of Chemical Releases

Type of Event	Assumption: Minimum Cost ($) to Company/Event	Number of Events Time Period: —/—/— through —/—/—	Estimated Minimum Cost ($) to Company
Release contained On-site			
Release with off-site consequence			
Transportation emergency on highway			
Transportation emergency at customer location			
Other:			
Other:			

■ APPENDIX C.2 Environmental Aspects Characterization

Facility: ______________________ Scope:

__ Activity __ Product __ Service
__ New __ Modification __ Existing (no modification)

EPA ID #: ______________________

	Ongoing	**Ongoing**	**New or Modification**	**New or Modification**	**Action**
ENVIRONMENTAL ASPECT(S)	Describe baseline environmental parameters	Proposed environmental objective for improvement	Describe baseline environmental parameters	Proposed environmental objective for improvement	Briefly describe how objective will be achieved
Emissions to atmosphere (controlled)					
Emissions to atmosphere (uncontrolled)					
Discharge to water (controlled)					
Discharge to water (uncontrolled)					
Produce solid waste (nonhazardous)					
Produce solid waste (hazardous, not commingled)					
Produce solid waste (hazardous, commingled)					

	Ongoing	Ongoing	New or Modification	New or Modification	Action
ENVIRONMENTAL ASPECT(S)	Describe baseline environmental parameters	Proposed environmental objective for improvement	Describe baseline environmental parameters	Proposed environmental objective for improvement	Briefly describe how objective will be achieved
Produce special waste (state variations, radioactive, biohazard)					
Discharge to land (controlled)					
Discharge to land (uncontrolled)					
Use of land, fuels, water, energy, and other natural resources					
Odor, noise, dust, vibration and/or visual impact					
Other ecosystem effects					
Other ecosystem effects					

Completed by: ______________
signature

____
print name

Date Completed: __/__/__

■ APPENDIX C.3 Audit Format

Facility: ________

Pollution Prevention Audit
Inspection Date: ________

INSTRUCTIONS: Check or complete as appropriate. Checking NI requires completion of Action Date, Responsible Party at time of audit. Complete Date Resolved after the action item has been corrected.

NA = not applicable, or none observed during inspection   OK = Acceptable   NI = Needs Improvement

Item of Environmental Concern	NA	OK	NI	Action Date	Responsible Party	Date Resolved	If NI, Identify Location	Other Comments
OFFICE								
1. General housekeeping: work areas uncluttered and clear for emergency egress								
2. Slip, trip, fall hazards eliminated								
3. Environmental policy posted								
4. Night/security lighting and timers functional								
5. Electrical outlet integrity/unblocked								
6. Extension cords/portable power tools in good condition								
7. Perimeter locks on building entrances in place/work								

Item of Environmental Concern	NA	OK	NI	Action Date	Responsible Party	Date Resolved	If NI, Identify Location	Other Comments
8. Visitor control in place/used								
9. Intrusion alarms tested								
10. Smoke detector tested								
11. Running, horseplay, unsafe work habits avoided								
12. Exit doors marked/egress unlocked during business hours								
13. Non-exit doors so marked								
14. Emergency phone numbers prominently posted								
15. Diagram of fire extinguishers/ emergency equipment prominently posted								
16. Fire extinguishers (20# A/B/C): accessible (75′ in office), tagged, inspected								
WAREHOUSE (building or zone)								
17. General housekeeping: work areas uncluttered and clear for emergency egress								
18. Slip, trip, fall hazards eliminated								
19. Rubbish containers emptied								
20. Night/security lighting and timers functional								

Item of Environmental Concern	NA	OK	NI	Action Date	Responsible Party	Date Resolved	If NI, Identify Location	Other Comments
21. Electrical outlet integrity/unblocked								
22. Extension cords/portable power tools in good condition								
23. Employees use proper lifting techniques								
24. Perimeter locks on building entrances in place/work								
25. Visitor control in place/used								
26. Intrusion alarms tested								
27. Fire alarms tested								
28. Running, horseplay, unsafe work habits avoided								
29. Exit doors marked/egress unlocked during business hours								
30. Non-exit doors so marked								
31. Emergency phone numbers prominently posted								
32. Diagram of fire extinguishers/ emergency equipment prominently posted								
33. Fire Extinguishers (20# A/B/C): accessible (50′ in warehouse), tagged, inspected								

Item of Environmental Concern	NA	OK	NI	Action Date	Responsible Party	Date Resolved	If NI, Identify Location	Other Comments
34. Mezzanine/balcony used for storage marked for load capacity with rails and toe boards								
35. Damaged bags/drums recovered/ repackaged ASAP								
36. Scales certified per state regs; date of last certification: ____________								
37. Hazardous chemical list available (full facility)								
38. Guard on compressor; warning sign in place for automatic starting								
39. Forklifts: working horn, fire extinguisher, overhead guard, good tires, rubber pads on pedals, no fluid leaks, hour meter, hand brake								
40. Emergency response equipment accessible, inspected								
41. PPE accessible and used								
42. Pallet racks marked for capacity and properly utilized								
43. No unmarked containers of material requiring analysis or identification for disposal								
44. No-smoking signs posted								

Item of Environmental Concern	NA	OK	NI	Action Date	Responsible Party	Date Resolved	If NI, Identify Location	Other Comments
45. Environmental policy posted								
46. Flammable liquids in drums stored outside in maximum groups of 40; away from buildings								
47. Chemical product segregated in storage by compatibility								
YARD AREA								
48. General housekeeping: work areas uncluttered and clear for emergency egress								
49. Slip, trip, fall hazards eliminated								
50. Rubbish containers emptied								
51. Night/security lighting and timers functional								
52. Electrical outlet integrity/unblocked								
53. Perimeter locks on fences in place/work								
54. Visitor control in place/used								
55. Damaged bags/drums recovered/repackaged ASAP								
56. Employees use proper lifting techniques								
57. Product segregated in storage by compatibility								

Item of Environmental Concern	NA	OK	NI	Action Date	Responsible Party	Date Resolved	If NI, Identify Location	Other Comments
58. Running, horseplay, unsafe work habits avoided								
59. PPE accessible and used								
60. Emergency phone numbers prominently posted								
61. Diagram of fire extinguishers/ emergency equipment prominently posted								
62. Fire extinguishers (20# A/B/C): accessible (50′), tagged, inspected								
63. No evidence of spillage on yard and paving leading to possible regulatory censure								
64. Pallets stacked safely; broken pallet accumulation limited; none against fence								
65. Flammable liquids in drums stored outside in maximum groups of 40; away from buildings								
66. Extension cords/portable power tools in good condition								
67. Trash receptacles away from buildings and dock								
68. Damaged bags/drums recovered/ repackaged ASAP								

Item of Environmental Concern	NA	OK	NI	Action Date	Responsible Party	Date Resolved	If NI, Identify Location	Other Comments
69. Cylinders secured								
70. No-smoking signs posted								
71. Emergency response equipment accessible, inspected								
72. No unmarked containers of material requiring analysis or identification for disposal								
73. Safety shower/eyewash unobstructed and functional								
74. Perimeter fence line walked; all sections functional								
TANK FARM/LIQUID REPACKAGING								
75. General housekeeping: work areas uncluttered and clear for emergency egress								
76. Slip, trip, fall hazards eliminated								
77. Rubbish containers emptied								
78. Night/security lighting and timers functional								
79. Electrical outlet integrity/unblocked/ explosion-proof if flammables in area								
80. Extension cords/portable power tools in good condition								
81. Locks on valves in place/work								

Item of Environmental Concern	NA	OK	NI	Action Date	Responsible Party	Date Resolved	If NI, Identify Location	Other Comments
82. Tank valves kept closed except as needed								
83. Tank ladders and catwalks in good condition								
84. Visitor control in place/used								
85. Employees use proper lifting techniques								
86. Running, horseplay, unsafe work habits avoided								
87. Safety shower/eyewash unobstructed and functional								
88. Emergency phone numbers prominently posted								
89. Diagram of fire extinguishers/ emergency equipment prominently posted								
90. Fire extinguishers (20# A/B/C): accessible (50′), tagged, inspected								
91. No-smoking signs posted								
92. Emergency response equipment accessible, inspected								
93. Metal tanks maintained and grounded								
94. Dikes in good condition without cracks								

Item of Environmental Concern	NA	OK	NI	Action Date	Responsible Party	Date Resolved	If NI, Identify Location	Other Comments
95. Product transfer lines identified								
96. No evidence of line or hose drainage to dikes								
97. No valve drippage								
98. Storage tank contents identified, using NFPA signals and product name								
99. Storage valves identified by name of product								
100. Tank gauges operable; gauge glasses clear, tapes, floats functioning								
101. Pump maintenance: no seal leakage, necessary drip from packed shafts contained; shaft guards in place								
102. No excessive noise from pumps or other power equipment								
103. Engineering controls for vapor/dust collection available, used, in good condition								
104. PPE accessible, used, in good condition								
105. Grounding system used for all liquid packaging/bulk transfers								
106. Scales certified per state regs; date of last certification: ____________								

Item of Environmental Concern	NA	OK	NI	Action Date	Responsible Party	Date Resolved	If NI, Identify Location	Other Comments
107. Damaged bags/drums recovered/ repackaged ASAP								
108. IBCs in test per DOT date or segregated for testing								
109. IBCs labeled/marked per DOT/OSHA								
110. Guard on compressor; warning sign in place for automatic starting.								
111. Transfer hoses inspected								
112. No unmarked containers of material requiring analysis or identification for disposal								
113. Waste tank disposal records maintained								
114. Retain samples managed								
LOADING/UNLOADING AREA								
115. Dock plates								
116. Safety shower/eyewash unobstructed and functional								
117. Edge of dock clearly marked								
118. PPE accessible, used, maintained								
119. No-smoking signs posted								
120. Employees use proper lifting techniques								
121. Cylinders secured								

Item of Environmental Concern	NA	OK	NI	Action Date	Responsible Party	Date Resolved	If NI, Identify Location	Other Comments
TRANSPORTATION EQUIPMENT								
122. Grounding system used for all liquid bulk transfers								
123. Transfer hoses inspected								
124. Forklifts: working horn, fire extinguisher, overhead guard, good tires, rubber pads on pedals, no fluid leaks, parking brake								
125. Wheel chocks: access/use								
126. Equipped with fire extinguisher/spill response kit/PPE								
127. Preventive maintenance scheduled								
128. Drivers complete daily written inspections/logs								
RAIL SIDING								
129. Rail siding clean of spilled product/trash; tracks/ties appear safe								
130. Derail devices in place where required; rail tank car flagged when connected; wheels chocked during bulk unloading								
131. Safety shower/eyewash function; fire extinguisher accessible; static lines in place/functional								

Item of Environmental Concern	NA	OK	NI	Action Date	Responsible Party	Date Resolved	If NI, Identify Location	Other Comments
CONTRACTOR PERFORMANCE								
132. Pre-startup reviews conducted								
133. General housekeeping: work areas uncluttered and clear for emergency egress								
134. Slip, trip, fall hazards eliminated								
135. No-smoking signs posted								
136. PPE accessible and used								
137. Emergency response equipment accessible, inspected								
138. Fire extinguishers (20# A/B/C): accessible (50′), tagged, inspected								
LABORATORY								
139. General housekeeping: work areas uncluttered and clear for emergency egress								
140. Slip, trip, fall hazards eliminated								
141. Rubbish containers emptied								
142. Night/security lighting and timers functional								
143. Fume hoods evaluated at installation and quarterly; hood face air velocity between 60–125 linear feet/minute								

Item of Environmental Concern	NA	OK	NI	Action Date	Responsible Party	Date Resolved	If NI, Identify Location	Other Comments
144. Fume hoods								
145. PPE accessible and used								
146. Emergency phone numbers prominently posted								
147. Diagram of fire extinguishers/ emergency equipment prominently posted								
148. Fire extinguishers (20# A/B/C): accessible (50′), tagged, inspected								
149. Product segregated in storage by compatibility								
MANUFACTURING (customize)								

AUDITOR(S) NAME(S): ____________________ AUDITOR(S) SIGNATURE(S) ____________________

■ APPENDIX C.4 Source Description

Company: ____________________
Facility Business ____________________
Product: ____________________
Environmental Impact: ____________________
Area: ____________________
Objective: ____________________
Target: ____________________

Medium of Stream	_Solid _Semisolid _Liquid _Gas
Specific generation point(s) in process	
How generated?	
Cause/possible cause	
Volumes at expected levels?	__Yes __No (describe)
Volume accuracy	__Measured __Inferred
Estimated cost of raw materials that comprise waste stream	
Source	__Point __Nonpoint
Creation agent	__Human __Mechanical
Condition at point of origin	__Hazardous __Nonhazardous
Transformation during production?	__Yes (describe) __No __Becomes hazardous __Becomes nonhazardous __Medium changes to: __solid __liquid __gas __Other
Production frequency	__Periodic at intervals of: ______ __Intermittent __One-time
Special characteristics	__Can be segregated __Commingled __Special waste per state __Radioactive
EPA classification	__Listed (Give list #) ______ __Characteristic __Ignitable __Corrosive __Reactive __Toxic

■ APPENDIX C.5 Mandatory Training: Assessment Checklist

Instructions:

The following table lists topics covered by OSHA regulations that have a training requirement.

- Place a check mark in column 1 if the organization has such an activity.
- In column 4, designate whether training is currently being conducted on this topic.
- If training is not currently being conducted, use column 5 to summarize actions to take to improve training compliance.

1. Organizational Activity	2. Regulatory Citation	3. Topic	4. Training Currently Conducted (Yes or No)	5. Improvement Plan Summary
	29 CFR	**GENERAL INDUSTRY**		
	Subpart E	Means of Egress		
	§1910.38	Employee emergency plans and fire prevention plans		
	Subpart F	Powered platforms, manlifts, and vehicle-mounted work platforms		
	§1910.66	Powered platforms for exterior building maintenance		
	§1910.68	Manlifts		
	Subpart G	Occupational Health and Environmental Control		
	§1910.94	Ventilation		
	§1910.95	Occupational noise exposure		
	§1910.96	Ionizing radiation		
	Subpart H	Hazardous Materials		
	§1910.103	Hydrogen		

1. Organizational Activity	2. Regulatory Citation	3. Topic	4. Training Currently Conducted (Yes or No)	5. Improvement Plan Summary
	§1910.106	Flammable and combustible liquids		
	§1910.109	Explosives and blasting agents		
	§1910.110	Storage and handling of liquefied petroleum gases		
	§1910.111	Storage and handling of anhydrous ammonia		
	§1910.119	Process safety management of highly hazardous chemicals		
	§1910.120	Hazardous waste operations and emergency response		
	Subpart I	Personal Protective Equipment		
	§1910.132	General requirements		
	§1910.133	Eye and face protection		
	§1910.134	Respiratory protection		
	§1910.135	Head protection		
	§1910.136	Foot protection		
	§1910.137	Electrical protective equipment		
	§1910.138	Hand protection		
	Subpart J	General Environmental Controls		
	§1910.142	Temporary labor camps		

1. Organizational Activity	2. Regulatory Citation	3. Topic	4. Training Currently Conducted (Yes or No)	5. Improvement Plan Summary
	§1910.145	Specifications for accident prevention signs and tags		
	§1910.146	Permit-required confined spaces		
	§1910.147	The control of hazardous energy (lock-out/tag-out)		
	Subpart K	Medical and First Aid		
	§1910.151	Medical services and first aid		
	Subpart L	Fire Protection		
	§1910.155	Scope, application and definitions applicable to this subpart		
	§1910.156	Fire brigades		
	§1910.157	Portable fire extinguishers		
	§1910.158	Standpipe and hose systems		
	§1910.160	Fixed extinguishing systems, general		
	§1910.164	Fire detection systems		
	Subpart N	Materials Handling and Storage		
	§1910.177	Servicing multi-piece and single-piece rim wheels		
	§1910.178	Powered industrial trucks		
	§1910.179	Overhead and gantry cranes		

1. Organizational Activity	2. Regulatory Citation	3. Topic	4. Training Currently Conducted (Yes or No)	5. Improvement Plan Summary
	§1910.180	Crawler locomotive and truck cranes		
	§1910.181	Derricks		
	§1910.184	Slings		
	Subpart O	Machinery and Machine Guarding		
	§1910.213	Woodworking machinery requirements		
	§1910.217	Mechanical power presses		
	§1910.218	Forging machines		
	Subpart Q	Welding, Cutting, and Brazing		
	§1910.252	General requirements		
	§1910.253	Oxygen-fuel gas welding and cutting		
	§1910.254	Arc welding and cutting		
	§1910.255	Resistance welding		
	Subpart R	Special Industries		
	§1910.261	Pulp, paper, and paperboard mills		
	§1910.264	Laundry machinery and operations		
	§1910.265	Sawmills		
	§1910.266	Pulpwood logging		
	§1910.268	Telecommunications		

1. Organizational Activity	2. Regulatory Citation	3. Topic	4. Training Currently Conducted (Yes or No)	5. Improvement Plan Summary
	§1910.269	Electric power generation, transmission, and distribution		
	§1910.272	Grain handling facilities		
	Subpart S	Electrical		
	§1910.332	Training		
	Subpart T	Commercial Diving Operations		
	§1910.410	Qualifications of dive team		
	Subpart Z	Toxic and Hazardous Substances		
	§1910.1001	Asbestos		
	§1910.1003	4-nitrobiphenyl		
	§1910.1004	alpha-Napthylamine		
	§1910.1006	Methyl chloromethyl ether		
	§1910.1007	3,3′-dichlorobenzidine (and its salts)		
	§1910.1008	bis-Chloromethyl ether		
	§1910.1009	Beta-naphthylamine		
	§1910.1010	Benzidine		
	§1910.1011	4-aminodiphenyl		
	§1910.1012	Ethyleneimine		
	§1910.1013	beta-Propiolactone		
	§1910.1014	2-acetylaminofluorene		
	§1910.1015	4-dimethylaminoazobenzeine		

1. Organizational Activity	2. Regulatory Citation	3. Topic	4. Training Currently Conducted (Yes or No)	5. Improvement Plan Summary
	§1910.1016	N-nitrosodimethylamine		
	§1910.1017	Vinyl chloride		
	§1910.1018	Inorganic arsenic		
	§1910.1025	Lead		
	§1910.1027	Cadmium		
	§1910.1028	Benzene		
	§1910.1029	Coke oven emissions		
	§1910.1030	Bloodborne pathogens		
	§1910.1043	Cotton dust		
	§1910.1044	1,2-dibromo-3-chloropropane		
	§1910.1045	Acrylonitrile		
	§1910.1047	Ethylene oxide		
	§1910.1048	Formaldehyde		
	§1910.1050	Methylenedianiline		
	§1910.1200	Hazard communication		
	§1910.1450	Occupational exposure to hazardous chemicals in laboratories		
	29 CFR	**CONSTRUCTION INDUSTRY**		
	Subpart C	General Safety and Health Provisions		

1. Organizational Activity	2. Regulatory Citation	3. Topic	4. Training Currently Conducted (Yes or No)	5. Improvement Plan Summary
	§1926.20	General safety and health provisions		
	§1926.21	Safety training and education		
	Subpart D	Occupational Health and Environmental Controls		
	§1926.50	Medical services and first aid		
	§1926.53	Ionizing radiation		
	§1926.54	Nonionizing radiation		
	§1926.55	Gases, vapors, fumes, dusts, and mists		
	§1926.58	Asbestos		
	Subpart E	Personal Protective and Life-Saving Equipment		
	§1926.101	Hearing protection		
	§1926.103	Respiratory protection		
	Subpart F	Fire Protection and Prevention		
	§1926.150	Fire protection		
	Subpart G	Signs, Signals, and Barricades		
	§1926.201	Signaling		
	Subpart I	Tools—Hand and Power		
	§1926.304	Woodworking tools		

1. Organizational Activity	2. Regulatory Citation	3. Topic	4. Training Currently Conducted (Yes or No)	5. Improvement Plan Summary
	Subpart J	Welding and Cutting		
	§1926.350	Gas welding and cutting		
	§1926.351	Arc welding and cutting		
	§1926.352	Fire prevention		
	§1926.354	Welding, cutting, and heating in way of preservative coatings		
	Subpart K	Electrical		
	§1926.404	Wiring design and protection		
	Subpart L	Ladders and Scaffolding		
	§1926.451	Scaffolding		
	Subpart M	Floors and Wall Openings, and Stairways		
	§1926.500	Guardrails, handrails, and covers		
	§1926.503	Training requirements		
	Subpart N	Cranes, Derricks, Hoists, Elevators, and Conveyors		
	§1926.550	Cranes and derricks		
	§1926.552	Material hoists, personnel hoists, and elevators		
	Subpart O	Motor Vehicles, Mechanized Equipment, and Marine Operations		

1. Organizational Activity	2. Regulatory Citation	3. Topic	4. Training Currently Conducted (Yes or No)	5. Improvement Plan Summary
	§1926.602	Material handling equipment		
	§1926.604	Site clearing		
	Subpart P	Excavations, Trenching, and Shoring		
	§1926.651	General requirements		
	Subpart Q	Concrete, Concrete Forms, and Shoring		
	§1926.701	General requirements		
	§1926.703	Requirements for cast-in-place concrete		
	Subpart R	Steel Erection		
	§1926.752	Bolting, riveting, fitting-up, and plumbing-up		
	Subpart S	Tunnels and Shafts, Caissons, Cofferdams, and Compressed Air		
	§1926.800	Underground Construction		
	§1926.803	Compressed air		
	Subpart T	Demolition		
	§1926.850	Preparatory operations		
	§1926.852	Chutes		
	§1926.859	Mechanical demolition		

1. Organizational Activity	2. Regulatory Citation	3. Topic	4. Training Currently Conducted (Yes or No)	5. Improvement Plan Summary
	Subpart U	Blasting and Use of Explosives		
	§1926.900	General provisions		
	§1926.901	Blaster qualifications		
	§1926.902	Surface transportation of explosives		
	§1926.909	Firing the blast		
	Subpart V	Power Transmission and Distribution		
	§1926.950	General requirements		
	§1926.955	Overhead lines		
	§1926.956	Underground lines		
	§1926.957	Construction in energized substations		
	Subpart X	Stairways and Ladders		
	§1926.1053	Ladders		
	§1926.1060	Training requirements		
	Subpart Z	Toxic and Hazardous Substances		
	§1926.1101	Asbestos		
	§1926.1127	Cadmium		

1. Organizational Activity	2. Regulatory Citation	3. Topic	4. Training Currently Conducted (Yes or No)	5. Improvement Plan Summary
	29 CFR	**MARITIME INDUSTRY**		
	Subpart A	General Provisions		
	§1915.7	Competent person		
	Subpart B	Explosive and Other Dangerous Atmospheres		
	§1915.12	Precautions before entering		
	§1915.13	Cleaning and other cold work		
	§1915.14	Certification before hot work is begun		
	§1915.15	Maintaining gas-free conditions		
	Subpart C	Surface Preparation and Preservation		
	§1915.35	Painting		
	§1915.36	Flammable liquids		
	Subpart D	Welding, Cutting and Heating		
	§1915.52	Fire prevention		
	§1915.53	Welding, cutting, and heating in way of preservative coatings		
	§1915.54	Welding, cutting and heating of hollow metal containers and structures not covered by §1915.12.		

1. Organizational Activity	2. Regulatory Citation	3. Topic	4. Training Currently Conducted (Yes or No)	5. Improvement Plan Summary
	§1915.55	Gas welding and cutting		
	§1915.56	Arc welding and cutting		
	§1915.57	Uses of fissionable material in ship repairing and shipbuilding		
	Subpart E	Scaffolds, Ladders, and Other Working Surfaces		
	§1915.71	Scaffolds or staging		
	Subpart G	Gear and Equipment for Rigging and Materials Handling		
	§1915.95	Ship repairing and shipbuilding work on or in the vicinity of radar and radio		
	§1915.97	Health and sanitation		
	§1915.98	First aid		
	§1915.112	Ropes, chains, and slings		
	§1915.116	Use of gear		
	§1915.117	Qualifications of operators		
	Subpart H	Tools and Related Equipment		
	§1915.135	Power-actuated fastening tools		
	§1915.136	Internal combustion engines, other than ship's equipment		
	Subpart I	Personal Protective Equipment		

1. Organizational Activity	2. Regulatory Citation	3. Topic	4. Training Currently Conducted (Yes or No)	5. Improvement Plan Summary
	§1915.152	Respiratory protection		
	Subpart K	Portable, Unfired Pressure Vessels, Drums and Containers, other than Ship's Equipment		
	§1915.172	Portable air receivers and other unfired pressure vessels.		
	Subpart Z	Toxic and Hazardous Substances		
	§1915.1001	Asbestos		
	§1915.1200	Hazard communication		
	Part 1917	Marine terminals		
	Subpart B	Marine Terminal Operations		
	§1917.25	Fumigants, pesticides, insecticides and hazardous preservations (see §1917.2(p))		
	§1917.27	Personnel		
	§1917.28	Hazard communication		
	Subpart C	Cargo Handling Gear and Equipment		
	§1917.44	General rules applicable to vehicles		
	Subpart D	Specialized Terminals		
	§1917.73	Terminal facilities handling menhaden and similar species of fish		

1. Organizational Activity	2. Regulatory Citation	3. Topic	4. Training Currently Conducted (Yes or No)	5. Improvement Plan Summary
	Subpart G	Related Terminal Operations and Equipment		
	§1917.152	Welding, cutting and heating (hot work)		
	Part 1918	Longshoring		
	Subpart I	General Working Conditions		
	§1918.90	Hazard communication		
	Subpart J	Personal Protective Equipment		
	§1918.102	Respiratory protection		
	40 CFR	**ENVIRONMENTAL**		
	§82.40	Servicing of motor vehicle air conditioners—technician training and certification		
	§265.16	(RCRA) Interim standards for TSDs: personnel training		
	§265.55	(RCRA) Interim standards for TSDs: contingency plan and emergency procedures		
	§721.63	Protection in the workplace		
	§721.72	Certain significant new uses: hazard communication		

1. Organizational Activity	2. Regulatory Citation	3. Topic	4. Training Currently Conducted (Yes or No)	5. Improvement Plan Summary
	§763.92	Training and periodic surveillances		
	§763.121	Asbestos abatement projects: regulatory requirements		
	Part 763 Subpart E Appendix C	EPA Model Contractor Accreditation Plan		
	49 CFR	**TRANSPORTATION**		
	Part 40	Procedures for Transportation Workplace Drug Testing Programs		
	§40.23	Preparation for testing		
	§40.27	Laboratory personnel		
	§40.51	The breath alcohol technician		
	Part 172	Subpart H Training		
	§173.7	Shippers—general requirements for shipments and packaging		
	§174.7	Carriage by rail: compliance and training		
	§175.20	Carriage by aircraft: compliance and training		
	§176.13	Carriage by vessel: responsibility for compliance and training		

1. Organizational Activity	2. Regulatory Citation	3. Topic	4. Training Currently Conducted (Yes or No)	5. Improvement Plan Summary
	§177.800	Carriage by public highway: responsibility for compliance and training		
	§177.816	Driver training		
	§177.818	Special instructions: Division 2.1 (flammable gas) materials that are cryogenic liquids		
	§177.825	Routing and training requirements for Class 7 (radioactive) materials		
	Part 199	Drug Testing		
	§199.21	Contractor employees		
	§199.25	Reporting of anti-drug testing results		
	§199.202	Alcohol misuse plan		
	§199.229	Reporting of alcohol testing results		
	§391.11	Qualifications of drivers		
	§391.121	EAP training program		
	§392.1	Driving of motor vehicles; scope of the rules in this part		
	§396.19	Inspector qualifications		
	§396.25	Qualifications of brake inspectors		
	§397.101	Requirements for motor carriers and drivers		

■ APPENDIX C.6 Organization: Pre-Startup Environmental Review Checklist

Job Site: ______________________ Contractor: ____________
Date of Conference: __/__/__ Contractor Supervisor: ____________

Checklist completed by: ______________________ ______________________
signature print name

Organization to provide:

__ 1. List of Organization's hazardous chemicals that may affect Contractor work area.

____

____

____

__ 2. Copies of MSDS(s) for appropriate hazardous chemicals supplied to Contractor.

__ 3. Identification of precautionary measures Contractor must take to protect its employees during Organization's normal operation:

__ 4. Identification of precautionary measures Contractor must take to protect its property during Organization's normal operation:

__ 5. Identification of precautionary measures Contractor must take in the event of an Organizational emergency:
__ alarm signals __ escape routes __ assembly points
__ emergency shutdown (describe) __ other (describe)

__ 6. Decision on whose confined space entry permit system:
__ Organization __ Contractor __ Not applicable

__ 7. Decision on whose hot work permit system:
__ Organization __ Contractor __ Not applicable

__ 8. Decision on whose lock-out/tag-out permit system:
__ Organization __ Contractor __ Not applicable

__ 9. Description of Organization's hazardous chemical labeling system at the site:

Contractor to provide:

__ 1. List of hazardous chemicals that Contractor may bring on site.

____

____

____

__ 2. Copies of MSDS(s) for Contractor hazardous chemicals.

__ 3. Identification of precautionary measures Organization must take to protect its employees during Contractor's normal operation:

__ 4. Identification of precautionary measures Organization must take to protect against accidental environmental releases during Contractor's normal operation:

__ 5. Identification of precautionary measures Organization must take in the event of a Contractor emergency:

__ alarm signals __ escape routes __ assembly points
__ emergency shutdown (describe) __ other (describe)

__ 6. Description of Contractor's hazardous chemical labeling system at the site:

__ rely on mfg/distrib labels __ NFPA signals __ HMIS __ Other (describe)

____

____

__ 7. Verification of appropriate Contractor employee training certifications:

Certification Description	Identification of Contractor Employees
A. OSHA HAZWOPER (29 CFR 1910.120) Effective Through: __/__/__ __ 40-Hour Level __ Emergency Response Level __ Awareness __ Operations __ HazMat Technician __ On-Scene Commander __ Manager/Supervisor of Operations __ Instructor/Trainer	
B. OSHA Process Safety (29 CFR 1910.119) Effective Through: __/__/__ __ Process Hazard Analysis __ Contractor Responsibilities	
C. OSHA Confined Space Entry (29 CFR 1910.146) Effective Through: __/__/__ __ Authorized Entrant __ Attendant __ Entry Supervisor __ Rescue and Emergency Service	
D. OSHA Marine Safety (29 CFR 1915.7) Effective Through: __/__/__ __ Competent Person (29 CFR 1915.7)	
E. DOT (49 CFR 172.704) Effective Through: __/__/__ __ General Awareness __ Safety __ Commercial Driver's License __ Function-Specific: __	
F. Other (describe)	

■ APPENDIX C.7 Regulatory Inspection Report

Facility: ______ **Date of Completion:** __/__/__

Instructions:

- Complete this report whenever the facility has had a regulatory inspection.
- File copies of this report with the __________ within ____ day(s) of the inspection.
- Completed reports are environmental records.

Date of Inspection: __/__/__ This Report Completed By: __________

Name of Inspector: __________ Title: __________

Agency: ______

Mailing Address: __________ Phone: __________

__________ Fax: __________

Reason for Inspection:

__ due to complaint __ routine/periodic

__ other (describe): __________

Scope/Focus of Inspection:

__ air __ noise, vibration, dust

__ land __ waste

__ water __ other (describe): ______

Environmental Records Reviewed:

Environmental Records Copied/Given to Inspector:

Were photographs taken? __ yes __ no

Inspection Results

Will this inspection require company follow-up? __ yes __ no

If yes, who is responsible for follow-up: __________

If yes, by when must follow-up be completed: __/__/__.

__ Attach copy of any inspection reports left at facility.

__ Send final inspection report to __________ after it is received.

Corrective Action

Will this inspection require corrective action? __yes __no

If yes, summarize the proposed corrective action:

If yes, who is responsible for corrective action: ____________________

If yes, by when must corrective action be completed: __/__/__.

Who is responsible for updating the facility environmental management program to reflect this corrective action? ________________________

Complete within _____ day(s) of inspection and forward copies to

____.

APPENDIX C.8 Environmental Incident Investigation

Facility: ______ Environmental Incident Investigation

Identity of Activity, Product, Service: ______

Instructions:

- Ensure that all parts of this report are completed for workers' compensation and other incidents, including near misses.
- If a section is not applicable, write N/A.
- Attach support documents (such as DOT 5800.1 forms) as they become available to make up an incident investigation file.
- Incident investigation must begin within 48 hours of the incident (or knowledge of the incident by management).
- Forward Parts 1–4 to the facility ______________ within 72 hours of the incident. Provide them with the status of Parts 5–7 within one week of the incident.

Part 1—Administration

Report prepared by: ____________________

Incident date: ____ Time: ____ am/pm Investigation began: Date: ____ Time: ____ am/pm

Investigation Team Members:

Part 2—Incident Summary

Specific location: ____________________

__Type of injuries (number of victims, names, initial description of injury):

Note: If injury is to employee of the company, log the injury on the OSHA 200 form.

__Type of vehicular accident (vehicles involved):

Note: If vehicular accident involved a company truck, it may be necessary to complete a DOT 5800.1 report. Verify need and make sure required reports are filed.

__Type of property damage (estimate amount, description, extent):

__Characteristics of environmental release (chemical product, estimated amount, areas of contamination):

Note: Release reporting to regulatory authorities may be required. Verify need and make sure required reports (verbal and written) are filed.

Was emergency response plan activated? __yes __no
__Other incident (describe)__Sequence of events

Part 3—Investigation Summary

Conditions at time of incident
Weather: ______________
Lighting: ______________
Traffic: ______________
Work activity in progress: ____________________________

Equipment in use:

Engineering controls in place:

Administrative controls in place:

Part 4—Conditions During Initial Response to Incident

Response assistance (police, fire, public, company):

Response activity:

Response equipment:

Public involvement: ____________ Were the media involved? __yes __no

Part 5—Disposition of Nonconforming Product Created by Incident

Description:

Who is responsible for this disposition? ______________________________

What is the time schedule for disposition? ______________________________

Part 6—Conclusion of Investigating Team

Factors contributing to incident:

- Work practices:

- Engineering controls:

- Equipment:

- Administrative controls:

Part 7—Corrective Actions Recommended by the Investigating Team

Date of Recommendations: ______

Item	Corrective Action	Responsible Party	Target Date for Implementation

■ APPENDIX C.9 Procedures Audit

Application: Referencing requirements of ISO 14001, the following matrix is used to audit existing procedures by assessing whether they are written to facilitate pollution prevention efforts.

Procedure (name/reference number): ______________________

Audit Date: __/__/__ Auditor: ______________________

With regard to environmental aspects, does the procedure:	OK	Needs Improvement	Not Applicable	Comments
Identify suitable work environment?				
Reference compliance with applicable federal/state regulations?				
Reference compliance with standards/codes or other documented procedures?				
Address monitoring of suitable process parameters and their environmental impacts?				
Specify approved processes and equipment to mitigate environmental impacts?				
State criteria for environmental performance?				
Address suitable maintenance of equipment to ensure continuing process capability, including mitigation of environmental impacts?				
Identify information used to verify compliance to procedure, specifying accuracy of results?				
Give schedule (locations, times) for compliance verification (facility-level audit)?				
Contain or provide reference to quality control procedures?				
State environmental compliance (acceptance) criteria?				
Address action to be taken if environmental compliance (acceptance) criteria are not met (reporting of findings)?				

With regard to environmental aspects, does the procedure:	OK	Needs Improvement	Not Applicable	Comments
Require the assessment of data validity when environmental compliance (acceptance) systems are found to malfunction?				
Address safeguarding of measurement and test facilities from unauthorized adjustments or damage?				
Specify training needs?				
Identify environmental records?				
Are internal environmental records produced on forms designed within the organization • Consistent with the procedure? • Understood and in use where needed in the organization? • When completed, retained per a recognized schedule?				
Apply to contractors? If yes, does it state how the procedure is communicated to the contractor?				

■ APPENDIX D.1 Registration for Offeror of Hazardous Materials

Exempted from Paperwork Reduction Act by 49 App. U.S.C. 1805(c)(13)

U. S. DEPARTMENT OF TRANSPORTATION
HAZARDOUS MATERIALS REGISTRATION STATEMENT
REGISTRATION YEAR 19____ - 19____

(Please Type or Print all Responses)

Initial Registration ____ Renewal of Registration ____ Amendment to Registration ____ Expedited Follow-up ____

Current Registration # _ _ _ _ _ _ _ _ _ _ _ _ _ _

1. Registrant______________________________
(Company Name)

(Place pre-printed label here if provided and if name and address are correct. Otherwise, provide correct information.)

2. Mailing Address of Principal Place of Business

Street or P.O. Box ______________________ City ______________

County ______________ State _ _ Zip Code _ _ _ _ _ - _ _ _ _ Country ______________

3. Carrier's US DOT ID Number, ICC Number, or Reporting Railroad Alphabetic Code (if applicable)

US DOT ID #______________ ICC # ______________ Railroad Alphabetic Code ______________

4. Mode(s) Used to Transport Hazardous Materials: Highway ____ Rail ____ Water ____ Air ____

5. Industrial Classification: Check the primary industry in which the registrant operates. **Mark Only One.**

Transportation
- ☐ Carrier
- ☐ Warehousing
- ☐ Freight Forwarding, Agent Services
- ☐ Transportation Repair & Service Facilities

Manufacturing
- ☐ Petroleum Refining & Related Industries
- ☐ Apparel & Other Textile Products
- ☐ Paper, Wood, & Allied Products
- ☐ Printing & Publishing
- ☐ Chemicals & Allied Products
- ☐ Food & Kindred Products
- ☐ Rubber & Miscellaneous Plastic Products
- ☐ Electric & Electronic Equipment
- ☐ Explosives
- ☐ Other Manufacturing Industries

Other
- ☐ Agriculture & Support Services
- ☐ Mining - Metal & Non-Metal
- ☐ Oil & Gas Extraction
- ☐ Gasoline, Fuel Oil, Propane Sales & Delivery
- ☐ Construction - All Types
- ☐ Wholesale or Retail Trade
- ☐ Non-Transportation Repair Facilities
- ☐ Hazardous Waste Services - Transportation, Disposal, Treatment
- ☐ Other ______________

6. Annual Registration Fee. The combined registration and processing fee is $300.00. (Complete only when submitting initial or renewal registration.)

Total Amount Enclosed:______________

Make check or money order in U.S. funds, drawn on a U.S. bank, and payable to "U.S. Department of Transportation," and identified as payment for the "Hazmat Registration Fee."

Method of Payment

Check ____ Money Order ____ Credit Card: *VISA* ____ *MasterCard* ____

Card Number: ☐☐☐☐☐☐☐☐☐☐☐☐☐☐☐☐ Expiration Date: ☐☐ ☐☐ (MO YR)

Name as it appears on the card ______________________

Authorized Signature ______________________

Cardholder acknowledges ordering goods or services in the amount of the total shown hereon and agrees to perform the obligations set forth in the Cardholder's agreement with the issuer. Credit card statement will list this payment as "US DOT Hazmat Regis."

NOTE: If completing an Expedited Registration, **do not** resubmit credit card information here.

Form DOT F 5800.2 (Revised 04/95) THIS FORM MAY BE REPRODUCED

7. PRIOR-YEAR SURVEY INFORMATION: Hazardous Materials Activities, and States in Which Activity was Conducted. Indicate those activities conducted by the registrant during the previous calendar year (e.g., 1994 for the 1995-96 Registration Year). Mark "A" through "E," as appropriate, to indicate the category or categories and the activity or activities (shipper, carrier, or other) in which the registrant acted. Check all categories and activities that apply. "Other" may be checked to indicate offeror activities not covered under the heading of shipper or carrier, such as freight forwarder or agent. Carriers should circle all states in which they operated as a hazardous materials carrier. Shippers and others engaged in offering hazardous materials should circle only those states from which they offered hazardous materials. They do not need to indicate to or through which states shipments were sent. A list of the states and their abbreviations appears in the accompanying materials. Circle "48 Contiguous States," if appropriate, to indicate that the activity was conducted in all of the 48 contiguous states. If the registrant did not engage in activities covered by "A" through "E" during the previous year, but plans to do so in the current registration year, mark only "F."

A. ____ Offered or transported in commerce any highway route-controlled quantity of a Class 7 (radioactive) material.

1. Shipper ____ 2. Carrier ____ 3. Other (Freight Forwarder, Agent, etc.) ____

AL AR AZ CA CO CT DE FL GA ID IL IN IA KS KY LA MA MD ME MI MN
MO MS MT NC ND NE NH NJ NM NV NY OH OK OR PA RI SC SD TN TX UT
VT VA WA WV WI WY 48 Contiguous States AK AS DC GU HI MP PR VI

B. ____ Offered or transported in commerce more than 25 kilograms (55 pounds) of a Division 1.1, 1.2, or 1.3 (explosive) material in a motor vehicle, rail car, or freight container.

1. Shipper ____ 2. Carrier ____ 3. Other (Freight Forwarder, Agent, etc.) ____

AL AR AZ CA CO CT DE FL GA ID IL IN IA KS KY LA MA MD ME MI MN
MO MS MT NC ND NE NH NJ NM NV NY OH OK OR PA RI SC SD TN TX UT
VT VA WA WV WI WY 48 Contiguous States AK AS DC GU HI MP PR VI

C. ____ Offered or transported in commerce more than 1 liter (1.06 quarts) per package of a material extremely toxic by inhalation (Division 2.3, Hazard Zone A, or Division 6.1, Packing Group I, Hazard Zone A).

1. Shipper ____ 2. Carrier ____ 3. Other (Freight Forwarder, Agent, etc.) ____

AL AR AZ CA CO CT DE FL GA ID IL IN IA KS KY LA MA MD ME MI MN
MO MS MT NC ND NE NH NJ NM NV NY OH OK OR PA RI SC SD TN TX UT
VT VA WA WV WI WY 48 Contiguous States AK AS DC GU HI MP PR VI

D. ____ Offered or transported in commerce a hazardous material or hazardous waste in a bulk packaging (see 49 CFR 171.8) having a capacity equal to or greater than 13,248 liters (3,500 gallons) for liquids or gases or more than 13.24 cubic meters (468 cubic feet) for solids.

1. Shipper ____ 2. Carrier ____ 3. Other (Freight Forwarder, Agent, etc.) ____

AL AR AZ CA CO CT DE FL GA ID IL IN IA KS KY LA MA MD ME MI MN
MO MS MT NC ND NE NH NJ NM NV NY OH OK OR PA RI SC SD TN TX UT
VT VA WA WV WI WY 48 Contiguous States AK AS DC GU HI MP PR VI

E. ____ Offered or transported in commerce a shipment, in other than a bulk packaging, of 2,268 kilograms (5,000 pounds) gross weight or more of one class of hazardous materials or hazardous waste for which placarding of a vehicle, rail car, or freight container is required.

1. Shipper ____ 2. Carrier ____ 3. Other (Freight Forwarder, Agent, etc.) ____

AL AR AZ CA CO CT DE FL GA ID IL IN IA KS KY LA MA MD ME MI MN
MO MS MT NC ND NE NH NJ NM NV NY OH OK OR PA RI SC SD TN TX UT
VT VA WA WV WI WY 48 Contiguous States AK AS DC GU HI MP PR VI

F. ____ Did not engage in any of the activities listed in A through E during the previous calendar year.

8. Certification of Information. I certify that, to the best of my knowledge, the above information is true, accurate, and complete.

Certifier's Name ________________________ (Print) Date ____________

Title ________________________ Phone (_ _ _) _ _ _ - _ _ _ _

Certifier's Signature ________________________

FALSE STATEMENTS MAY VIOLATE 18 U.S.C. 1001.

MAIL COMPLETED FORM TO :

U.S. Department of Transportation
Hazardous Materials Registration
P.O. Box 740188
Atlanta, GA 30374-0188

Please retain a copy of this form for your records.

■ APPENDIX D.2 MCS-90 Form

POLICY NUMBER:

COMMERCIAL AUTO
ISSUE DATE:

THIS ENDORSEMENT CHANGES THE POLICY. PLEASE READ IT CAREFULLY.

ENDORSEMENT FOR MOTOR CARRIER POLICIES OF INSURANCE FOR PUBLIC LIABILITY UNDER SECTIONS 29 AND 30 OF THE MOTOR CARRIER ACT OF 1980

DEFINITIONS AS USED IN THIS ENDORSEMENT

ACCIDENT includes continuous or repeated exposure to conditions which results in bodily injury, property damage, or environmental damage which the insured neither expected nor intended.

MOTOR VEHICLE means a land vehicle, machine, truck, tractor, trailer, or semitrailer propelled or drawn by mechanical power and used on a highway for transporting property, or any combination thereof.

BODILY INJURY means injury to the body, sickness, or disease to any person, including death resulting from any of these.

ENVIRONMENTAL RESTORATION means restitution for the loss, damage, or destruction of natural resources arising out of the accidental discharge, dispersal, release or escape into or upon the land, atmosphere, watercourse, or body of water, of any commodity transported by a motor carrier. This shall include the cost of removal and the cost of necessary measures taken to minimize or mitigate damage to human health, the natural environment, fish, shellfish, and wildlife.

PROPERTY DAMAGE means damage to or loss of use of tangible property.

PUBLIC LIABILITY means liability for bodily injury, property damage, and environmental restoration.

The insurance policy to which this endorsement is attached provides automobile liability insurance and is amended to assure compliance by the insured, within the limits stated herein, as a motor carrier of property, with Sections 29 and 30 of the Motor Carrier Act of 1980 and the rules and regulations of the Federal Highway Administration (FHWA) and the Interstate Commerce Commission (ICC).

In consideration of the premium stated in the policy to which this endorsement is attached, the insurer (the company) agrees to pay, within the limits of liability described herein, any final judgment recovered against the insured for public liability resulting from negligence in the operation, maintenance or use of motor vehicles subject to the financial responsibility requirements of Sections 29 and 30 of the Motor Carrier Act of 1980 regardless of whether or not each motor vehicle is specifically described in the policy and whether or not such negligence occurs on any route or in any territory authorized to be served by the insured or elsewhere. Such insurance as is afforded, for public liability, does not apply to injury to or death of the insured's employees while engaged in the course of their employment, or property transported by the insured, designated as cargo.

It is understood and agreed that no condition, provision, stipulation, or limitation contained in the policy, this endorsement, or any other endorsement thereon, or violation thereof, shall relieve the company from liability or from the payment of any final judgment, within the limits of liability herein described, irrespective of the financial condition, insolvency or bankruptcy of the insured. However, all terms, conditions, and limitations in the policy to which the endorsement is attached shall remain in full force and effect as binding between the insured and the company. The insured agrees to reimburse the company for any payment made by the company on account of any accident, claim, or suit involving a breach of the terms of the policy, and for any payment that the company would not have been obligated to make under the provisions of the policy except for the agreement contained in this endorsement.

It is further understood and agreed that, upon failure of the company to pay any final judgment recovered against the insured as provided herein, the judgment creditor may maintain an action in any court of competent jurisdiction against the company to compel such payment.

The limits of the company's liability for the amounts prescribed in this endorsement apply separately to each accident and any payment under the policy because of any one accident shall not operate to reduce the liability of the company for the payment of final judgments resulting from any other accident.

MCS-90 (5-91)

Page 1 of 2

COMMERCIAL AUTO

The policy to which this endorsement is attached provides primary or excess insurance, as indicated by "X," for the limits shown:

☐ This insurance is primary and the company shall not be liable for amounts in excess of $ for each accident.

☐ This insurance is excess and the company shall not be liable for amounts in excess of $ for each accident in excess of the underlying limit of $ for each accident.

Whenever required by the FHWA or the ICC the company agrees to furnish the FHWA or the ICC a duplicate of said policy and all its endorsements. The company also agrees, upon telephone request by an authorized representative of the FHWA or the ICC, to verify that the policy is in force as of a particular date.

Cancellation of this endorsement may be effected by the company or the insured by giving (1) thirty-five (35) days notice in writing to the other party (said 35 days notice to commence from the date the notice is mailed, proof of mailing shall be sufficient proof of notice), and (2) if the insured is subject to the ICC's jurisdiction, by providing thirty (30) days notice to the ICC (said 30 days notice to commence from the date the notice is received by the ICC at its office in Washington, D.C.).

Issued to

of

Dated:

Amending Policy Number: Effective Date:

Countersigned by: ______________________________
Authorized Company Representative

Name of Insurance Company:

The Motor Carrier Act of 1980 requires limits of financial responsibility according to type of carriage and commodity transported by the motor carrier.

It is the MOTOR CARRIER'S obligation to obtain the required limits of financial responsibility.

THE SCHEDULE OF LIMITS SHOWN BELOW DOES NOT PROVIDE COVERAGE.

The limits shown in the schedule are for information purposes only.

SCHEDULE OF LIMITS

Public Liability

Type of Carriage*	Commodity Transported	Minimum Insurance
(1) For-hire (in interstate or foreign commerce).	Property (nonhazardous).	$ 750,000
(2) For-hire and Private (in interstate, foreign, or intrastate commerce).	Hazardous substances, as defined in 49 CFR 171.8, transported in cargo tanks, portable tanks, or hopper-type vehicles with capacities in excess of 3,500 water gallons; or in bulk Class A or B explosives, poison gas (Poison A), liquefied compressed gas or compressed gas; or highway route controlled quantity radioactive materials as defined in 49 CFR 173.403.	$5,000,000
(3) For-hire and Private (in interstate or foreign commerce in any quantity) or (in intrastate commerce: in bulk only).	Oil listed in 49 CFR 172.101; hazardous waste, hazardous materials and hazardous substances defined in 49 CFR 171.8 and listed in 49 CFR 172.101, but not mentioned in (2) above or (4) below.	$1,000,000
(4) For-hire and Private (in interstate or foreign commerce).	Any quantity of Class A or B explosives; any quantity of poison gas (Poison A); or highway route controlled quantity radioactive materials as defined in 49 CFR 173.403.	$5,000,000

*NOTE: The type of carriage listed under numbers (1), (2) and (3) apply to vehicles with a gross vehicle weight rating of 10,000 pounds or more. The type of carriage listed under number (4) applies to all vehicles with a gross vehicle wright rating of less than 10,000 pounds.

Page 2 of 2 MCS-90 (5-91)

■ APPENDIX E.1 Environmental Protection Agency 40 CFR Waste Streams Exempted from Hazardous Waste Management

261.4
Exclusions.
261.4(a)
(a) Materials which are not solid wastes. The following materials are not solid wastes for the purpose of this part:
261.4(a)(1)(i)
(1)(i) Domestic sewage; and
261.4(a)(1)(ii)
(ii) Any mixture of domestic sewage and other wastes that passes through a sewer system to a publicly-owned treatment works for treatment. “Domestic sewage” means untreated sanitary wastes that pass through a sewer system.
261.4(a)(2)
(2) Industrial wastewater discharges that are point source discharges subject to regulation under section 402 of the Clean Water Act, as amended.

[*Comment:* This exclusion applies only to the actual point source discharge. It does not exclude industrial wastewaters while they are being collected, stored or treated before discharge, nor does it exclude sludges that are generated by industrial wastewater treatment.]

261.4(a)(3)
(3) Irrigation return flows.
261.4(a)(4)
(4) Source, special nuclear or by-product material as defined by the Atomic Energy Act of 1954, as amended, 42 U.S.C. 2011 et seq.
261.4(a)(5)
(5) Materials subjected to in-situ minings techniques which are not removed from the ground as part of the extraction process.
261.4(a)(6)
(6) Pulping liquors (i.e., black liquor) that are reclaimed in a pulping liquor recovery furnace and then reused in the pulping process, unless it is accumulated speculatively as defined in §261.1(c) of this chapter.
261.4(a)(7)
(7) Spent sulfuric acid used to produce virgin sulfuric acid, unless it is accumulated speculatively as defined in §261.1(c) of this chapter.
261.4(a)(8)
(8) Secondary materials that are reclaimed and returned to the original process or processes in which they were generated where they are reused in the production process provided:
261.4(a)(8)(i)
(i) Only tank storage is involved, and the entire process through completion of reclamation is closed by being entirely connected with pipes or other comparable enclosed means of conveyance;
261.4(a)(8)(ii)
(ii) Reclamation does not involve controlled flame combustion (such as occurs in boilers, industrial furnaces, or incinerators);

261.4(a)(8)(iii)

(iii) The secondary materials are never accumulated in such tanks for over twelve months without being reclaimed; and

261.4(a)(8)(iv)

(iv) The reclaimed material is not used to produce a fuel, or used to produce products that are used in a manner constituting disposal.

261.4(a)(9)(i)

(9)(i) Spent wood preserving solutions that have been reclaimed and are reused for their original intended purpose; and

261.4(a)(9)(ii)

(ii) Wastewaters from the wood preserving process that have been reclaimed and are reused to treat wood.

261.4(a)(10)

(10) EPA Hazardous Waste Nos. K060, K087, K141, K142, K143, K144, K145, K147, and K148, and any wastes from the coke by-products processes that are hazardous only because they exhibit the Toxicity Characteristic (TC) specified in section 261.24 of this part when, subsequent to generation, these materials are recycled to coke ovens, to the tar recovery process as a feedstock to produce coal tar, or mixed with coal tar prior to the tar's sale or refining. This exclusion is conditioned on there being no land disposal of the wastes from the point they are generated to the point they are recycled to coke ovens or tar recovery or refining processes, or mixed with coal tar.

261.4(a)11

(11) Nonwastewater splash condenser dross residue from the treatment of K061 in high temperature metals recovery units, provided it is shipped in drums (if shipped) and not land disposed before recovery.

261.4(a)(12)

(12) Recovered oil from petroleum refining, exploration and production, and from transportation incident thereto, which is to be inserted into the petroleum refining process (SIC Code 2911) at or before a point (other than direct insertion into a coker) where contaminants are removed. This exclusion applies to recovered oil stored or transported prior to insertion, except that the oil must not be stored in a manner involving placement on the land, and must not be accumulated speculatively, before being so recycled. Recovered oil is oil that has been reclaimed from secondary materials (such as wastewater) generated from normal petroleum refining, exploration and production, and transportation practices. Recovered oil includes oil that is recovered from refinery wastewater collection and treatment systems, oil recovered from oil and gas drilling operations, and oil recovered from wastes removed from crude oil storage tanks. Recovered oil does not include (among other things) oil-bearing hazardous waste listed in 40 CFR part 261 D (e.g., K048-K052, F037, F038). However, oil recovered from such wastes may be considered recovered oil. Recovered oil also does not include used oil as defined in 40 CFR 279.1.

261.4(b)

(b) Solid wastes which are not hazardous wastes. The following solid wastes are not hazardous wastes:

261.4(b)(1)

(1) Household waste, including household waste that has been collected, transported, stored, treated, disposed, recovered (e.g., refuse-derived fuel) or reused. "Household

waste" means any material (including garbage, trash and sanitary wastes in septic tanks) derived from households (including single and multiple residences, hotels and motels, bunkhouses, ranger stations, crew quarters, campgrounds, picnic grounds and day-use recreation areas). A resource recovery facility managing municipal solid waste shall not be deemed to be treating, storing, disposing of, or otherwise managing hazardous wastes for the purposes of regulation under this subtitle, if such facility:

261.4(b)(1)(i)

(i) Receives and burns only

261.4(b)(1)(i)(A)

(A) Household waste (from single and multiple dwellings, hotels, motels, and other residential sources) and

261.4(b)(1)(i)(B)

(B) Solid waste from commercial or industrial sources that does not contain hazardous waste; and

261.4(b)(1)(ii)

(ii) Such facility does not accept hazardous wastes and the owner or operator of such facility has established contractual requirements or other appropriate notification or inspection procedures to assure that hazardous wastes are not received at or burned in such facility.

261.4(b)(2)

(2) Solid wastes generated by any of the following and which are returned to the soils as fertilizers:

261.4(b)(2)(i)

(i) The growing and harvesting of agricultural crops.

261.4(b)(2)(ii)

(ii) The raising of animals, including animal manures.

261.4(b)(3)

(3) Mining overburden returned to the mine site.

261.4(b)(4)

(4) Fly ash waste, bottom ash waste, slag waste, and flue gas emission control waste, generated primarily from the combustion of coal or other fossil fuels, except as provided by §266.112 of this chapter for facilities that burn or process hazardous waste.

261.4(b)(5)

(5) Drilling fluids, produced waters, and other wastes associated with the exploration, development, or production of crude oil, natural gas or geothermal energy.

261.4(b)(6)(i)

(6)(i) Wastes which fail the test for the Toxicity Characteristic because chromium is present or are listed in subpart D due to the presence of chromium, which do not fail the test for the Toxicity Characteristic for any other constituent or are not listed due to the presence of any other constituent, and which do not fail the test for any other characteristic, if it is shown by a waste generator or by waste generators that:

261.4(b)(6)(i)(A)

(A) The chromium in the waste is exclusively (or nearly exclusively) trivalent chromium; and

261.4(b)(6)(i)(B)

(B) The waste is generated from an industrial process which uses trivalent chromium exclusively (or nearly exclusively) and the process does not generate hexavalent chromium; and

261.4(b)(6)(i)(C)

(C) The waste is typically and frequently managed in non-oxidizing environments.

261.4(b)(6)(ii)

(ii) Specific waste which meet the standard in paragraphs (b)(6)(i)(A), (B), and (C) (so long as they do not fail the test for the toxicity characteristic for any other constituent, and do not exhibit any other characteristic) are: ***

261.4(b)(6)(ii)(A)

(A) Chrome (blue) trimmings generated by the following subcategories of the leather tanning and finishing industry; hair pulp/chrome tan/retan/wet finish; hair save/chrome tan/retan/wet finish; retan/wet finish; no beamhouse; through-the-blue; and shearling.

261.4(b)(6)(ii)(B)

(B) Chrome (blue) shavings generated by the following subcategories of the leather tanning and finishing industry: Hair pulp/chrome tan/retan/wet finish; hair save/chrome tan/retan/wet finish; retan/wet finish; no beamhouse; through-the-blue; and shearling.

261.4(b)(6)(ii)(C)

(C) Buffing dust generated by the following subcategories of the leather tanning and finishing industry; hair pulp/chrome tan/retan/wet finish; hair save/chrome tan/retan/wet finish; retan/wet finish; no beamhouse; through-the-blue.

261.4(b)(6)(ii)(D)

(D) Sewer screenings generated by the following subcategories of the leather tanning and finishing industry: Hair pulp/chrome tan/retan/wet finish; hair save/chrome tan/retan/wet finish; retan/wet finish; no beamhouse; through-the-blue; and shearling.

261.4(b)(6)(ii)(E)

(E) Wastewater treatment sludges generated by the following subcategories of the leather tanning and finishing industry: Hair pulp/chrome tan/retan/wet finish; hair save/chrome tan/retan/wet finish; retan/wet finish; no beamhouse; through-the-blue; and shearling.

261.4(b)(6)(ii)(F)

(F) Wastewater treatment sludges generated by the following subcategories of the leather tanning and finishing industry: Hair pulp/chrome tan/retan/wet finish; hair save/chrome tan/retan/wet finish; and through-the-blue.

261.4(b)(6)(ii)(G)

(G) Waste scrap leather from the leather tanning industry, the shoe manufacturing industry, and other leather product manufacturing industries.

261.4(b)(6)(ii)(H)

(H) Wastewater treatment sludges from the production of TiO_2 pigment using chromium-bearing ores by the chloride process.

261.4(b)(7)

(7) Solid waste from the extraction, beneficiation, and processing of ores and minerals (including coal, phosphate rock and overburden from the mining of uranium ore), except as provided by §266.112 of this chapter for facilities that burn or process hazardous waste. For purposes of §261.4(b)(7), beneficiation of ores and minerals is restricted to the following activities: Crushing; grinding; washing; dissolution; crystallization; filtration; sorting; sizing; drying; sintering; pelletizing; briquetting; calcining to remove water and/or carbon dioxide; roasting, autoclaving, and/or

chlorination in preparation for leaching (except where the roasting (and/or autoclaving and/or chlorination)/leaching sequence produces a final or intermediate product that does not undergo further beneficiation or processing); gravity concentration; magnetic separation; electrostatic separation; flotation; ion exchange; solvent extraction; electrowinning; precipitation; amalgamation; and heap, dump, vat, tank, and in situ leaching. For the purpose of §261.4(b)(7), solid waste from the processing of ores and minerals includes only the following wastes:

261.4(b)(7)(i)

(i) Slag from primary copper processing;

261.4(b)(7)(ii)

(ii) Slag from primary lead processing;

261.4(b)(7)(iii)

(iii) Red and brown muds from bauxite refining;

261.4(b)(7)(iv)

(iv) Phosphogypsum from phosphoric acid production;

261.4(b)(7)(v)

(v) Slag from elemental phosphorus production;

261.4(b)(7)(vi)

(vi) Gasifier ash from coal gasification;

261.4(b)(7)(vii)

(vii) Process wastewater from coal gasification;

261.4(b)(7)(viii)

(viii) Calcium sulfate wastewater treatment plant sludge from primary copper processing;

261.4(b)(7)(ix)

(ix) Slag tailings from primary copper processing;

261.4(b)(7)(x)

(x) Fluorogypsum from hydrofluoric acid production;

261.4(b)(7)(xi)

(xi) Process wastewater from hydrofluoric acid production;

261.4(b)(7)(xii)

(xii) Air pollution control dust/sludge from iron blast furnaces;

261.4(b)(7)(xiii)

(xiii) Iron blast furnace slag;

261.4(b)(7)(xiv)

(xiv) Treated residue from roasting/leaching of chrome ore;

261.4(b)(7)(xv)

(xv) Process wastewater from primary magnesium processing by the anhydrous process;

261.4(b)(7)(xvi)

(xvi) Process wastewater from phosphoric acid production;

261.4(b)(7)(xvii)

(xvii) Basic oxygen furnace and open hearth furnace air pollution control dust/sludge from carbon steel production;

261.4(b)(7)(xviii)

(xviii) Basic oxygen furnace and open hearth furnace slag from carbon steel production;

261.4(b)(7)(xix)
(xix) Chloride process waste solids from titanium tetrachloride production;
261.4(b)(7)(xx)
(xx) Slag from primary zinc processing.
261.4(b)(8)
(8) Cement kiln dust waste, except as provided by §266.112 of this chapter for facilities that burn or process hazardous waste.
261.4(b)(9)
(9) Solid waste which consists of discarded arsenical-treated wood or wood products which fails the test for the Toxicity Characteristic for Hazardous Waste Codes D004 through D017 and which is not a hazardous waste for any other reason if the waste is generated by persons who utilize the arsenical-treated wood and wood product for these materials' intended end use.
261.4(b)(10)
(10) Petroleum-contaminated media and debris that fail the test for the Toxicity Characteristic of §261.24 (Hazardous Waste Codes D018 through D043 only) and are subject to the corrective action regulations under part 280 of this chapter.
261.4(b)(11)
(11) Injected groundwater that is hazardous only because it exhibits the Toxicity Characteristic (Hazardous Waste Codes D018 through D043 only) in §261.24 of this part that is reinjected through an underground injection well pursuant to free phase hydrocarbon recovery operations undertaken at petroleum refineries, petroleum marketing terminals, petroleum bulk plants, petroleum pipelines, and petroleum transportation spill sites until January 25, 1993. This extension applies to recovery operations in existence, or for which contracts have been issued, on or before March 25, 1991. For groundwater returned through infiltration galleries from such operations at petroleum refineries, marketing terminals, and bulk plants, until [insert date six months after publication]. New operations involving injection wells (beginning after March 25, 1991) will qualify for this compliance date extension (until January 25, 1993) only if:
261.4(b)(11)(i)
(i) Operations are performed pursuant to a written state agreement that includes a provision to assess the groundwater and the need for further remediation once the free phase recovery is completed; and
261.4(b)(11)(ii)
(ii) A copy of the written agreement has been submitted to: Characteristics Section (OS-333), U.S. Environmental Protection Agency, 401 M Street, SW., Washington, DC 20460.
261.4(b)(12)
(12) Used chlorofluorocarbon refrigerants from totally enclosed heat transfer equipment, including mobile air conditioning systems, mobile refrigeration, and commercial and industrial air conditioning and refrigeration systems that use chlorofluorocarbons as the heat transfer fluid in a refrigeration cycle, provided the refrigerant is reclaimed for further use.
261.4(b)(13)
(13) Non-terne plated used oil filters that are not mixed with wastes listed in Subpart

D of this part if these oil filters have been gravity hot-drained using one of the following methods:

261.4(b)(13)(i)

(i) Puncturing the filter anti-drain back valve or the filter dome end and hot-draining;

261.4(b)(13)(ii)

(ii) Hot-draining and crushing;

261.4(b)(13)(iii)

(iii) Dismantling and hot-draining; or

261.4(b)(13)(iv)

(iv) Any other equivalent hot-draining method that will remove used oil.

261.4(b)(14)

(14) Used oil re-refining distillation bottoms that are used as feedstock to manufacture asphalt products.

261.4(c)

(c) Hazardous wastes which are exempted from certain regulations. A hazardous waste which is generated in a product or raw material storage tank, a product or raw material transport vehicle or vessel, a product or raw material pipeline, or in a manufacturing process unit or an associated non-waste-treatment-manufacturing unit, is not subject to regulation under parts 262 through 265, 268, 270, 271 and 124 of this chapter or to the notification requirements of section 3010 of RCRA until it exits the unit in which it was generated, unless the unit is a surface impoundment, or unless the hazardous waste remains in the unit more than 90 days after the unit ceases to be operated for manufacturing, or for storage or transportation of product or raw materials.

261.4(d)

(d) Samples.

261.4(d)(1)

(1) Except as provided in paragraph (d)(2) of this section, a sample of solid waste or a sample of water, soil, or air, which is collected for the sole purpose of testing to determine its characteristics or composition, is not subject to any requirements of this part or parts 262 through 268 or part 270 or part 124 of this chapter or to the notification requirements of section 3010 of RCRA, when:

261.4(d)(1)(i)

(i) The sample is being transported to a laboratory for the purpose of testing; or

261.4(d)(1)(ii)

(ii) The sample is being transported back to the sample collector after testing; or

261.4(d)(1)(iii)

(iii) The sample is being stored by the sample collector before transport to a laboratory for testing; or

261.4(d)(1)(iv)

(iv) The sample is being stored in a laboratory before testing; or

261.4(d)(1)(v)

(v) The sample is being stored in a laboratory after testing but before it is returned to the sample collector; or

261.4(d)(1)(vi)

(vi) The sample is being stored temporarily in the laboratory after testing for a specific purpose (for example, until conclusion of a court case or enforcement action where further testing of the sample may be necessary).

261.4(d)(2)

(2) In order to qualify for the exemption in paragraphs (d)(1) (i) and (ii) of this section, a sample collector shipping samples to a laboratory and a laboratory returning samples to a sample collector must:

261.4(d)(2)(i)

(i) Comply with U.S. Department of Transportation (DOT), U.S. Postal Service (USPS), or any other applicable shipping requirements; or

261.4(d)(2)(ii)

(ii) Comply with the following requirements if the sample collector determines that DOT, USPS, or other shipping requirements do not apply to the shipment of the sample:

261.4(d)(2)(ii)(A)

(A) Assure that the following information accompanies the sample:

261.4(d)(2)(ii)(A)(1)

(1) The sample collector's name, mailing address, and telephone number;

261.4(d)(2)(ii)(A)(2)

(2) The laboratory's name, mailing address, and telephone number;

261.4(d)(2)(ii)(A)(3)

(3) The quantity of the sample;

261.4(d)(2)(ii)(A)(4)

(4) The date of shipment; and

261.4(d)(2)(ii)(A)(5)

(5) A description of the sample.

261.4(d)(2)(ii)(B)

(B) Package the sample so that it does not leak, spill, or vaporize from its packaging.

261.4(d)(3)

(3) This exemption does not apply if the laboratory determines that the waste is hazardous but the laboratory is no longer meeting any of the conditions stated in paragraph (d)(1) of this section.

261.4(e)

(e) Treatability Study Samples.

261.4(e)(1)

(1) Except as provided in paragraph (e)(2) of this section, persons who generate or collect samples for the purpose of conducting treatability studies as defined in section 260.10, are not subject to any requirement of parts 261 through 263 of this chapter or to the notification requirements of Section 3010 of RCRA, nor are such samples included in the quantity determinations of §261.5 and §262.34 (d) when:

261.4(e)(1)(i)

(i) The sample is being collected and prepared for transportation by the generator or sample collector; or

261.4(e)(1)(ii)

(ii) The sample is being accumulated or stored by the generator or sample collector prior to transportation to a laboratory or testing facility; or

261.4(e)(1)(iii)

(iii) The sample is being transported to the laboratory or testing facility for the purpose of conducting a treatability study.

261.4(e)(2)

(2) The exemption in paragraph (e)(1) of this section is applicable to samples of hazardous waste being collected and shipped for the purpose of conducting treatability studies provided that:

261.4(e)(2)(i)

(i) The generator or sample collector uses (in "treatability studies") no more than 1000 kg of any non-acute hazardous waste, 1 kg of acute hazardous waste, or 250 kg of soils, water, or debris contaminated with acute hazardous waste for each process being evaluated for each generated waste stream; and

261.4(e)(2)(ii)

(ii) The mass of each sample shipment does not exceed 1000 kg of non-acute hazardous waste, 1 kg of acute hazardous waste, or 250 kg of soils, water, or debris contaminated with acute hazardous waste; and

261.4(e)(2)(iii)

(iii) The sample must be packaged so that it will not leak, spill, or vaporize from its packaging during shipment and the requirements of paragraph A or B of this subparagraph are met.

261.4(e)(2)(iii)(A)

(A) The transportation of each sample shipment complies with U.S. Department of Transportation (DOT), U.S. Postal Service (USPS), or any other applicable shipping requirements; or

261.4(e)(2)(iii)(B)

(B) If the DOT, USPS, or other shipping requirements do not apply to the shipment of the sample, the following information must accompany the sample:

261.4(e)(2)(iii)(B)(1)

(1) The name, mailing address, and telephone number of the originator of the sample;

261.4(e)(2)(iii)(B)(2)

(2) The name, address, and telephone number of the facility that will perform the treatability study;

261.4(e)(2)(iii)(B)(3)

(3) The quantity of the sample;

261.4(e)(2)(iii)(B)(4)

(4) The date of shipment; and

261.4(e)(2)(iii)(B)(5)

(5) A description of the sample, including its EPA Hazardous Waste Number.

261.4(e)(2)(iv)

(iv) The sample is shipped to a laboratory or testing facility which is exempt under §261.4(f) or has an appropriate RCRA permit or interim status.

261.4(e)(2)(v)

(v) The generator or sample collector maintains the following records for a period ending 3 years after completion of the treatability study:

261.4(e)(2)(v)(A)

(A) Copies of the shipping documents;

261.4(e)(2)(v)(B)

(B) A copy of the contract with the facility conducting the treatability study;

261.4(e)(2)(v)(C)

(C) Documentation showing:

261.4(e)(2)(v)(C)(1)

(1) The amount of waste shipped under this exemption;

261.4(e)(2)(v)(C)(2)

(2) The name, address, and EPA identification number of the laboratory or testing facility that received the waste;

261.4(e)(2)(v)(C)(3)

(3) The date the shipment was made; and

261.4(e)(2)(v)(C)(4)

(4) Whether or not unused samples and residues were returned to the generator.

261.4(e)(2)(vi)

(vi) The generator reports the information required under paragraph (e)(v)(C) of this section in its biennial report.

261.4(e)(3)

(3) The Regional Administrator, or State Director (if located in an authorized State), may grant requests, on a case-by-case basis, for quantity limits in excess of those specified in paragraph (e)(2)(i) of this section, for up to an additional 500 kg of non-acute hazardous waste, 1 kg of acute hazardous waste, and 250 kg of soils, water, or debris contaminated with acute hazardous waste, to conduct further treatability study evaluation when: There has been an equipment or mechanical failure during the conduct of a treatability study; there is a need to verify the results of a previously conducted treatability study; there is a need to study and analyze alternative techniques within a previously evaluated treatment process; or there is a need to do further evaluation of an ongoing treatability study to determine final specifications for treatment. The additional quantities allowed are subject to all the provisions in paragraphs (e)(1) and (e)(2)(ii)(vi) of this section. The generator or sample collector must apply to the Regional Administrator in the Region where the sample is collected and provide in writing the following information:

261.4(e)(3)(i)

(i) The reason why the generator or sample collector requires additional quantity of sample for the treatability study evaluation and the additional quantity needed;

261.4(e)(3)(ii)

(ii) Documentation accounting for all samples of hazardous waste from the waste stream which have been sent for or undergone treatability studies including the data each previous sample from the waste stream was shipped, the quantity of each previous shipment, the laboratory or testing facility to which it was shipped, what treatability study processes were conducted on each sample shipped, and the available results of each treatability study;

261.4(e)(3)(iii)

(iii) A description of the technical modifications or change in specifications which will be evaluated and the expected results;

261.4(e)(3)(iv)

(iv) If such further study is being required due to equipment or mechanical failure, the applicant must include information regarding the reason for the failure or breakdown and also include what procedures or equipment improvements have been made to protect against further breakdowns; and

261.4(e)(3)(v)

(v) Such other information that the Regional Administrator considers necessary.

261.4(f)

(f) Samples Undergoing Treatability Studies at Laboratories and Testing Facilities. Samples undergoing treatability studies and the laboratory or testing facility conducting such treatability studies (to the extent such facilities are not otherwise subject to RCRA requirements) are not subject to any requirement of this part, part 124, parts 262-266, 268, and 270, or to the notification requirements of Section 3010 of RCRA provided that the conditions of paragraphs (f) (1) through (11) of this section are met. A mobile treatment unit (MTU) may qualify as a testing facility subject to paragraphs (f) (1) through (11) of this section. Where a group of MTUs are located at the same site, the limitations specified in (f) (1) through (11) of this section apply to the entire group of MTUs collectively as if the group were one MTU.

261.4(f)(1)

(1) No less than 45 days before conducting treatability studies, the facility notifies the Regional Administrator, or State Director (if located in an authorized State), in writing that it intends to conduct treatability studies under this paragraph.

261.4(f)(2)

(2) The laboratory or testing facility conducting the treatability study has an EPA identification number.

261.4(f)(3)

(3) No more than a total of 250 kg of "as received" hazardous waste is subjected to initiation of treatment in all treatability studies in any single day. "As received" waste refers to the waste as received in the shipment from the generator or sample collector.

261.4(f)(4)

(4) The quantity of "as received" hazardous waste stored at the facility for the purpose of evaluation in treatability studies does not exceed 1000 kg, the total of which can include 500 kg of soils, water, or debris contaminated with acute hazardous waste or 1 kg of acute hazardous waste. This quantity limitation does not include:

261.4(f)(4)(i)

(i) Treatability study residues; and

261.4(f)(4)(ii)

(ii) Treatment materials (including nonhazardous solid waste) added to "as received" hazardous waste.

261.4(f)(5)

(5) No more than 90 days have elapsed since the treatability study for the sample was completed, or no more than one year has elapsed since the generator or sample collector shipped the sample to the laboratory or testing facility, whichever date first occurs.

261.4(f)(6)

(6) The treatability study does not involve the placement of hazardous waste on the land or open burning of hazardous waste.

261.4(f)(7)

(7) The facility maintains records for 3 years following completion of each study that show compliance with the treatment rate limits and the storage time and quantity limits. The following specific information must be included for each treatability study conducted:

261.4(f)(7)(i)

(i) The name, address, and EPA identification number of the generator or sample collector of each waste sample;

261.4(f)(7)(ii)
(ii) The date the shipment was received;
261.4(f)(7)(iii)
(iii) The quantity of waste accepted;
261.4(f)(7)(iv)
(iv) The quantity of "as received" waste in storage each day;
261.4(f)(7)(v)
(v) The date the treatment study was initiated and the amount of "as received" waste introduced to treatment each day;
261.4(f)(7)(vi)
(vi) The date the treatability study was concluded;
261.4(f)(7)(vii)
(vii) The date any unused sample or residues generated from the treatability study were returned to the generator or sample collector or, if sent to a designated facility, the name of the facility and the EPA identification number.
261.4(f)(8)
(8) The facility keeps, on-site, a copy of the treatability study contract and all shipping papers associated with the transport of treatability study samples to and from the facility for a period ending 3 years from the completion date of each treatability study.
261.4(f)(9)
(9) The facility prepares and submits a report to the Regional Administrator, or State Director (if located in an authorized State), by March 15 of each year that estimates the number of studies and the amount of waste expected to be used in treatability studies during the current year, and includes the following information for the previous calendar year:
261.4(f)(9)(i)
(i) The name, address, and EPA identification number of the facility conducting the treatability studies;
261.4(f)(9)(ii)
(ii) The types (by process) of treatability studies conducted;
261.4(f)(9)(iii)
(iii) The names and addresses of persons for whom studies have been conducted (including their EPA identification numbers);
261.4(f)(9)(iv)
(iv) The total quantity of waste in storage each day;
261.4(f)(9)(v)
(v) The quantity and types of waste subjected to treatability studies;
261.4(f)(9)(vi)
(vi) When each treatability study was conducted;
261.4(f)(9)(vii)
(vii) The final disposition of residues and unused sample from each treatability study.
261.4(f)(10)
(10) The facility determines whether any unused sample or residues generated by the treatability study are hazardous waste under §261.3 and, if so, are subject to parts 261 through 268, and part 270 of this chapter, unless the residues and unused samples are returned to the sample originator under the §261.4 (e) exemption.

261.4(f)(11)

(11) The facility notifies the Regional Administrator, or State Director (if located in an authorized State), by letter when the facility is no longer planning to conduct any treatability studies at the site.

(Approved by the Office of Management and Budget under control number 2050-0088) [45 Fed. Reg. 33119, May 19, 1980; 56 Fed. Reg. 41177, Aug. 19, 1991; 57 Fed. Reg. 21534, May 20, 1992; 57 Fed. Reg. 27888, June 22, 1992; 57 Fed. Reg. 37281, Aug. 18, 1992; 58 Fed. Reg. 26420, May 3, 1993; 58 Fed. Reg. 8362, Feb. 18, 1994; 59 Fed. Reg. 38536, July 28, 1994; 61 Fed. Reg. 13103, Mar. 26, 1996]

■ APPENDIX E.2 Environmental Protection Agency 40 CFR Listed Hazardous Wastes

261.31

Hazardous wastes from non-specific sources.

261.31(a)

(a) The following solid wastes are listed hazardous wastes from non-specific sources unless they are excluded under §§260.20 and 260.22 and listed in appendix IX.

Industry and EPA Hazardous Waste No.	Hazardous Waste	Hazard Code
Generic:		
F001	The following spent halogenated solvents used in degreasing: Tetrachloroethylene, trichloroethylene, methylene chloride, 1,1,1-trichloroethane, carbon tetrachloride, and chlorinated fluorocarbons; all spent solvent mixtures/blends used in degreasing containing, before use, a total of ten percent or more (by volume) of one or more of the above halogenated solvents or those solvents listed in F002, F004, and F005; and still bottoms from the recovery of these spent solvents and spent solvent mixtures	(T)
F002	The following spent halogenated solvents: Tetrachloroethylene, methylene chloride, trichloroethylene, 1,1,1-trichloroethane, chlorobenzene, 1,1,2-trichloro-1,2,2-trifluoroethane, ortho-dichlorobenzene, trichlorofluoromethane, and 1,1,2-trichloroethane; all spent solvent mixtures/blends containing, before use, a total of ten percent or more (by volume) of one or more of the above halogenated solvents or those listed in F001, F004, or F005; and still bottoms from the recovery of these spent solvents and spent solvent mixtures	(T)
F003	The following spent non-halogenated solvents: Xylene, acetone, ethyl acetate, ethyl benzene, ethyl ether, methyl isobutyl ketone, n-butyl alcohol, cyclohexanone, and methanol; all spent solvent mixtures/blends containing, before use, only the above spent non-halogenated solvents; and all spent solvent mixtures/blends containing, before use, one or more of the above non-halogenated solvents, and, a total of ten	(I)*

* (I,T) should be used to specify mixtures containing ignitable and toxic constituents.

Industry and EPA Hazardous Waste No.	Hazardous Waste	Hazard Code
	percent or more (by volume) of one or more of those solvents listed in F001, F002, F004, and F005; and still bottoms from the recovery of these spent solvents and spent solvent mixtures	
F004	The following spent non-halogenated solvents: Cresols and cresylic acid, and nitrobenzene; all spent solvent mixtures/blends containing, before use, a total of ten percent or more (by volume) of one or more of the above non-halogenated solvents or those solvents listed in F001, F002, and F005; and still bottoms from the recovery of these spent solvents and spent solvent mixtures	(T)
F005	The following spent non-halogenated solvents: Toluene, methyl ethyl ketone, carbon disulfide, isobutanol, pyridine, benzene, 2-ethoxyethanol, and 2-nitropropane; all spent solvent mixtures/blends containing, before use, a total of 10 percent or more (by volume) of one or more of the above non-halogenated solvents or those solvents listed in F001, F002, or F004; and still bottoms from the recovery of these spent solvents and spent solvent mixtures	(I, T)
F006	Wastewater treatment sludges from electroplating operations except from the following processes: (1) Sulfuric acid anodizing of aluminum; (2) tin plating on carbon steel; (3) zinc plating (segregated basis) on carbon steel; (4) aluminum or zinc-aluminum plating on carbon steel; (5) cleaning/stripping associated with tin, zinc and aluminum plating on carbon steel; and (6) chemical etching and milling of aluminum	(T)
F007	Spent cyanide plating bath solutions from electroplating operations	(R, T)
F008	Plating bath residues from the bottom of plating baths from electroplating operations where cyanides are used in the process	(R, T)
F009	Spent stripping and cleaning bath solutions from electroplating operations where cyanides are used in the process	(R, T)
F010	Quenching bath residues from oil baths from metal heat treating operations where cyanides are used in the process	(R, T)
F011	Spent cyanide solutions from salt bath pot cleaning from metal heat treating operations	(R, T)

Industry and EPA Hazardous Waste No.	Hazardous Waste	Hazard Code
F012	Quenching waste water treatment sludges from metal heat treating operations where cyanides are used in the process	(T)
F019	Wastewater treatment sludges from the chemical conversion coating of aluminum except from zirconium phosphating in aluminum can washing when such phosphating is an exclusive conversion coating process	(T)
F020	Wastes (except wastewater and spent carbon from hydrogen chloride purification) from the production or manufacturing use (as a reactant, chemical intermediate, or component in a formulating process) of tri- or tetrachlorophenol, or of intermediates used to produce their pesticide derivatives. (This listing does not include wastes from the production of Hexachlorophene from highly purified 2,4,5-trichlorophenol.)	(H)
F021	Wastes (except wastewater and spent carbon from hydrogen chloride purification) from the production or manufacturing use (as a reactant, chemical intermediate, or component in a formulating process) of pentachlorophenol, or of intermediates used to produce its derivatives	(H)
F022	Wastes (except wastewater and spent carbon from hydrogen chloride purification) from the manufacturing use (as a reactant, chemical intermediate, or component in a formulating process) of tetra-, penta-, or hexachlorobenzenes under alkaline conditions	(H)
F023	Wastes (except wastewater and spent carbon from hydrogen chloride purification) from the production of materials on equipment previously used for the production or manufacturing use (as a reactant, chemical intermediate, or component in a formulating process) of tri- and tetrachlorophenols. (This listing does not include wastes from equipment used only for the production or use of Hexachlorophene from highly purified 2,4,5-trichlorophenol.)	(H)
F024	Process wastes, including but not limited to, distillation residues, heavy ends, tars, and reactor clean-out wastes, from the production of certain chlorinated aliphatic hydrocarbons by free radical catalyzed processes. These chlorinated aliphatic hydrocarbons are those having carbon chain lengths ranging from one to and including five, with varying amounts and positions of chlorine substitution. (This listing does not	(T)

Industry and EPA Hazardous Waste No.	Hazardous Waste	Hazard Code
	include wastewaters, wastewater treatment sludges, spent catalysts, and wastes listed in §261.31 or §261.32.)	
F025	Condensed light ends, spent filters and filter aids, and spent desiccant wastes from the production of certain chlorinated aliphatic hydrocarbons, by free radical catalyzed processes. These chlorinated aliphatic hydrocarbons are those having carbon chain lengths ranging from one to and including five, with varying amounts and positions of chlorine substitution	(T)
F026	Wastes (except wastewater and spent carbon from hydrogen chloride purification) from the production of materials on equipment previously used for the manufacturing use (as a reactant, chemical intermediate, or component in a formulating process) of tetra-, penta-, or hexachlorobenzene under alkaline conditions	(H)
F027	Discarded unused formulations containing tri-, tetra-, or pentachlorophenol or discarded unused formulations containing compounds derived from these chlorophenols. (This listing does not include formulations containing Hexachlorophene synthesized from prepurified 2,4,5-trichlorophenol as the sole component.)	(H)
F028	Residues resulting from the incineration or thermal treatment of soil contaminated with EPA Hazardous Waste Nos. F020, F021, F022, F023, F026, and F027	(T)
F032[1]	Wastewaters (except those that have not come into contact with process contaminants), process residuals, preservative drippage, and spent formulations from wood preserving processes generated at plants that currently use or have previously used chlorophenolic formulations (except potentially cross-contaminated wastes that have had the F032 waste code deleted in accordance with §261.35 of this chapter or	(T)

[1] The F032, F034, and F035 listings are administratively stayed with respect to the process area receiving drippage of these wastes provided persons desiring to continue operating notify EPA by August 6, 1991 of their intent to upgrade or install drip pads, and by November 6, 1991 provide evidence to EPA that they have adequate financing to pay for drip pad upgrades or installation, as provided in the administrative stay. The stay of the listings will remain in effect until February 6, 1992 for existing drip pads and until May 6, 1992 for new drip pads.

Industry and EPA Hazardous Waste No.	Hazardous Waste	Hazard Code
	potentially cross-contaminated wastes that are otherwise currently regulated as hazardous wastes (i.e., F034 or F035), and where the generator does not resume or initiate use of chlorophenolic formulations). This listing does not include K001 bottom sediment sludge from the treatment of wastewater from wood preserving processes that use creosote and/or pentachlorophenol	(T)
F034[1]	Wastewaters (except those that have not come into contact with process contaminants), process residuals, preservative drippage, and spent formulations from wood preserving processes generated at plants that use creosote formulations. This listing does not include K001 bottom sediment sludge from the treatment of wastewater from wood preserving processes that use creosote and/or pentachlorophenol	(T)
F035[1]	Wastewaters (except those that have not come into contact with process contaminants), process residuals, preservative drippage, and spent formulations from wood preserving processes generated at plants that use inorganic preservatives containing arsenic or chromium. This listing does not include K001 bottom sediment sludge from the treatment of wastewater from wood preserving processes that use creosote and/or pentachlorophenol	(T)
F037	Petroleum refinery primary oil/water/solids separation sludge—Any sludge generated from the gravitational separation of oil/water/solids during the storage or treatment of process wastewaters and oily cooling wastewaters from petroleum refineries. Such sludges include, but are not limited to, those generated in: oil/water/solids separators; tanks and impoundments; ditches and other conveyances; sumps; and stormwater units receiving dry weather flow. Sludge generated in stormwater units that do not receive dry weather flow, sludges generated from non-contact once-through cooling waters segregated for treatment from other process or oily cooling waters, sludges generated in aggressive biological treatment units as defined in §261.31 (b)(2) (including sludges generated in one or more additional units after wastewaters have been treated in aggressive biological treatment units) and K051 wastes are not included in this listing	(T)

Industry and EPA Hazardous Waste No.	Hazardous Waste	Hazard Code
F038	Petroleum refinery secondary (emulsified) oil/water/solids separation sludge—Any sludge and/or float generated from the physical and/or chemical separation of oil/water/solids in process wastewaters and oily cooling wastewaters from petroleum refineries. Such wastes include, but are not limited to, all sludges and floats generated in: induced air flotation (IAF) units, tanks and impoundments, and all sludges generated in DAF units. Sludges generated in stormwater units that do not receive dry weather flow, sludges generated from non-contact once-through cooling waters segregated for treatment from other process or oily cooling waters, sludges and floats generated in aggressive biological treatment units as defined in §261.31 (b)(2) (including sludges and floats generated in one or more additional units after wastewaters have been treated in aggressive biological treatment units) and F037, K048, and K051 wastes and not included in this listing	(T)
F039	Leachate (liquids that have percolated through land disposed wastes) resulting from the disposal of more than one restricted waste classified as hazardous under subpart D of this part. (Leachate resulting from the disposal of one or more of the following EPA Hazardous Wastes and no other Hazardous Wastes retains its EPA Hazardous Waste Number(s): F020, F021, F022, F026, F027, and/or F028.)	(T)

261.31(b)

(b) Listing Specific Definitions:

261.31(b)(1)

(1) For the purposes of the F037 and F038 listings, oil/water/solids is defined as oil and/or water and/or solids.

261.31(b)(2)(i)

(2)(i) For the purposes of the F037 and F038 listings, aggressive biological treatment units are defined as units which employ one of the following four treatment methods: activated sludge; trickling filter; rotating biological contactor for the continuous accelerated biological oxidation of wastewaters; or high-rate aeration. High-rate aeration is a system of surface impoundments or tanks, in which intense mechanical aeration is used to completely mix the wastes, enhance biological activity, and (A) the units employs a minimum of 6 hp per million gallons of treatment volume; and either (B) the hydraulic retention time of the unit is no longer than 5 days; or (C) the hydraulic retention time is no longer than 30 days and the unit does not generate a sludge that is a hazardous waste by the Toxicity Characteristic.

261.31(b)(2)(ii)

(ii) Generators and treatment, storage and disposal facilities have the burden of proving that their sludges are exempt from listing as F037 and F038 wastes under this definition. Generators and treatment, storage and disposal facilities must maintain, in their operating or other onsite records, documents and data sufficient to prove that: (A) the unit is an aggressive biological treatment unit as defined in this subsection; and (B) the sludges sought to be exempted from the definitions of F037 and/or F038 were actually generated in the aggressive biological treatment unit.

261.31(b)(3)(i)

(3)(i) For the purposes of the F037 listing, sludges are considered to be generated at the moment of deposition in the unit, where deposition is defined as at least a temporary cessation of lateral particle movement.

261.31(b)(3)(ii)

(ii) For the purposes of the F038 listing,

261.31(b)(3)(ii)(A)

(A) sludges are considered to be generated at the moment of deposition in the unit, where deposition is defined as at least a temporary cessation of lateral particle movement and

261.31(b)(3)(ii)(B)

(B) floats are considered to be generated at the moment they are formed in the top of the unit.

[46 Fed. Reg. 4617, Jan. 16, 1981; 57 Fed. Reg. 61492, Dec. 24, 1992]

261.32

Hazardous wastes from specific sources.

The following solid wastes are listed hazardous wastes from specific sources unless they are excluded under §§260.20 and 260.22 and listed in appendix IX.

Industry and EPA Hazardous Waste No.	Hazardous Waste	Hazard Code
Wood preservation:		
K001	Bottom sediment sludge from the treatment of wastewaters from wood preserving processes that use creosote and/or pentachlorophenol	(T)

Industry and EPA Hazardous Waste No.	Hazardous Waste	Hazard Code
Inorganic pigments:		
K002	Wastewater treatment sludge from the production of chrome yellow and orange pigments	(T)
K003	Wastewater treatment sludge from the production of molybdate orange pigments	(T)
K004	Wastewater treatment sludge from the production of zinc yellow pigments	(T)
K005	Wastewater treatment sludge from the production of chrome green pigments	(T)
K006	Wastewater treatment sludge from the production of chrome oxide green pigments (anhydrous and hydrated)	(T)
K007	Wastewater treatment sludge from the production of iron blue pigments	(T)
K008	Oven residue from the production of chrome oxide green pigments	(T)
Organic chemicals:		
K009	Distillation bottoms from the production of acetaldehyde from ethylene	(T)
K010	Distillation side cuts from the production of acetaldehyde from ethylene	(T)
K011	Bottom stream from the wastewater stripper in the production of acrylonitrile	(R, T)
K013	Bottom stream from the acetonitrile column in the production of acrylonitrile	(R, T)
K014	Bottoms from the acetonitrile purification column in the production of acrylonitrile	(T)
K015	Still bottoms from the distillation of benzyl chloride	(T)
K016	Heavy ends or distillation residues from the production of carbon tetrachloride	(T)
K017	Heavy ends (still bottoms) from the purification column in the production of epichlorohydrin	(T)
K018	Heavy ends from the fractionation column in ethyl chloride production	(T)
K019	Heavy ends from the distillation of ethylene dichloride in ethylene dichloride production	(T)
K020	Heavy ends from the distillation of vinyl chloride in vinyl chloride monomer production	(T)

Industry and EPA Hazardous Waste No.	Hazardous Waste	Hazard Code
K021	Aqueous spent antimony catalyst waste from fluoromethanes production	(T)
K022	Distillation bottom tars from the production of phenol/acetone from cumene	(T)
K023	Distillation light ends from the production of phthalic anhydride from naphthalene	(T)
K024	Distillation bottoms from the production of phthalic anhydride from naphthalene	(T)
K025	Distillation bottoms from the production of nitrobenzene by the nitration of benzene	(T)
K026	Stripping still tails from the production of methyl ethyl pyridines	(T)
K027	Centrifuge and distillation residues from toluene diisocyanate production	(R, T)
K028	Spent catalyst from the hydrochlorinator reactor in the production of 1,1,1-trichloroethane	(T)
K029	Waste from the product steam stripper in the production of 1,1,1-trichloroethane	(T)
K030	Column bottoms or heavy ends from the combined production of trichloroethylene and perchloroethylene	(T)
K083	Distillation bottoms from aniline production	(T)
K085	Distillation or fractionation column bottoms from the production of chlorobenzenes	(T)
K093	Distillation light ends from the production of phthalic anhydride from ortho-xylene	(T)
K094	Distillation bottoms from the production of phthalic anhydride from ortho-xylene	(T)
K095	Distillation bottoms from the production of 1,1,1-trichloroethane	(T)
K096	Heavy ends from the heavy ends column from the production of 1,1,1-trichloroethane	(T)
K103	Process residues from aniline extraction from the production of aniline	(T)
K104	Combined wastewater streams generated from nitrobenzene/aniline production	(T)
K105	Separated aqueous stream from the reactor product washing step in the production of chlorobenzenes	(T)
K107	Column bottoms from product separation from the production of 1,1-dimethyl-hydrazine (UDMH) from carboxylic acid hydrazines	(C, T)

Industry and EPA Hazardous Waste No.	Hazardous Waste	Hazard Code
K108	Condensed column overheads from product separation and condensed reactor vent gases from the production of 1,1-dimethylhydrazine (UDMH) from carboxylic acid hydrazides	(I, T)
K109	Spent filter cartridges from product purification from the production of 1,1-dimethylhydrazine (UDMH) from carboxylic acid hydrazides	(T)
K110	Condensed column overheads from intermediate separation from the production of 1,1-dimethylhydrazine (UDMH) from carboxylic acid hydrazides	(T)
K111	Product washwaters from the production of dinitrotoluene via nitration of toluene	(C, T)
K112	Reaction by-product water from the drying column in the production of toluenediamine via hydrogenation of dinitrotoluene	(T)
K113	Condensed liquid light ends from the purification of toluenediamine in the production of toluenediamine via hydrogenation of dinitrotoluene	(T)
K114	Vicinals from the purification of toluenediamine in the production of toluenediamine via hydrogenation of dinitrotoluene	(T)
K115	Heavy ends from the purification of toluenediamine in the production of toluenediamine via hydrogenation of dinitrotoluene	(T)
K116	Organic condensate from the solvent recovery column in the production of toluene diisocyanate via phosgenation of toluenediamine	(T)
K117	Wastewater from the reactor vent gas scrubber in the production of ethylene dibromide via bromination of ethene	(T)
K118	Spent adsorbent solids from purification of ethylene dibromide in the production of ethylene dibromide via bromination of ethene	(T)
K136	Still bottoms from the purification of ethylene dibromide in the production of ethylene dibromide via bromination of ethene	(T)

Industry and EPA Hazardous Waste No.	Hazardous Waste	Hazard Code
K149	Distillation bottoms from the production of alpha- (or methyl-) chlorinated toluenes, ring-chlorinated toluenes, benzoyl chlorides, and compounds with mixtures of these functional groups (This waste does not include still bottoms from the distillation of benzyl chloride.)	(T)
K150	Organic residuals, excluding spent carbon adsorbent, from the spent chlorine gas and hydrochloric acid recovery processes associated with the production of alpha- (or methyl-) chlorinated toluenes, ring-chlorinated toluenes, benzoyl chlorides, and compounds with mixtures of these functional groups	(T)
K151	Wastewater treatment sludges, excluding neutralization and biological sludges, generated during the treatment of wastewaters from the production of alpha- (or methyl-) chlorinated toluenes, ring-chlorinated toluenes, benzoyl chlorides, and compounds with mixtures of these functional groups.	(T)
K156	Organic waste (including heavy ends, still bottoms, light ends, spent solvents, filtrates, and decantates) from the production of carbamates and carbamoyl oximes	(T)
K157	Wastewaters (including scrubber waters, condenser waters, washwaters, and separation waters) from the production of carbamates and carbamoyl oximes	(T)
K158	Bag house dusts and filter/separation solids from the production of carbamates and carbamoyl oximes	(T)
K159	Organics from the treatment of thiocarbamate wastes	(T)
K160	Solids (including filter wastes, separation solids, and spent catalysts) from the production of thiocarbamates and solids from the treatment of thiocarbamate wastes	(T)
K161	Purification solids (including filtration, evaporation, and centrifugation solids), bag house dust and floor sweepings from the production of dithiocarbamate acids and their salts (This listing does not include K125 or K126.)	(R,T)
Inorganic chemicals:		
K071	Brine purification muds from the mercury cell process in chlorine production, where separately prepurified brine is not used	(T)

Industry and EPA Hazardous Waste No.	Hazardous Waste	Hazard Code
K073	Chlorinated hydrocarbon waste from the purification step of the diaphragm cell process using graphite anodes in chlorine production	(T)
K106	Wastewater treatment sludge from the mercury cell process in chlorine production	(T)
Pesticides:		
K031	By-product salts generated in the production of MSMA and cacodylic acid	(T)
K032	Wastewater treatment sludge from the production of chlordane	(T)
K033	Wastewater and scrub water from the chlorination of cyclopentadiene in the production of chlordane	(T)
K034	Filter solids from the filtration of hexachlorocyclopentadiene in the production of chlordane	(T)
K035	Wastewater treatment sludges generated in the production of creosote	(T)
K036	Still bottoms from toluene reclamation distillation in the production of disulfoton	(T)
K037	Wastewater treatment sludges from the production of disulfoton	(T)
K038	Wastewater from the washing and stripping of phorate production	(T)
K039	Filter cake from the filtration of diethylphosphorodithioic acid in the production of phorate	(T)
K040	Wastewater treatment sludge from the production of phorate	(T)
K041	Wastewater treatment sludge from the production of toxaphene	(T)
K042	Heavy ends or distillation residues from the distillation of tetrachlorobenzene in the production of 2,4,5-T	(T)
K043	2,6-Dichlorophenol waste from the production of 2,4-D	(T)
K097	Vacuum stripper discharge from the chlordane chlorinator in the production of chlordane	(T)
K098	Untreated process wastewater from the production of toxaphene	(T)
K099	Untreated wastewater from the production of 2,4-D	(T)

Industry and EPA Hazardous Waste No.	Hazardous Waste	Hazard Code
K123	Process wastewater (including supernates, filtrates, and washwaters) from the production of ethylenebisdithiocarbamic acid and its salts	(T)
K124	Reactor vent scrubber water from the production of ethylenebisdithiocarbamic acid and its salts	(C,T)
K125	Filtration, evaporation, and centrifugation solids form the production of ethylenebisdithiocarbamic acid and its salts	(T)
K126	Baghouse dust and floor sweepings in milling and packaging operations from the production or formulation of ethylenebisdithiocarbamic acid and its salts	(T)
K131	Wastewater from the reactor and spent sulfuric acid from the acid dryer from the production of methyl bromide	(C,T)
K132	Spent absorbent and wastewater separator solids from the production of methyl bromide	(T)
Explosives:		
K044	Wastewater treatment sludges from the manufacturing and processing of explosives	(R)
K045	Spent carbon from the treatment of wastewater containing explosives	(R)
K046	Wastewater treatment sludges from the manufacturing, formulation and loading of lead-based initiating compounds	(T)
K047	Pink/red water from TNT operations	(R)
Petroleum refining:		
K048	Dissolved air flotation (DAF) float from the petroleum refining industry	(T)
K049	Slop oil emulsion solids from the petroleum refining industry	(T)
K050	Heat exchanger bundle cleaning sludge from the petroleum refining industry	(T)
K051	API separator sludge from the petroleum refining industry	(T)
K052	Tank bottoms (leaded) from the petroleum refining industry	(T)

Industry and EPA Hazardous Waste No.	Hazardous Waste	Hazard Code
Iron and steel:		
K061	Emission control dust/sludge from the primary production of steel in electric furnaces	(T)
K062	Spent pickle liquor generated by steel finishing operations of facilities within the iron and steel industry (SIC Codes 331 and 332)	(C,T)
Primary copper:		
K064	Acid plant blowdown slurry/sludge resulting from the thickening of blowdown slurry from primary copper production	(T)
Primary lead:		
K065	Surface impoundment solids contained in and dredged from surface impoundments at primary lead smelting facilities	(T)
Primary zinc:		
K066	Sludge from treatment of process wastewater and/or acid plant blowdown from primary zinc production	(T)
Primary aluminum:		
K088	Spent potliners from primary aluminum reduction	(T)
Ferroalloys:		
K090	Emission control dust or sludge from ferrochromiumsilicon production	(T)
K091	Emission control dust or sludge from ferrochromium production	(T)
Secondary lead:		
K069	Emission control dust/sludge from secondary lead smelting (Note: This listing is stayed administratively for sludge generated from secondary acid scrubber systems. The stay will remain in effect until further administrative action is taken. If EPA takes further action effecting this stay, EPA will publish a notice of the action in the Federal Register.)	(T)
K100	Waste leaching solution from acid leaching of emission control dust/sludge from secondary lead smelting	(T)

Industry and EPA Hazardous Waste No.	Hazardous Waste	Hazard Code
Veterinary pharmaceuticals:		
K084	Wastewater treatment sludges generated during the production of veterinary pharmaceuticals from arsenic or organo-arsenic compounds	(T)
K101	Distillation tar residues from the distillation of aniline-based compounds in the production of veterinary pharmaceuticals from arsenic or organo-arsenic compounds	(T)
K102	Residue form the use of activated carbon for decolorization in the production of veterinary pharmaceuticals from arsenic or organo-arsenic compounds	(T)
Ink formulation:		
K086	Solvent washes and sludges, caustic washes and sludges, or water washes and sludges from cleaning tubs and equipment used in the formulation of ink from pigments, driers, soaps, and stabilizers containing chromium and lead	(T)
Coking:		
K060	Ammonia still lime sludge from coking operations	(T)
K087	Decanter tank tar sludge from coking operations	(T)
K141	Process residues from the recovery of coal tar, including, but not limited to, collecting sump residues from the production of coke from coal or the recovery of coke by-products produced from coal. This listing does not include K087 (decanter tank tar sludges from coking operations).	(T)
K142	Tar storage tank residues from the production of coke from coal or from the recovery of coke by-products produced from coal	(T)
K143	Process residues from the recovery of light oil, including, but not limited to, those generated in stills, decanters, and wash oil recovery units from the recovery of coke by-products produced from coal	(T)
K144	Wastewater sump residues from light oil refining, including, but not limited to, intercepting or contamination sump sludges from the recovery of coke by-products produced from coal	(T)

Industry and EPA Hazardous Waste No.	Hazardous Waste	Hazard Code
K145	Residues from naphthalene collection and recovery operations from the recovery of coke by-products produced from coal	(T)
K147	Tar storage tank residues from coal tar refining	(T)
K148	Residues from coal tar distillation, including but not limited to, still bottoms	(T)

[46 Fed. Reg. 4618, Jan. 16, 1981; 57 Fed. Reg. 37281, Aug. 18, 1992; 57 Fed. Reg. 47376, Oct. 15, 1992; 60 Fed. Reg. 7824, Feb. 9, 1995]

261.33

Discarded commercial chemical products, off-specification species, container residues, and spill residues thereof.

The following materials or items are hazardous wastes if and when they are discarded or intended to be discarded as described in §261.2 (a)(2)(i), when they are mixed with waste oil or used oil or other material and applied to the land for dust suppression or road treatment, when they are otherwise applied to the land in lieu of their original intended use or when they are contained in products that are applied to the land in lieu of their original intended use, or when, in lieu of their original intended use, they are produced for use as (or as a component of) a fuel, distributed for use as a fuel, or burned as a fuel.

261.33(a)

(a) Any commercial chemical product, or manufacturing chemical intermediate having the generic name listed in paragraph (e) or (f) of this section.

261.33(b)

(b) Any off-specification commercial chemical product or manufacturing chemical intermediate which, if it met specifications, would have the generic name listed in paragraph (e) or (f) of this section.

261.33(c)

(c) Any residue remaining in a container or in an inner liner removed from a container that has held any commercial chemical product or manufacturing chemical intermediate having the generic name listed in paragraphs (e) or (f) of this section, unless the container is empty as defined in §261.7(b) of this chapter.

[*Comment:* Unless the residue is being beneficially used or reused, or legitimately recycled or reclaimed; or being accumulated, stored, transported or treated prior to such use, reuse, recycling or reclamation, EPA considers the residue to be intended for discard, and thus, a hazardous waste. An example of a legitimate reuse of the residue would be where the residue remains in the container and the container is used to hold the same commercial chemical product or manufacturing chemical intermediate it previously held. An example of the discard of the residue would be where the drum is sent to a drum reconditioner who reconditions the drum but discards the residue.]

261.33(d)

(d) Any residue or contaminated soil, water or other debris resulting from the cleanup of a spill into or on any land or water of any commercial chemical product or manufacturing chemical intermediate having the generic name listed in paragraph (e) or (f) of this section, or any residue or contaminated soil, water or other debris resulting from the cleanup of a spill, into or on any land or water, of any off-specification chemical product and manufacturing chemical intermediate which, if it met specifications, would have the generic name listed in paragraph (e) or (f) of this section.

[*Comment:* The phrase "commercial chemical product or manufacturing chemical intermediate having the generic name listed in . . ." refers to a chemical substance which is manufactured or formulated for commercial or manufacturing use which consists of the commercially pure grade of the chemical, any technical grades of the chemical that are produced or marketed, and all formulations in which the chemical is the sole active ingredient. It does not refer to a material, such as a manufacturing process waste, that contains any of the substances listed in paragraph (e) or (f). Where a manufacturing process waste is deemed to be a hazardous waste because it contains a substance listed in paragraph (e) or (f), such waste will be listed in either §261.31 or §261.32 or will be identified as a hazardous waste by the characteristics set forth in subpart C of this part.]

261.33(e)

(e) The commercial chemical products, manufacturing chemical intermediates or off-specification commercial chemical products or manufacturing chemical intermediates referred to in paragraphs (a) through (d) of this section, are identified as acute hazardous wastes (H) and are subject to be the small quantity exclusion defined in §261.5(e).

[*Comment:* For the convenience of the regulated community the primary hazardous properties of these materials have been indicated by the letters T (Toxicity), and R (Reactivity). Absence of a letter indicates that the compound only is listed for acute toxicity.]

These wastes and their corresponding EPA Hazardous Waste Numbers are:

Hazardous Waste No.	**Chemical Abstracts No.**	**Substance**
P023	107-20-0	Acetaldehyde, chloro-
P002	591-08-2	Acetamide, N-aminothioxomethyl)-
P057	640-17-7	Acetamide, 2-fluoro-

Hazardous Waste No.	Chemical Abstracts No.	Substance
P058	62-74-8	Acetic acid, fluoro-, sodium salt
P002	591-08-2	1-Acetyl-2-thiourea
P003	107-02-8	Acrolein
P070	116-06-3	Aldicarb
P203	1646-88-4	Aldicarb sulfone
P004	309-00-2	Aldrin
P005	107-18-6	Allyl alcohol
P006	20859-73-8	Aluminum phosphide (R,T)
P007	2763-96-4	5-(Aminomethyl)-3-isoxazolol
P008	504-24-5	4-Aminopyridine
P009	131-74-8	Ammonium picrate (R)
P119	7803-55-6	Ammonium vanadate
P099	506-61-6	Argentate(1-), bis(cyano-C)-, potassium
P010	7778-39-4	Arsenic acid H_3AsO_4
P012	1327-53-3	Arsenic oxide As_2O_3
P011	1303-28-2	Arsenic oxide As_2O_5
P011	1303-28-2	Arsenic pentoxide
P012	1327-53-3	Arsenic trioxide
P038	692-42-2	Arsine, diethyl-
P036	696-28-6	Arsonous dichloride, phenyl-
P054	151-56-4	Aziridine
P067	75-55-8	Aziridine, 2-methyl-
P013	542-62-1	Barium cyanide
P024	106-47-8	Benzenamine, 4-chloro-
P077	100-01-6	Benzenamine, 4-nitro-
P028	100-44-7	Benzene, (chloromethyl)-
P042	51-43-4	1,2-Benzenediol, 4-[1-hydroxy-2-(methylamino)ethyl]-, (R)-
P046	122-09-8	Benzeneethanamine, alpha, alpha-dimethyl-
P014	108-98-5	Benzenethiol

Hazardous Waste No.	Chemical Abstracts No.	Substance
P127	1563-66-2	7-Benzofuranol, 2,3-dihydro-2,2-dimethyl-, methylcarbamate
P188	57-64-7	Benzoic acid, 2-hydroxy-, compd. with (3aS-cis)-1,2,3,3a,8,8a- hexahydro-1,3a, 8-trimethylpyrrolo [2,3-b]indol-5-ylmethylcarbamate ester (1:1)
P001	1 81-81-2	2H-1-Benzopyran-2-one, 4-hydroxy-3-(3-oxo-1-phenylbutyl)-, and salts, when present at concentrations greater than 0.3%
P028	100-44-7	Benzyl chloride
P015	7440-41-7	Beryllium powder
P017	598-31-2	Bromoacetone
P018	357-57-3	Brucine
P045	39196-18-4	2-Butanone, 3,3-dimethyl-1-(methylthio)-,
P021	592-01-8	Calcium cyanide
P021	592-01-8	Calcium cyanide $Ca(CN)_2$
P189	55285-14-8	Carbamic acid, [(dibutylamino)-thio]methyl-, 2,3-dihydro-2,2- dimethyl-7-benzofuranyl ester
P191	644-64-4	Carbamic acid, dimethyl-, 1-[(dimethyl-amino)carbonyl]-5-methyl-1H-pyrazol-3-yl ester
P192	119-38-0	Carbamic acid, dimethyl-, 3-methyl-1-(1-methylethyl)-1H-pyrazol-5-yl ester
P190	1129-41-5	Carbamic acid, methyl-, 3-methylphenyl ester
P127	1563-66-2	Carbofuran
P022	75-15-0	Carbon disulfide
P095	75-44-5	Carbonic dichloride
P189	55285-14-8	Carbosulfan
P023	107-20-0	Chloroacetaldehyde
P024	106-47-8	p-Chloroaniline
P026	5344-82-1	1-(o-Chlorophenyl)thiourea
P027	542-76-7	3-Chloropropionitrile
P029	544-92-3	Copper cyanide
P029	544-92-3	Copper cyanide Cu(CN)
P202	64-00-6	m-Cumenyl methylcarbamate
P030		Cyanides (soluble cyanide salts), not otherwise specified

Hazardous Waste No.	Chemical Abstracts No.	Substance
P031	460-19-5	Cyanogen
P033	506-77-4	Cyanogen chloride
P033	506-77-4	Cyanogen chloride (CN)Cl
P034	131-89-5	2-Cyclohexyl-4,6-dinitrophenol
P016	542-88-1	Dichloromethyl ether
P036	696-28-6	Dichlorophenylarsine
P037	60-57-1	Dieldrin
P038	692-42-2	Diethylarsine
P041	311-45-5	Diethyl-p-nitrophenyl phosphate
P040	297-97-2	O,O-Diethyl O-pyrazinyl phosphorothioate
P043	55-91-4	Diisopropylfluorophosphate (DFP)
P004	309-00-2	1,4,5,8-Dimethanonaphthalene, 1,2,3,4,10, 10-hexa- chloro-1,4,4a,5,8,8a, -hexahydro-, (1alpha, 4alpha, 4abeta, 5alpha, 8alpha, 8abeta)-
P060	465-73-6	1,4,5,8-Dimethanonaphthalene, 1,2,3,4,10, 10-hexa- chloro-1,4,4a,5,8,8a-hexahydro-, (1alpha, 4alpha, 4abeta, 5beta, 8beta, 8abeta)-
P037	60-57-1	2,7:3,6-Dimethanonaphth[2,3-b]oxirene, 3,4,5,6,9,9-hexachloro-1a,2,2a,3,6,6a,7,7a-octahydro-, (1aalpha, 2beta, 2aalpha, 3beta, 6beta, 6aalpha, 7beta, 7aalpha)-
P051	1 72-20-8	2,7:3,6-Dimethanonaphth [2,3-b]oxirene, 3, 4,5,6,9,9-hexachloro-1a,2,2a,3,6,6a,7,7a-octahydro-, (1aalpha, 2beta, 2abeta, 3alpha, 6alpha, 6abeta, 7beta, 7aalpha)-, and metabolites
P044	60-51-5	Dimethoate
P046	122-09-8	alpha, alpha-Dimethylphenethylamine
P191	644-64-4	Dimetilan
P047	1 534-52-1	4,6-Dinitro-o-cresol, and salts
P048	51-28-5	2,4-Dinitrophenol
P020	88-85-7	Dinoseb
P085	152-16-9	Diphosphoramide, octamethyl-
P111	107-49-3	Diphosphoric acid, tetraethyl ester

Hazardous Waste No.	Chemical Abstracts No.	Substance
P039	298-04-4	Disulfoton
P049	541-53-7	Dithiobiuret
P185	26419-73-8	1,3-Dithiolane-2-carboxaldehyde, 2,4-dimethyl-, O-[(methylamino)- carbonyl] oxime
P050	115-29-7	Endosulfan
P088	145-73-3	Endothall
P051	72-20-8	Endrin
P051	72-20-8	Endrin, & metabolites
P042	51-43-4	Epinephrine
P031	460-19-5	Ethanedinitrile
P194	23135-22-0	Ethanimidothioc acid, 2-(dimethylamino)-N-[[(methylamino) carbonyl]oxy]-2-oxo-, methyl ester
P066	16752-77-5	Ethanimidothioic acid, N-[[(methylamino)carbonyl]oxy]-, methyl ester
P101	107-12-0	Ethyl cyanide
P054	151-56-4	Ethyleneimine
P097	52-85-7	Famphur
P056	7782-41-4	Fluorine
P057	640-19-7	Fluoroacetamide
P058	62-74-8	Fluoroacetic acid, sodium salt
P198	23422-53-9	Formetanate hydrochloride
P197	17702-57-7	Formparanate.
P065	628-86-4	Fulminic acid, mercury (2+) salt (R,T)
P059	76-44-8	Heptachlor
P062	757-58-4	Hexaethyl tetraphosphate
P116	79-19-6	Hydrazinecarbothioamide
P068	60-34-4	Hydrazine, methyl-
P063	74-90-8	Hydrocyanic acid
P063	74-90-8	Hydrogen cyanide
P096	7803-51-2	Hydrogen phosphide

Hazardous Waste No.	Chemical Abstracts No.	Substance
P060	465-73-6	Isodrin
P192	119-38-0	Isolan
P202	64-00-6	3-Isopropylphenyl N-methylcarbamate
P007	2763-96-4	3(2H)-Isoxazolone, 5-(aminomethyl)-
P196	15339-36-3	Manganese, bis(dimethylcarbamodithioato-S,S′)-
P196	15339-36-3	Manganese dimethyldithiocarbamate
P092	62-38-4	Mercury, (acetato-O)phenyl-
P065	628-86-4	Mercury fulminate (R,T)
P082	62-75-9	Methanamine, N-methyl-N-nitroso-
P064	624-83-9	Methane, isocyanato-
P016	542-88-1	Methane, oxybis[chloro-
P112	509-14-8	Methane, tetranitro- (R)
P118	75-70-7	Methanethiol, trichloro-
	23422-53-9	Methanimidamide, N,N-dimethyl-N′-[3-[[(methylamino)- carbonyl]oxy]phenyl]-, monohydrochloride.
P198	17702-57-7	Methanimidamide, N,N-dimethyl-N′-[2-methyl-4-[[(methylamino)carbonyl]oxy]phenyl]-
P199	2032-65-7	Methiocarb
P050	115-29-7	6,9-Methano-2,4,3-benzodioxathiepin, 6,7,8,9,10,10-hexachloro-1,5,5a,6,9,9a-hexahydro-, 3-oxide
P059	76-44-8	4,7-Methano-1H-indene, 1,4,5,6,7,8,8-heptachloro- 3a,4,7,7a-tetrahydro-
P066	16752-77-5	Methomyl
P068	60-34-4	Methyl hydrazine
P064	624-83-9	Methyl isocyanate
P069	75-86-5	2-Methyllactonitrile
P071	298-00-0	Methyl parathion
P190	1129-41-5	Metolcarb
P128	315-18-4	Mexacarbate
P072	86-88-4	alpha-Naphthylthiourea

Hazardous Waste No.	Chemical Abstracts No.	Substance
P073	13463-39-3	Nickel carbonyl
P073	13463-39-3	Nickel carbonyl $Ni(CO)_4$, (T-4)-
P074	557-19-7	Nickel cyanide
P074	557-19-7	Nickel cyanide $Ni(CN)_2$
P075	1 54-11-5	Nicotine, and salts
P076	10102-43-9	Nitric oxide
P077	100-01-6	p-Nitroaniline
P078	10102-44-0	Nitrogen dioxide
P076	10102-43-9	Nitrogen oxide NO
P078	10102-44-0	Nitrogen oxide NO_2
P081	55-63-0	Nitroglycerine (R)
P082	62-75-9	N-Nitrosodimethylamine
P084	4549-40-0	N-Nitrosomethylvinylamine
P085	152-16-9	Octamethylpyrophosphoramide
P087	20816-12-0	Osmium oxide OsO_4, (T-4)-
P087	20816-12-0	Osmium tetroxide
P088	145-73-3	7-Oxabicyclo[2.2.1]heptane-2,3-dicarboxylic acid
P194	23135-22-0	Oxamyl
P089	56-38-2	Parathion
P034	131-89-5	Phenol, 2-cyclohexyl-4,6-dinitro-
P128	315-18-4	Phenol, 4-(dimethylamino)-3,5-dimethyl-, methylcarbamate (ester)
P199	2032-65-7	Phenol, (3,5-dimethyl-4-(methylthio)-, methylcarbamate
P048	51-28-5	Phenol, 2,4-dinitro-
P047	1 534-52-1	Phenol, 2-methyl-4,6-dinitro-, and salts
P202	64-00-6	Phenol, 3-(1-methylethyl)-, methyl carbamate
P201	2631-37-0	Phenol, 3-methyl-5-(1-methylethyl)-, methyl carbamate
P020	88-85-7	Phenol, 2-(1-methylpropyl)-4,6-dinitro-
P009	131-74-8	Phenol, 2,4,6-trinitro-, ammonium salt (R)
P092	62-38-4	Phenylmercury acetate

Hazardous Waste No.	Chemical Abstracts No.	Substance
P093	103-85-5	Phenylthiourea
P094	298-02-2	Phorate
P095	75-44-5	Phosgene
P096	7803-51-2	Phosphine
P041	311-45-5	Phosphoric acid, diethyl 4-nitrophenyl ester
P039	298-04-4	Phosphorodithioic acid, O,O-diethyl S-[2-(ethylthio)ethyl] ester
P094	298-02-2	Phosphorodithioic acid, O,O-diethyl S-[(ethylthio)methyl] ester
P044	60-51-5	Phosphorodithioic acid, O,O-dimethyl S-[2-(methylamino)-2-oxoethyl] ester
P043	55-91-4	Phosphorofluoridic acid, bis(1-methylethyl) ester
P089	56-38-2	Phosphorothioic acid, O,O-diethyl O-(4-nitrophenyl) ester
P040	297-97-2	Phosphorothioic acid, O,O-diethyl O-pyrazinyl ester
P097	52-85-7	Phosphorothioic acid, O-[4-[(dimethylamino)sulfonyl]phenyl] O,O-dimethyl ester
P071	298-00-0	Phosphorothioic acid, O,O-dimethyl O-(4-nitrophenyl) ester
P204	57-47-6	Physostigmine
P188	57-64-7	Physostigmine salicylate
P110	78-00-2	Plumbane, tetraethyl-
P098	151-50-8	Potassium cyanide
P098	151-50-8	Potassium cyanide K(CN)
P099	506-61-6	Potassium silver cyanide
P201	2631-37-0	Promecarb
P203	1646-88-4	Propanal, 2-methyl-2-(methyl-sulfonyl)-, O-[(methylamino)carbonyl] oxime
P070	116-06-3	Propanal, 2-methyl-2-(methylthio)-, O-[(methylamino)carbonyl]oxime
P101	107-12-0	Propanenitrile
P027	542-76-7	Propanenitrile, 3-chloro-
P069	75-86-5	Propanenitrile, 2-hydroxy-2-methyl-
P081	55-63-0	1,2,3-Propanetriol, trinitrate (R)
P017	598-31-2	2-Propanone, 1-bromo-
P102	107-19-7	Propargyl alcohol
P003	107-02-8	2-Propenal

Hazardous Waste No.	Chemical Abstracts No.	Substance
P005	107-18-6	2-Propen-1-ol
P067	75-55-8	1,2-Propylenimine
P102	107-19-7	2-Propyn-1-ol
P008	504-24-5	4-Pyridinamine
P075	1 54-11-5	Pyridine, 3-(1-methyl-2-pyrrolidinyl)-, (S)-, and salts
P204	57-47-6	Pyrrolo[2,3-b]indol-5-ol, 1,2,3,3a,8,8a-hexahydro-1,3a,8-trimethyl-, methylcarbamate (ester), (3aS-cis)-
P114	12039-52-0	Selenious acid, dithallium (1+) salt
P103	630-10-4	Selenourea
P104	506-64-9	Silver cyanide
P104	506-64-9	Silver cyanide Ag(CN)
P105	26628-22-8	Sodium azide
P106	143-33-9	Sodium cyanide
P106	143-33-9	Sodium cyanide Na(CN)
P108	1 57-24-9	Strychnidin-10-one, and salts
P018	357-57-3	Strychnidin-10-one, 2,3-dimethoxy-
P108	1 57-24-9	Strychnine, & salts
P115	7446-18-6	Sulfuric acid, dithallium(1+) salt
P109	3689-24-5	Tetraethyldithiopyrophosphate
P110	78-00-2	Tetraethyl lead
P111	107-49-3	Tetraethyl pyrophosphate
P112	509-14-8	Tetranitromethane (R)
P062	757-58-4	Tetraphosphoric acid, hexaethyl ester
P113	1314-32-5	Thallic oxid
P113	1314-32-5	Thallium oxide Tl_2O_3
P114	12039-52-0	Thallium(I) selenite
P115	7446-18-6	Thallium(I) sulfate
P109	3689-24-5	Thiodiphosphoric acid, tetraethyl ester

Hazardous Waste No.	Chemical Abstracts No.	Substance
P045	39196-18-4	Thiofanox
P049	541-53-7	Thioimidodicarbonic diamide $[(H_2N)C(S)]_2NH$
P014	108-98-5	Thiophenol
P116	79-19-6	Thiosemicarbazide
P026	5344-82-1	Thiourea, (2-chlorophenyl)-
P072	86-88-4	Thiourea, 1-naphthalenyl-
P093	103-85-5	Thiourea, phenyl-
P123	8001-35-2	Toxaphene
P118	75-70-7	Trichloromethanethiol
P185	26419-73-8	Tirpate
P119	7803-55-6	Vanadic acid, ammonium salt
P120	1314-62-1	Vanadium oxide V_2O_5
P120	1314-62-1	Vanadium pentoxide
P084	4549-40-0	Vinylamine, N-methyl-N-nitroso-
P001	1 81-81-2	Warfarin, and salts, when present at concentrations greater than 0.3%
P205	137-30-4	Zinc, bis(dimethylcarbamodithioato-S,S′)-,
P121	557-21-1	Zinc cyanide
P121	557-21-1	Zinc cyanide $Zn(CN)_2$
P122	1314-84-7	Zinc phosphide Zn_3P_2, when present at concentrations greater than 10% (R,T)
P205	137-30-4	Ziram. 1 CAS Number given for parent compound only.

261.33(f)

(f) The commercial chemical products, manufacturing chemical intermediates, or off-specification commercial chemical products referred to in paragraphs (a) through (d) of this section, are identified as toxic wastes (T), unless otherwise designated and are subject to the small quantity generator exclusion defined in §261.5 (a) and (g).

[*Comment:* For the convenience of the regulated community, the primary hazardous properties of these materials have been indicated by the letters T (Toxicity), R (Reactivity), I (Ignitability) and C (Corrosivity). Absence of a letter indicates that the compound is only listed for toxicity.]

These wastes and their corresponding EPA Hazardous Waste Numbers are:

Hazardous waste No.	Chemical abstracts No.	Substance
U394	30558-43-1	A2213.
U001	75-07-0	Acetaldehyde (I)
U034	75-87-6	Acetaldehyde, trichloro-
U187	62-44-2	Acetamide, N-(4-ethoxyphenyl)-
U005	53-96-3	Acetamide, N-9H-fluoren-2-yl-
U240	1 94-75-7	Acetic acid, (2,4-dichlorophenoxy)-, salts and esters
U112	141-78-6	Acetic acid ethyl ester (I)
U144	301-04-2	Acetic acid, lead(2+) salt
U214	563-68-8	Acetic acid, thallium(1+) salt
see F027	93-76-5	Acetic acid, (2,4,5-trichlorophenoxy)-
U002	67-64-1	Acetone (I)
U003	75-05-8	Acetonitrile (I,T)
U004	98-86-2	Acetophenone
U005	53-96-3	2-Acetylaminofluorene
U006	75-36-5	Acetyl chloride (C,R,T)
U007	79-06-1	Acrylamide
U008	79-10-7	Acrylic acid (I)
U009	107-13-1	Acrylonitrile
U011	61-82-5	Amitrole
U012	62-53-3	Aniline (I,T)
U136	75-60-5	Arsinic acid, dimethyl-
U014	492-80-8	Auramine
U015	115-02-6	Azaserine
U365	2212-67-1	H-Azepine-1-carbothioic acid, hexahydro-, S-ethyl ester
U010	50-07-7	Azirino[2,3:3,4]pyrrolo[1,2-a]indole-4,7-dione,6-amino-8-[[(aminocarbonyl)oxy] methyl]-1, 1a,2,8,8a,8b-hexahydro-8a-methoxy-5-methyl-, [1aS-(1aalpha,8beta, 8aalpha,8balpha)]-

Hazardous waste No.	Chemical abstracts No.	Substance
U280	101-27-9	Barban
U278	22781-23-3	Bendiocarb
U364	22961-82-6	Bendiocarb phenol
U271	17804-35-2	Benomyl
U157	56-49-5	Benz[j]aceanthrylene, 1,2-dihydro-3-methyl-
U016	225-51-4	Benz[c]acridine
U017	98-87-3	Benzal chloride
U192	23950-58-5	Benzamide, 3,5-dichloro-N-(1,1-dimethyl-2-propynyl)-
U018	56-55-3	Benz[a]anthracene
U094	57-97-6	Benz[a]anthracene, 7,12-dimethyl-
U012	62-53-3	Benzenamine (I,T)
U014	492-80-8	Benzenamine, 4,4-carbonimidoylbis[N,N-dimethyl-
U049	3165-93-3	Benzenamine, 4-chloro-2-methyl-, hydrochloride
U093	60-11-7	Benzenamine, N,N-dimethyl-4-(phenylazo)-
U328	95-53-4	Benzenamine, 2-methyl-
U353	106-49-0	Benzenamine, 4-methyl-
U158	101-14-4	Benzenamine, 4,4-methylenebis[2-chloro-
U222	636-21-5	Benzenamine, 2-methyl-, hydrochloride
U181	99-55-8	Benzenamine, 2-methyl-5-nitro-
U019	71-43-2	Benzene (I,T)
U038	510-15-6	Benzeneacetic acid, 4-chloro-alpha-(4-chlorophenyl)-alpha-hydroxy-, ethyl ester
U030	101-55-3	Benzene, 1-bromo-4-phenoxy-
U035	305-03-3	Benzenebutanoic acid, 4-[bis(2-chloro-ethyl)amino]-
U037	108-90-7	Benzene, chloro-
U221	25376-45-8	Benzenediamine, ar-methyl-
U028	117-81-7	1,2-Benzenedicarboxylic acid, bis(2-ethyl-hexyl) ester
U069	84-74-2	1,2-Benzenedicarboxylic acid, dibutyl ester
U088	84-66-2	1,2-Benzenedicarboxylic acid, diethyl ester
U102	131-11-3	1,2-Benzenedicarboxylic acid, dimethyl ester

Hazardous waste No.	Chemical abstracts No.	Substance
U107	117-84-0	1,2-Benzenedicarboxylic acid, dioctyl ester
U070	95-50-1	Benzene, 1,2-dichloro-
U071	541-73-1	Benzene, 1,3-dichloro-
U072	106-46-7	Benzene, 1,4-dichloro-
U060	72-54-8	Benzene, 1,1-(2,2-dichloroethylidene)bis[4-chloro-
U017	98-87-3	Benzene, (dichloromethyl)-
U223	26471-62-5	Benzene, 1,3-diisocyanatomethyl- (R,T)
U239	1330-20-7	Benzene, dimethyl- (I,T)
U201	108-46-3	1,3-Benzenediol
U127	118-74-1	Benzene, hexachloro-
U056	110-82-7	Benzene, hexahydro- (I)
U220	108-88-3	Benzene, methyl-
U105	121-14-2	Benzene, 1-methyl-2,4-dinitro-
U106	606-20-2	Benzene, 2-methyl-1,3-dinitro-
U055	98-82-8	Benzene, (1-methylethyl)- (I)
U169	98-95-3	Benzene, nitro-
U183	608-93-5	Benzene, pentachloro-
U185	82-68-8	Benzene, pentachloronitro-
U020	98-09-9	Benzenesulfonic acid chloride (C,R)
U020	989-09-9	Benzenesulfonyl chloride (C,R)
U207	95-94-3	Benzene, 1,2,4,5-tetrachloro-
U061	50-29-3	Benzene, 1,1-(2,2,2-trichloroethylidene)bis[4-chloro-
U247	72-43-5	Benzene, 1,1-(2,2,2-trichloroethylidene)bis[4- methoxy-
U023	98-07-7	Benzene, (trichloromethyl)-
U234	99-35-4	Benzene, 1,3,5-trinitro-
U021	92-87-5	Benzidine
U202	1 81-07-2	1,2-Benzisothiazol-3(2H)-one, 1,1-dioxide, and salts
U203	94-59-7	1,3-Benzodioxole, 5-(2-propenyl)-
U141	120-58-1	1,3-Benzodioxole, 5-(1-propenyl)-

Hazardous waste No.	Chemical abstracts No.	Substance
U090	94-58-6	1,3-Benzodioxole, 5-propyl-
U064	189-55-9	Benzo[rst]pentaphene
U248	1 81-81-2	2H-1-Benzopyran-2-one, 4-hydroxy-3-(3-oxo-1-phenyl-butyl)-, and salts, when present at concentrations of 0.3% or less
U022	50-32-8	Benzo[a]pyrene
U278	22781-23-3	1,3-Benzodioxol-4-ol, 2,2-dimethyl-, methyl carbamate
U364	22961-82-6	1,3-Benzodioxol-4-ol, 2,2-dimethyl-,
U367	1563-38-8	7-Benzofuranol, 3-dihydro-2,2-dimethyl-
U197	106-51-4	p-Benzoquinone
U023	98-07-7	Benzotrichloride (C,R,T)
U085	1464-53-5	2,2-Bioxirane
U021	92-87-5	[1,1-Biphenyl]-4,4-diamine
U073	91-94-1	[1,1′-Biphenyl]-4,4′-diamine, 3′-dichloro-
U091	119-90-4	[1,1′-Biphenyl]-4,4′-diamine, 3,3′-dimethoxy-
U095	119-93-7	[1,1′-Biphenyl]-4,4′-diamine, 3′-dimethyl-
U401	97-74-5	Bis(dimethylthiocarbamoyl) sulfide
U400	120-54-7	Bis(pentamethylene)thiuram tetrasulfide
U225	75-25-2	Bromoform
U030	101-55-3	4-Bromophenyl phenyl ether
U128	87-68-3	1,3-Butadiene, 1,1,2,3,4,4-hexachloro-
U172	924-16-3	1-Butanamine, N-butyl-N-nitroso-
U031	71-36-3	1-Butanol (I)
U159	78-93-3	2-Butanone (I,T)
U160	1338-23-4	2-Butanone, peroxide (R,T)
U053	4170-30-3	2-Butenal
U074	764-41-0	2-Butene, 1,4-dichloro- (I,T)
U143	303-34-4	2-Butenoic acid, 2-methyl-, 7-[[2,3-dihydroxy-2-(1-methoxyethyl)-3-methyl-1-oxobutoxy]methyl]-2,3,5,7a-tetrahydro-1H-pyrrolizin-1-ylester, [1S-[1alpha(Z),7(2S*,3R*),7aalpha]]-

Hazardous waste No.	Chemical abstracts No.	Substance
U031	71-36-3	n-Butyl alcohol (I)
U392	2008-41-5	Butylate
U136	75-60-5	Cacodylic acid
U032	13765-19-0	Calcium chromate
U372	10605-21-7	Carbamic acid, 1H-benzimidazol-2-yl, methyl ester
U271	17804-35-2	Carbamic acid, [1-[(butylamino)carbonyl]-1H-benz imidazol-2-yl]-, methyl ester.
U375	55406-53-6	Carbamic acid, butyl-, 3-iodo-2-propynyl ester
U280	101-27-9	Carbamic acid, (3-chlorophenyl)-, 4-chloro-2-butynyl ester
U373	122-42-9	Carbamic acid, phenyl-, 1-methylethyl ester
U409	23564-05-8	Carbamic acid, [1,2-phenylenebis (iminocarbonothioyl)]bis-, dimethyl ester.
U238	51-79-6	Carbamic acid, ethyl ester
U178	615-53-2	Carbamic acid, methylnitroso-, ethyl ester
U097	79-44-7	Carbamic chloride, dimethyl-
U379	136-30-1	Carbamodithioic acid, dibutyl, sodium salt
U277	95-06-7	Carbamodithioic acid, diethyl-, 2-chloro-2-propenyl ester
U381	148-18-5	Carbamodithioic acid, diethyl-, sodium salt
U383	128-03-0	Carbamodithioic acid, dimethyl, potassium salt
U382	128-04-1	Carbamodithioic acid, dimethyl-, sodium salt
U376	144-34-3	Carbamodithioic acid, dimethyl-, tetraanhydrosulfide with orthothioselenious acid
U114	1 111-54-6	Carbamodithioic acid, 1,2-ethanediylbis-, salts and esters
U378	51026-28-9	Carbamodithioic acid, (hydroxymethyl)methyl-, monopotassium salt
U384	137-42-8	Carbamodithioic acid, methyl-, monosodium salt
U377	137-41-7	Carbamodithioic acid, methyl,-mono-potassium salt
U062	2303-16-4	Carbamothioic acid, bis(1-methylethyl)-, S-(2,3-dichloro-2-propenyl) ester
U389	2303-17-5	Carbamothioic acid, bis(1-methylethyl)-, S-(2,3,3-trichloro-2-propenyl) ester
U392	2008-41-5	Carbamothioic acid, bis(2-methylpropyl)-, S-ethyl ester
U391	1114-71-2	Carbamothioic acid, butylethyl-, S-propyl ester
U386	1134-23-2	Carbamothioic acid, cyclohexylethyl-, S-ethyl ester
U390	759-94-4	Carbamothioic acid, dipropyl-, S-ethyl ester

Hazardous waste No.	Chemical abstracts No.	Substance
U387	52888-80-9	Carbamothioic acid, dipropyl-, S-(phenylmethyl) ester
U385	1929-77-7	Carbamothioic acid, dipropyl-, S-propyl ester
U279	63-25-2	Carbaryl
U372	10605-21-7	Carbendazim
U367	1563-38-8	Carbofuran phenol
U215	6533-73-9	Carbonic acid, dithallium(1+) salt
U033	353-50-4	Carbonic difluoride
U156	79-22-1	Carbonochloridic acid, methyl ester (I,T)
U033	353-50-4	Carbon oxyfluoride (R,T)
U211	56-23-5	Carbon tetrachloride
U034	75-87-6	Chloral
U035	305-03-3	Chlorambucil
U036	57-74-9	Chlordane, alpha and gamma isomers
U026	494-03-1	Chlomaphazin
U037	108-90-7	Chlorobenzene
U038	510-15-6	Chlorobenzilate
U039	59-50-7	p-Chloro-m-cresol
U042	110-75-8	2-Chloroethyl vinyl ether
U044	67-66-3	Chloroform
U046	107-30-2	Chloromethyl methyl ether
U047	91-58-7	beta-Chloronaphthalene
U048	95-57-8	o-Chlorophenol
U049	3165-93-3	4-Chloro-o-toluidine, hydrochloride
U032	13765-19-0	Chromic acid H_2CrO_4, calcium salt
U050	218-01-9	Chrysene
U393	137-29-1	Copper,bis(dimethylcarbamodithioato-S,S′)-,
U393	137-29-1	Copper dimethyldithiocarbamate
U051		Creosote
U052	1319-77-3	Cresol (Cresylic acid)

Hazardous waste No.	Chemical abstracts No.	Substance
U053	4170-30-3	Crotonaldehyde
U055	98-82-8	Cumene (I)
U246	506-68-3	Cyanogen bromide (CN)Br
U197	106-51-4	2,5-Cyclohexadiene-1,4-dione
U386	1134-23-2	Cycloate
U056	110-82-7	Cyclohexane (I)
U129	58-89-9	Cyclohexane, 1,2,3,4,5,6-hexachloro-, (1alpha,2alpha,3beta,4alpha,5alpha,6beta)-
U057	108-94-1	Cyclohexanone (I)
U130	77-47-4	1,3-Cyclopentadiene, 1,2,3,4,5,5-hexachloro-
U058	50-18-0	Cyclophosphamide
U240	1 94-75-7	2,4-D, salts and esters
U059	20830-81-3	Daunomycin
U366	533-74-4	Dazomet
U060	72-54-8	DDD
U061	50-29-3	DDT
U062	2303-16-4	Diallate
U063	53-70-3	Dibenz[a,h]anthracene
U064	189-55-9	Dibenzo[a,i]pyrene
U066	96-12-8	1,2-Dibromo-3-chloropropane
U069	84-74-2	Dibutyl phthalate
U070	95-50-1	o-Dichlorobenzene
U071	541-73-1	m-Dichlorobenzene
U072	106-46-7	p-Dichlorobenzene
U073	91-94-1	3,3′-Dichlorobenzidine
U074	764-41-0	1,4-Dichloro-2-butent (I,T)
U075	75-71-8	Dichlorodifluoromethane
U078	75-35-4	1,1-Dichloroethylene
U079	156-60-5	1,2-Dichloroethylene
U025	111-44-4	Dichloroethyl ether

Hazardous waste No.	Chemical abstracts No.	Substance
U027	108-60-1	Dichloroisopropyl ether
U024	111-91-1	Dichloromethoxy ethane
U081	120-83-2	2,4-Dichlorophenol
U082	87-65-0	2,6-Dichlorophenol
U084	542-75-6	1,3-Dichloropropene
U085	1464-53-5	1,2:3,4-Diepoxybutane (I,T)
U395	5952-26-1	Diethylene glycol, dicarbamate
U108	123-91-1	1,4-Diethyleneoxide
U028	117-81-7	Diethylhexyl phthalate
U086	1615-80-1	N,N′-Diethylhydrazine
U087	3288-58-2	O,O-Diethyl S-methyl dithiophosphate
U088	84-66-2	Diethyl phthalate
U089	56-53-1	Diethylstilbesterol
U090	94-58-6	Dihydrosafrole
U091	119-90-4	3,3′-Dimethoxybenzidine
U092	124-40-3	Dimethylamine (I)
U093	60-11-7	p-Dimethylaminoazobenzene
U094	57-97-6	7,12-Dimethylbenz[a]anthracene
U095	119-93-7	3,3′-Dimethylbenzidine
U096	80-15-9	alpha,alpha-Dimethylbenzylhydroperoxide (R)
U097	79-44-7	Dimethylcarbamoyl chloride
U098	57-14-7	1,1-Dimethylhydrazine
U099	540-73-8	1,2-Dimethylhydrazine
U101	105-67-9	2,4-Dimethylphenol
U102	131-11-3	Dimethyl phthalate
U103	77-78-1	Dimethyl sulfate
U105	121-14-2	2,4-Dinitrotoluene
U106	606-20-2	2,6-Dinitrotoluene
U107	117-84-0	Di-n-octyl phthalate

Hazardous waste No.	Chemical abstracts No.	Substance
U108	123-91-1	1,4-Dioxane
U109	122-66-7	1,2-Diphenylhydrazine
U110	142-84-7	Dipropylamine (I)
U111	621-64-7	Di-n-propylnitrosamine
U403	97-77-8	Disulfiram
U041	106-89-8	Epichlorohydrin
U390	759-94-4	EPTC
U001	75-07-0	Ethanal (I)
U174	55-18-5	Ethanamine, N-ethyl-N-nitroso-
U404	101-44-8	Ethanamine, N,N-diethyl-
U155	91-80-5	1,2-Ethanediamine, N,N-dimethyl-N′–2-pyridinyl-N′-(2-thienylmethyl)-
U067	106-93-4	Ethane, 1,2-dibromo-
U076	75-34-3	Ethane, 1,1-dichloro-
U077	107-06-2	Ethane, 1,2-dichloro-
U131	67-72-1	Ethane, hexachloro-
U024	111-91-1	Ethane, 1,1′-[methylenebis(oxy)]bis[2-chloro-
U117	60-29-7	Ethane, 1,1′-oxybis-(I)
U025	111-44-4	Ethane, 1,1′-oxybis[2-chloro-
U184	76-01-7	Ethane, pentachloro-
U208	630-20-6	Ethane, 1,1,1,2-tetrachloro-
U209	79-34-5	Ethane, 1,1,2-tetrachloro-
U218	62-55-5	Ethanethioamide
U226	71-55-6	Ethane, 1,1,1-trichloro-
U227	79-00-5	Ethane, 1,1,2-trichloro-
U410	59669-26-0	Ethanimidothioic acid, N,N′-[thiobis[(methylimino) carbonyloxy]] bis-, dimethyl ester
U394	30558-43-1	Ethanimidothioic acid, 2-(dimethylamino)-N-hydroxy-2-oxo-, methyl ester
U359	110-80-5	Ethanol, 2-ethoxy-
U173	1116-54-7	Ethanol, 2,2′-(nitrosoimino)bis-

Hazardous waste No.	Chemical abstracts No.	Substance
U395	5952-26-1	Ethanol,2,2′-oxybis-, dicarbamate
U004	98-86-2	Ethanone, 1-phenyl-
U043	75-01-4	Ethene, chloro-
U042	110-75-8	Ethene, (2-chloroethoxy)-
U078	75-35-4	Ethene, 1,1-dichloro-
U079	156-60-5	Ethene, 1,2-dichloro-, (E)-
U210	127-18-4	Ethene, tetrachloro-
U228	79-01-6	Ethene, trichloro-
U112	141-78-6	Ethyl acetate (I)
U113	140-88-5	Ethyl acrylate (I)
U238	51-79-6	Ethyl carbamate (urethane)
U117	60-29-7	Ethyl ether (I)
U114	1 111-54-6	Ethylenebisdithiocarbamic acid, salts and esters
U067	106-93-4	Ethylene dibromide
U077	107-06-2	Ethylene dichloride
U359	110-80-5	Ethylene glycol monoethyl ether
U115	75-21-8	Ethylene oxide (I,T)
U116	96-45-7	Ethylenethiourea
U076	75-34-3	Ethylidene dichloride
U118	97-63-2	Ethyl methacrylate
U119	62-50-0	Ethyl methanesulfonate
U407	14324-55-1	Ethyl Ziram
U396	14484-64-1	Ferbam
U120	206-44-0	Fluoranthene
U122	50-00-0	Formaldehyde
U123	64-18-6	Formic acid (C,T)
U124	110-00-9	Furan (I)
U125	98-01-1	2-Furancarboxaldehyde (I)
U147	108-31-6	2,5-Furandione

Hazardous waste No.	Chemical abstracts No.	Substance
U213	109-99-9	Furan, tetrahydro-(I)
U125	98-01-1	Furfural (I)
U124	110-00-9	Furfuran (I)
U206	18883-66-4	Glucopyranose, 2-deoxy-2-(3-methyl-3-nitro soureido)-, D-
U206	18883-66-4	D-Glucose, 2-deoxy-2-[[(methylnitrosoamino)-carbonyl]amino]-
U126	765-34-4	Glycidylaldehyde
U163	70-25-7	Guanidine, N-methyl-N′-nitro-N-nitroso-
U127	118-74-1	Hexachlorobenzene
U128	87-68-3	Hexachlorobutadiene
U130	77-47-4	Hexachlorocyclopentadiene
U131	67-72-1	Hexachloroethane
U132	70-30-4	Hexachlorophene
U243	1888-71-7	Hexachloropropene
U133	302-01-2	Hydrazine (R,T)
U086	1615-80-1	Hydrazine, 1,2-diethyl-
U098	57-14-7	Hydrazine, 1,1-dimethyl-
U099	540-73-8	Hydrazine, 1,2-dimethyl-
U109	122-66-7	Hydrazine, 1,2-diphenyl-
U134	7664-39-3	Hydrofluoric acid (C,T)
U134	7664-39-3	Hydrogen fluoride (C,T)
U135	7783-06-4	Hydrogen sulfide
U135	7783-06-4	Hydrogen sulfide H2S
U096	80-15-9	Hydroperoxide, 1-methyl-1-phenylethyl- (R)
U116	96-45-7	2-Imidazolidinethionew
U137	193-39-5	Ideno[1,2,3-cd]pyrene
U375	55406-53-6	3-Iodo-2-propynyl n-butylcarbamate.
U396	14484-64-1	Iron, tris(dimethylcarbamodithioato-S,S′)-,
U190	85-44-9	1,3-Isobenzofurandione
U140	78-83-1	Isobutyl alcohol (I,T)
U141	120-58-1	Isosafrole

Hazardous waste No.	Chemical abstracts No.	Substance
U142	143-50-0	Kepone
U143	303-34-4	Lasiocarpine
U144	301-04-2	Lead acetate
U146	1335-32-6	Lead, bis(acetato-O)tetrahydroxytri-
U145	7446-27-7	Lead phosphate
U146	1335-32-6	Lead subacetate
U129	58-89-9	Lindane
U163	70-25-7	MNNG
U147	108-31-6	Maleic anhydride
U148	123-33-1	Maleic hydrazide
U149	109-77-3	Malononitrile
U150	148-82-3	Melphalan
U151	7439-97-6	Mercury
U384	137-42-8	Metam Sodium
U152	126-98-7	Methacrylonitrile (I,T)
U092	124-40-3	Methanamine, N-methyl- (I)
U029	74-83-9	Methane, bromo-
U045	74-87-3	Methane, chloro- (I,T)
U046	107-30-2	Methane, chloromethoxy-
U068	74-95-3	Methane, dibromo-
U088	75-09-2	Methane, dichloro-
U075	75-71-8	Methane, dichlorodifluoro-
U138	74-88-4	Methane, iodo-
U119	62-50-0	Methanesulfonic acid, ethyl ester
U211	56-23-5	Methane, tetrachloro-
U153	74-93-1	Methanethiol (I,T)
U225	75-25-2	Methane, tribromo-
U044	67-77-3	Methane, trichloro-
U121	75-69-4	Methane, trichlorofluoro-

Hazardous waste No.	Chemical abstracts No.	Substance
U036	57-74-9	4,7-Methano-1H-indene, 1,2,4,5,6,7,8,8-octachloro-2,3,3a,4,7,7a-hexahydro-
U154	67-56-1	Methanol (I)
U155	91-80-5	Methapyrilene
U142	143-50-0	1,3,4-Metheno-2H-cyclobuta[cd]pentalen-2-one, 1,1a,3,3a,4,5,5,5a,5b,6-decachloro-octahydro-
U247	72-43-5	Methoxychlor
U154	67-56-1	Methyl alcohol (I)
U029	74-83-9	Methyl bromide
U186	504-60-9	1-Methylbutadiene (I)
U045	74-87-3	Methyl chloride (I,T)
U156	79-22-1	Methyl chlorocarbonate (I,T)
U226	71-55-6	Methyl chloroform
U157	56-49-5	3-Methylcholanthrene
U158	101-14-4	4,4′-Methylenebis(2-chloraniline)
U068	74-95-3	Methyl bromide
U080	75-09-2	Methyl chloride
U159	78-93-3	Methyl ethyl ketone (MEK) (I,T)
U160	1338-23-4	Methyl ethyl ketone peroxide (R,T)
U138	74-88-4	Methyl iodide
U161	108-10-1	Methyl isobutyl ketone (I)
U162	80-62-6	Methyl methacrylate (I,T)
U161	108-10-1	4-Methyl-2-pentanone (I)
U164	56-04-2	Methylthiouracil
U010	50-07-7	Mitomycin C
U365	2212-67-1	Molinate
U059	20830-81-3	5,12-Naphthacenedione, 8-acetyl-10-[(3-amino-2,3,6-trideoxy)-alpha-L-lyxo-hexo pyranosyl)oxy]-7,8,9,10-tetrahydro-6,8,11-trihydroxy-1-methoxy-, (8S-cis)-
U167	134-32-7	1-Naphthalenamine
U168	91-59-8	2-Naphthalenamine

Hazardous waste No.	Chemical abstracts No.	Substance
U026	494-03-1	Naphthalenamine, N,N′-bis(2-chloroethyl)-
U165	91-20-3	Naphthalene
U047	91-58-7	Naphthalene, 2-chloro-
U166	130-15-4	1,4-Naphthalenedione
U236	72-57-1	2,7-Naphthalenedisulfonic acid, 3,3′-[(3,3′-dimethyl[1,1′-biphenyl]-4,4′-diyl)bis(azo)(bis[5-amino-4-hydroxy]-l, tetrasodium salt
U279	63-25-2	1-Naphthalenol, methylcarbamate
U166	130-15-4	1,4-Naphthoquinone
U167	134-32-7	alpha-Naphthylamine
U168	91-59-8	beta-Naphthylamine
U217	10102-45-1	Nitric acid, thallium(1+) salt
U169	98-95-3	Nitrobenzene (I,T)
U170	100-02-7	p-Nitrophenol
U171	79-46-9	2-Nitropropane (I,T)
U172	924-16-3	N-Nitrosodi-n-butylamine
U173	1116-54-7	N-Nitrosodiethanolamine
U174	55-18-5	N-Nitrosodiethylamine
U176	759-73-9	N-Nitroso-N-ethylurea
U177	684-93-5	N-Nitroso-N-methylurea
U178	615-53-2	N-Nitroso-N-methylurethane
U179	100-75-4	N-Nitrosopiperidine
U180	930-55-2	N-Nitrosopyrrolidine
U181	99-55-8	5-Nitro-o-toluidine
U193	1120-71-4	1,2-Oxathiolane, 2,2-dioxide
U058	50-18-0	2H-1,3,2-Oxazaphosphorin-2-amine, N,N-bis(2-chloroethyl)tetrahydro-, 2-oxide
U115	75-21-8	Oxirane (I,T)
U126	765-34-4	Oxiranecarboxyaldehyde
U041	106-89-8	Oxirane, (chloromethyl)-

Hazardous waste No.	Chemical abstracts No.	Substance
U182	123-63-7	Paraldehyde
U391	1114-71-2	Pebulate
U183	608-93-5	Pentachlorobenzene
U184	76-01-7	Pentachloroethane
U185	82-68-8	Pentachloronitrobenzene (PCNB)
See F027	87-86-5	Pentachlorophenol
U161	108-10-1	Pentanol, 4-methyl-
U186	504-60-9	1,3-Pentadiene (I)
U187	62-44-2	Phenacetin
U188	108-95-2	Phenol
U048	95-57-8	Phenol, 2-chloro-
U039	59-50-7	Phenol, 4-chloro-3-methyl-
U081	120-83-2	Phenol, 2,4-dichloro-
U082	87-65-0	Phenol, 2,6-dichloro-
U089	56-53-1	Phenol, 4,4′-(1,2-diethyl-1,2-ethenediyl)bis-, (E)-
U101	105-67-9	Phenol, 2,4-dimethyl-
U052	1319-77-3	Phenol, methyl-
U132	70-30-4	Phenol, 2,2′-methylenebis[3,4,6-trichloro-
U411	114-26-1	Phenol, 2-(1-methylethoxy)-, methylcarbamate
U170	100-02-7	Phenol, 4-nitro-
See F027	87-86-5	Phenol, pentachloro-
See F027	58-90-2	Phenol, 2,3,4,6-tetrachloro-
See F027	95-95-4	Phenol, 2,4,5-trichloro-
See F027	88-06-2	Phenol, 2,4,6-trichloro-
U150	148-82-3	L-Phenylalanine, 4-[bis(2-chloroethyl)amino]-
U145	7446-27-7	Phosphoric acid, lead(2+) salt (2:3)
U087	3288-58-2	Phosphorodithioic acid, O,O-diethyl S-methyl ester
U189	1314-80-3	Phosphorus sulfide (R)
U190	85-44-9	Phthalic anhydride

Hazardous waste No.	Chemical abstracts No.	Substance
U191	109-06-8	2-Picoline
U179	100-75-4	Piperidine, 1-nitroso-
U400	120-54-7	Piperidine, 1,1′-(tetrathiodicarbonothioyl)-bis-
U383	128-03-0	Potassium dimethyldithiocarbamate
U378	51026-28-9	Potassium n-hydroxymethyl-n-methyldithio-carbamate
U377	137-41-7	Potassiun n-methyldithiocarbamate
U192	23950-58-5	Pronamide
U194	107-10-8	1-Propanamine (I,T)
U111	621-64-7	1-Propanamine, N-nitroso-N-propyl-
U110	142-84-7	1-Propanamine, N-propyl- (I)
U066	96-12-8	Propane, 1,2-dibromo-3-chloro-
U083	78-87-5	Propane, 1,2-dichloro-
U149	109-77-3	Propanedinitrile
U171	79-46-9	Propane, 2-nitro- (I,T)
U027	108-60-1	Propane, 2,2′-oxybis[2-chloro-
U193	1120-71-4	1,3-Propane sultone
See F027	93-72-1	Propanoic acid, 2-(2,4,5-trichlorophenoxy)-
U235	126-72-7	1-Propanol, 2,3-dibromo-, phosphate (3:1)
U140	78-83-1	1-Propanol, 2-methyl- (I,T)
U002	67-64-1	2-Propanone (I)
U007	79-06-1	2-Propenamide
U084	542-75-6	1-Pro-pene, 1,3-dichloro-
U243	1888-71-7	1-Propene, 1,1,2,3,3,3-hexachloro-
U009	107-13-7	2-Propenenitrile
U152	126-98-7	2-Propenenitrile, 2-methyl- (I,T)
U008	79-10-7	2-Propenoic acid (I)
U113	140-88-5	2-Propenoic acid, ethyl ester (I)
U118	97-63-2	2-Propenoic acid, 2-methyl-, ethyl ester
U162	80-62-6	2-Propenoic acid, 2-methyl-, methyl ester (I,T)

Hazardous waste No.	Chemical abstracts No.	Substance
U373	122-42-9	Propham
U411	114-26-1	Propoxur
U194	107-10-8	n-Propylamine (I,T)
U083	78-87-5	Propylene dichloride
U387	52888-880-9	Prosulocarb
U148	123-33-1	3,6-Pyridazinedione, 1,2-dihydro-
U196	110-06-8	Pyridine, 2-methyl-
U191	109-06-8	Pyridine, 2-methyl-
U237	66-75-1	2,4-(1H,3H)-Pyrimidinedione, 5-[bis(2-chloroethyl)amino]-
U164	56-04-2	4(1H)-Pyrimidinone, 2,3-dihydro-6-methyl-2-thyioxo-
U180	930-55-2	Pyrrolidine, 1-nitroso-
U200	50-55-5	Reserine
U201	108-46-3	Resorcinol
U202	1 81-07-2	Saccharin, and salts
U203	94-59-7	Safrole
U204	7783-00-8	Selenious acid
U204	7783-00-8	Selenium dioxide
U205	7488-56-4	Selenium sulfide
U205	7488-56-4	Selenium sulfide SeS_2 (R,T)
U376	144-34-3	Selenium, tetrakis(dimethyldithiocarbamate).
U015	115-02-6	L-Serine, diazoacetate (ester)
See F027	93-72-1	Silvex (2,4,5-TP)
U379	136-30-1	Sodium dibutyldithiocarbamate
U381	148-18-5	Sodium diethyldithiocarbamate
U382	128-04-1	Sodium dimethyldithiocarbamate
U206	18883-66-4	Streptozotocin
U277	95-06-7	Sulfallate
U103	77-78-1	Sulfuric acid, dimethyl ester
U189	1314-80-3	Sulfur phosphide (R)
See F027	93-76-5	2,4,5-T

Hazardous waste No.	Chemical abstracts No.	Substance
U402	1634-02-2	Tetrabutylthiuram disulfide
U207	95-94-3	1,2,4,5-Tetrachlorobenzene
U208	630-20-6	1,1,1,2-Tetrachloroethane
U209	79-34-5	1,1,2,2-Tetrachloroethane
U210	127-18-4	Tetrachloroethylene
See F027	58-90-2	2,3,4,6-Tetrachlorophenol
U213	109-99-9	Tetrahydrofuran (I)
U401	97-74-5	Tetramethylthiuram monosulfide
U366	533-74-4	2H-1,3,5-Thiadiazine-2-thione, tetrahydro-3,5-dimethyl-
U214	563-68-8	Thallium(I) acetate
U215	6533-73-9	Thallium(I) carbonate
U216	7791-12-0	Thallium(I) chloride
U216	7791-12-0	Thallium chloride TlCl
U217	10102-45-1	Thallium(I) nitrate
U218	62-55-5	Thioacetamide
U410	59669-26-0	Thiodicarb
U153	74-93-1	Thiomethanol (I,T)
U244	137-26-8	Thioperoxydicarbonic diamide [(H2N)C(S)]2S2, tetramethyl-
U402	1634-02-2	Thioperoxydicarbonic diamide, tetrabutyl
U403	97-77-8	Thioperoxydicarbonic diamide, tetraethyl
U409	23564-05-8	Thiophanate-methyl
U219	62-56-6	Thiourea
U244	137-26-8	Thiram
U220	108-88-3	Toluene
U221	25376-45-8	Toluenediamine
U223	26471-62-5	Toluene diisocyanate (R,T)
U328	95-563-4	o-Toluidine
U353	106-49-0	p-Toluidine

Hazardous waste No.	Chemical abstracts No.	Substance
U222	636-21-5	o-Toluidine hydrochloride
U389	2303-17-5	Triallate
U011	61-82-5	1H-1,2,4-Triazol-3-amine
U227	79-00-5	1,1,2-Trichloroethane
U228	79-01-6	Trichloroethylene
U121	75-69-4	Trichloromonofluoromethane
See F027	95-95-4	2,4,5-Trichlorophenol
See F027	88-06-2	2,4,6-Trichlorophenol
U404	121-44-8	Triethylamine
U234	99-35-4	1,3,5-Trinitrobenzene (R,T)
U182	123-63-7	1,3,5-Trioxane, 2,4,6-trimethyl-
U235	126-72-7	Tris(2,3-dibromopropyl) phosphate
U236	72-57-1	Trypan blue
U237	66-75-1	Uracil mustard
U176	759-73-9	Urea, N-ethyl-N-nitroso-
U177	684-93-5	Urea, N-methyl-N-nitroso-
U385	1929-77-7	Vernolate
U043	75-01-4	Vinyl chloride
U248	1 81-81-2	Warfarin, and salts, when present at concentrations of 0.3% or less
U239	1330-20-7	Xylene (I)
U200	50-55-5	Yohimban-16-carboxylic acid, 11,17-dimethoxy-18-[(3,4,5-trimethoxybenzoyl) oxy]-, methyl ester, (3beta, 16beta, 17alpha, 18beta, 20alpha)-
U407	14324-55-1	Zinc, bis(diethylcarbamodithioato-S,S′)-
U249	1314-84-7	Zinc phosphide Zn_3P_2, when present at concentrations of 10% or less 1 CAS Number given for parent compound only

Approved by the Office of Management and Budget under control number 2050-0047) [45 Fed. Reg. 78529, 78541, Nov. 25, 1980; 59 Fed. Reg. 31551, June 20, 1994; 60 Fed. Reg. 7824, Feb. 9, 1995; 60 Fed. Reg. 19165, April 17, 1995; 60 Fed. Reg. 25619, May 12, 1995]

■ APPENDIX E.3 List of Regulated Toxic Substances, Flammable Substances and Threshold Quantities for Accidental Release Prevention (40 CFR 68.130)

Basis for Listing:

- a Mandated for listing by Congress
- b On EHS list, vapor pressure 10 mm Hg or greater
- c Toxic gas
- d Toxicity of hydrogen chloride, potential to release hydrogen chloride, and history of accidents
- e Toxicity of sulfur trioxide and sulfuric acid, potential to release sulfur trioxide, and history of accidents
- f Flammable gas
- g Volatile flammable liquid

Toxic Substances

Chemical Name	CAS Number	Threshold Quantity (lbs)	Basis for Listing
Acrolein [2-Propenal]	107-02-8	5,000	b
Acrylonitrile [2-Propenenitrile]	107-13-1	20,000	b
Acrylyl chloride [2-Propenoyl chloride]	814-68-6	5,000	b
Allyl alcohol [2-Propen-l-ol]	107-18-61	15,000	b
Allylamine [2-Propen-l-amine]	107-11-9	10,000	b
Ammonia (anhydrous)	7664-41-7	10,000	a,b
Ammonia (conc. 20% or greater)	7664-41-7	20,000	a,b
Arsenous trichloride	7784-34-1	15,000	b
Arsine	7784-42-1	1,000	b
Boron trichloride [Borane, trichloro-]	10294-34-5	5,000	b
Boron trifluoride [Borane, trifluoro-]	7637-07-2	5,000	b
Boron trifluoride compound with methyl ether (1:1) [Boron, trifluoro[oxybis[metane]]-, T-4-	353-42-4	15,000	b
Bromine	7726-95-6	10,000	a,b
Carbon disulfide	75-15-0	20,000	b
Chlorine	7782-50-5	2,500	a,b
Chlorine dioxide [Chlorine oxide (ClO_2)]	10049-04-4	1,000	c
Chloroform [Methane, trichloro-]	67-66-3	20,000	b
Chloromethyl ether [Methane, oxybis[chloro-]	542-88-1	1,000	b

Chemical Name	CAS Number	Threshold Quantity (lbs)	Basis for Listing
Chloromethyl methyl ether [Methane, chloromethoxy-]	107-30-2	5,000	b
Crotonaldehyde [2-Butenal]	4170-30-3	20,000	b
Crotonaldehyde, (E)- [2-Butenal, (E)-]	123-73-9	20,000	b
Cyanogen chloride	506-77-4	10,000	c
Cyclohexylamine [Cyclohexanamine]	108-91-8	15,000	b
Diborane	19287-45-7	2,500	b
Dimethyldichlorosilane [Silane, dichlorodimethyl-]1,1-	75-78-5	5,000	b
Dimethylhydrazine [Hydrazine, 1,1-dimethyl-]	57-14-7	15,000	b
Epichlorohydrin [Oxirane, (chloromethyl)-]	106-89-8	20,000	b
Ethylenediamine [1,2-Ethanediamine]	107-15-3	20,000	b
Ethyleneimine [Aziridine]	151-56-4	10,000	b
Ethylene oxide [Oxirane]	75-21-8	10,000	a,b
Fluorine	7782-41-4	1,000	b
Formaldehyde (solution)	50-00-0	15,000	b
Furan	110-00-9	5,000	b
Hydrazine	302-01-2	15,000	b
Hydrochloric acid (conc. 30% or greater)	7647-01-0	15,000	d
Hydrocyanic acid	74-90-8	2,500	a,b
Hydrogen chloride (anhydrous) [Hydrochloric acid]	7647-01-0	5,000	a
Hydrogen fluoride/Hydrofluoric acid (conc. 50% or greater) [Hydrofluoric acid]	7664-39-3	1,000	a,b
Hydrogen selenide	7783-07-5	500	b
Hydrogen sulfide	7783-06-4	10,000	a,b
Iron, pentacarbonyl- [Iron carbonyl $(Fe(CO)_5$, (TB-5-11)-]	13463-40-6	2,500	b
Isobutyronitrile [Propanenitrile, 2-methyl-]	78-82-0	20,000	b
Isopropyl chloroformate [Carbonochloridic acid, 1-methylethyl ester]	108-23-6	15,000	b

Chemical Name	CAS Number	Threshold Quantity (lbs)	Basis for Listing
Methacrylonitrile [2-Propenenitrile, 2-methyl-]	126-98-7	10,000	b
Methyl chloride [Methane, chloro-]	74-87-3	10,000	a
Methyl chloroformate [Carbonochloridic acid, methylester]	79-22-1	5,000	b
Methyl hydrazine [Hydrazine, methyl-]	60-34-4	15,000	b
Methyl isocyanate [Methane, isocyanato-]	624-83-9	10,000	a,b
Methyl mercaptan [Methanethiol]	74-93-1	10,000	b
Methyl thiocyanate [Thiocyanic acid, methyl ester]	556-64-9	20,000	b
Methyltrichloro silane [Silane, trichloromethyl-]	75-79-6	5,000	b
Nickel carbonyl	13463-39-3	1,000	b
Nitric acid (conc. 80% or greater)	7697-37-2	15,000	b
Nitric oxide [Nitrogen oxide (NO)]	10102-43-9	10,000	b
Oleum (Fuming sulfuric acid) [Sulfuric acid, mixture with sulfur trioxide] 1	8014-95-7	10,000	e
Peracetic acid [Ethaneperoxoic acid]	79-21-0	10,000	b
Perchloromethyl mercaptan [Methanesulfenyl chloride, trichloro-]	594-42-3	10,000	b
Phosgene [Carbonic dichloride]	75-44-5	500	a,b
Phosphine	7803-51-2	5,000	b
Phosphorus oxychloride [Phosphoryl chloride]	10025-87-3	5,000	b
Phosphorus trichloride [Phosphorous trichloride]	7719-12-2	15,000	b
Piperidine	110-89-4	15,000	b
Propionitrile [Propanenitrile]	107-12-0	10,000	b
Propyl chloroformate [Carbonochloridic acid, propylester]	109-61-5	15,000	b

Chemical Name	CAS Number	Threshold Quantity (lbs)	Basis for Listing
Propyleneimine [Aziridine, 2-methyl-]	75-55-8	10,000	b
Propylene oxide [Oxirane, methyl-]	75-56-9	10,000	b
Sulfur dioxide (anhydrous)	7446-09-5	5,000	a,b
Sulfur tetrafluoride [Sulfur fluoride (SF_4), (T-4)-]	7783-60-0	2,500	b
Sulfur trioxide	7446-11-9	10,000	a,b
Tetramethyl lead [Plumbane, tetramethyl-]	75-74-1	10,000	b
Tetranitromethane [Methane, tetranitro-]	509-14-8	10,000	b
Titanium tetrachloride [Titanium chloride ($TiCl_4$) (T-4)-]	7550-45-0	2,500	b
Toluene 2,4-diisocyanate [Benzene, 2,4-diisocyanato-1-methyl-] 1	584-84-9	10,000	a
Toluene 2,6-diisocyanate [Benzene, 1,3-diisocyanato-2-methyl-] 1	91-08-7	10,000	a
Toluene diisocyanate (unspecified isomer) [Benzene, 1,3-diisocyanatomethyl-] 1	26471-62-5	10,000	a
Trimethyl chlorosilane [Silane, chlorotrimethyl-]	75-77-4	10,000	b
Vinyl acetate monomer [Acetic acid ethenyl ester]	108-05-4	15,000	b

Flammable Substances

Chemical Name	CAS Number	Threshold Quantity (lbs)	Basis for Listing
Acetaldehyde	75-07-0	10,000	g
Acetylene [Ethyne]	74-86-2	10,000	f
Bromotrifluoroethylene [Ethene, bromotrifluoro-]	598-73-2	10,000	f
1,3-Butadiene	106-99-0	10,000	f
Butane	106-97-8	10,000	f
1-Butene	106-98-9	10,000	f
2-Butene	107-01-7	10,000	f
Butene	25167-67-3	10,000	f

Chemical Name	CAS Number	Threshold Quantity (lbs)	Basis for Listing
2-Butene-trans [2-Butene, (E)]	624-64-6	10,000	f
Carbon oxysulfide [Carbon oxide sulfide (COS)]	463-58-1	10,000	f
Chlorine monoxide [Chlorine oxide]	7791-21-1	10,000	f
2-Chloropropylene [1-Propene, 2-chloro-]	557-98-2	10,000	g
1-Chloropropylene [1-Propene, 1-chloro-]	590-21-6	10,000	g
Cyanogen [Ethanedinitrile]	460-19-5	10,000	f
Cyclopropane	75-19-4	10,000	f
Dichlorosilane [Silane, dichloro-]	4109-96-0	10,000	f
Difluoroethane [Ethane, 1,1-difluoro-]	75-37-6	10,000	f
Dimethylamine [Methanamine, N-methyl-]	124-40-3	10,000	f
2,2-Dimethylpropane [Propane, 2,2-dimethyl-]	463-82-1	10,000	f
Ethane	74-84-0	10,000	f
Ethyl acetylene [1-Butyne]	107-00-6	10,000	f
Ethylamine [Ethanamine]	75-04-7	10,000	f
Ethyl chloride [Ethane, chloro-]	75-00-3	10,000	f
Ethylene [Ethene]	74-85-1	10,000	f
Ethyl ether [Ethane, 1,1′-oxybis-]	60-29-7	10,000	g
Ethyl mercaptan [Ethanethiol]	75-08-1	10,000	g
Ethyl nitrite [Nitrous acid, ethyl ester]	109-95-5	10,000	f
Hydrogen	1333-74-0	10,000	f
Isobutane [Propane, 2-methyl]	75-28-5	10,000	f
Isopentane [Butane, 2-methyl-]	78-78-4	10,000	g
Isoprene [1,3-Butadinene, 2-methyl-]	78-79-5	10,000	g
Isopropylamine [2-Propanamine]	75-31-0	10,000	g
Isopropyl chloride [Propane, chloro-]	275-29-6	10,000	g
Methane	74-82-8	10,000	f
Methylamine [Methanamine]	74-89-5	10,000	f
3-Methyl-1-butene	563-45-1	10,000	f
2-Methyl-1-butene	563-46-2	10,000	g

Chemical Name	CAS Number	Threshold Quantity (lbs)	Basis for Listing
Methyl ether [Methane, oxybis-]	115-10-6	10,000	f
Methyl formate [Formic acid, methyl ester]	107-31-3	10,000	g
2-Methylpropene [1-Propene, 2-methyl-]	115-11-7	10,000	f
1,3-Pentadinene	504-60-9	10,000	f
Pentane	109-66-0	10,000	g
1-Pentene	109-67-1	10,000	g
2-Pentene, (E)-	646-04-8	10,000	g
2-Pentene, (Z)-	627-20-3	10,000	g
Propadiene [1,2-Propadiene]	463-49-0	10,000	f
Propane	74-98-6	10,000	f
Propylene [1-Propene]	115-07-1	10,000	f
Propyne [1-Propyne]	74-99-7	10,000	f
Silane	7803-62-5	10,000	f
Tetrafluoroethylene [Ethene, tetrafluoro-]	116-14-3	10,000	f
Tetramethylsilane [Silane, tetramethyl-]	75-76-3	10,000	g
Trichlorosilane [Silane, trichloro-]	10025-78-2	10,000	g
Trifluorochloroethylene [Ethene, chlorotrifluoro-]	79-38-9	10,000	f
Trimethylamine [Methanamine, N, N-dimethyl-]	75-50-3	10,000	f
Vinyl acetylene [1-Buten-3 yne]	689-97-4	10,000	f
Vinyl chloride [Ethene, chloro-]	75-01-4	10,000	a,f
Vinyl ethyl ether [Ethene, ethoxy-]	109-92-2	10,000	g
Vinyl fluoride [Ethene, fluoro-]	75-02-5	10,000	f
Vinylidene chloride [Ethene, 1,1-dichloro-]	75-35-4	10,000	g
Vinylidene fluoride [Ethene, 1,1-difluoro-]	75-38-7	10,000	f
Vinyl methyl ether [Ethene, methoxy-]	107-25-5	10,000	f

■ APPENDIX E.4 A Summary of Federal Regulations for Underground Storage Tank Systems

United States Environmental Protection Agency
Solid Waste And Emergency Response 5403W
EPA 510-K-95-002
July 1995

EPA

Musts For USTs

A Summary Of Federal Regulations For Underground Storage Tank Systems

CONTENTS

WHAT ARE THESE REGULATIONS ABOUT?

Over one million underground storage tank systems (USTs) in the United States contain petroleum or hazardous substances regulated by the U.S. Environmental Protection Agency (EPA). Many of these USTs have leaked or are currently leaking. More USTs will leak unless owners and operators make sure their USTs meet the requirements described in this booklet.

Releases from USTs can threaten human health and safety. UST releases can also contaminate soil and drinking water supplies. As of April 1995, more than 287,000 UST releases had been confirmed. EPA estimates that about half of these releases reached groundwater.

Releases from USTs--from spills, overfills, or leaking tanks and piping--can cause fires or explosions that threaten human safety. Releases from USTs can also contaminate the groundwater that many of us depend on for the water we drink.

Properly managed, USTs will not threaten our health or environment. Federal legislation, therefore, directed EPA to develop the UST regulations described in this booklet. Regulations require owners and operators of USTs to:

- **Prevent** releases from USTs (see pages 7, 12-17, and 19-21);
- **Detect** releases from USTs (see pages 8-11 and 19); and
- **Correct** the problems created by releases from USTs (see pages 22-24).

In addition, the regulations require owners and operators of USTs to demonstrate their ability to pay for correcting the problems created if their USTs do leak (see pages 2 and 5).

How Will These Regulations Affect You?

If your USTs do not meet these requirements by the indicated deadlines, you can be cited for violations and fined.

The regulations describe the steps you, as an UST owner or operator, need to take to help protect our health and environment. These steps will also help you avoid the high cost of cleaning up the environment and involving yourself in legal actions that can result if your tank or its piping leaks.

You must take these steps:

- If you install an UST after December 22, 1988, it must meet the **requirements for new USTs** concerning correct installation, leak detection, and spill, overfill, and corrosion protection (see pages 6-17).

- If you have an UST that was installed before December 22, 1988, it must meet two major requirements:

  1) By December 1993, your UST must have **leak detection** (see pages 8-11 and 19).

  2) By December 1998, your UST must have **spill, overfill, and corrosion protection** (see pages 12-17 and 19-21).

- You must take **corrective action** in response to leaks (see pages 22-24).

- You must follow **closure requirements** for tanks you temporarily or permanently close (see page 25).

EPA Dollars And Sense

Financial Responsibility Requirements For Underground Storage Tanks

- You must demonstrate your **financial responsibility** for the cost of cleaning up a leak and compensating other people for bodily injury and property damage. The financial responsibility requirements appear in the **Code of Federal Regulations, 40 CFR Part 280.** These requirements are briefly summarized in a free EPA booklet, **Dollars And Sense**, which you can order by calling EPA's toll-free Hotline at 800 424-9346.

  Basically, as an owner or operator of **USTs storing petroleum**, you must be able to demonstrate your ability to pay for damage costs that could occur if your USTs leaked. You need either $500,000 or $1 million in "per occurrence" coverage and either $1 million or $2 million in "aggregate" coverage, depending on your ownership category. Compliance deadlines were phased in according to categories of UST owners, but by February 1994 virtually every UST owner should have met the financial responsibility requirements. The table on page 5 displays UST categories, coverage amounts, and deadlines. See **Dollars And Sense** for details.

Who Is The "Regulatory Authority"?

If you don't know how to reach your state agency, see the list of state contacts starting on page 35.

This booklet describes EPA's basic requirements for USTs, but your state or local regulatory authority may have requirements that are somewhat different or more stringent. Contact your regulatory authority for its specific UST requirements. If you are not sure who your regulatory authority is, see the list of state UST contacts starting on page 35.

What's An "UST"?

An UST is a tank and any underground piping connected to the tank that has at least 10 percent of its combined volume underground. The federal regulations apply only to USTs storing either petroleum or certain hazardous substances.

The "For Hazardous Substance USTs Only" section starting on page 28 identifies hazardous substances and special requirements for USTs storing them. Generally, the requirements for both petroleum and hazardous substance USTs are very similar.

Requirements and definitions of USTs are found in the Code of Federal Regulations, 40 CFR Part 280.

Some kinds of tanks are **not** covered by these regulations:

- Farm and residential tanks of 1,100 gallons or less capacity holding motor fuel used for noncommercial purposes.
- Tanks storing heating oil used on the premises where it is stored.
- Tanks on or above the floor of underground areas, such as basements or tunnels.
- Septic tanks and systems for collecting storm water and wastewater.
- Flow-through process tanks.
- Emergency spill and overfill tanks.

Other storage sites, such as surface impoundments, are not covered by the federal requirements. Some tanks, such as field-constructed tanks, have been deferred from most of the regulations. (For details see the **Code of Federal Regulations, 40 CFR Part 280.**)

✔ ***The tables on the next two pages display basic UST requirements and deadlines...***

When Do You Have To Act?

TYPE OF TANK & PIPING	LEAK DETECTION	SPILL & OVERFILL PROTECTION	CORROSION PROTECTION
New Tanks & Piping (installed *after* December 22, 1988)	At installation	At installation (Does not apply to piping)	At installation
Existing Tanks & Piping (installed *before* December 22, 1988)	No later than December 1993	No later than December 22, 1998 (Does not apply to piping)	No later than December 22, 1998

What Do You Have To Do?

LEAK DETECTION *(see pages 8-11 and 19)*	
NEW TANKS	♦ Monthly Monitoring*; *or* ♦ Inventory Control Plus Tank Tightness Testing** (only for 10 years after installation)
EXISTING TANKS	♦ Monthly Monitoring*; *or* ♦ Inventory Control Plus Tank Tightness Testing** (only for 10 years after adding spill, overfill, and corrosion protection); *or* ♦ Inventory Control Plus Annual Tank Tightness Testing (only until December 1998)
NEW & EXISTING PRESSURIZED PIPING	♦ Automatic Shutoff Device *or* Flow Restrictor *or* Continuous Alarm System; AND ♦ Annual Line Tightness Test *or* Monthly Monitoring* [except Automatic Tank Gauging]
NEW & EXISTING SUCTION PIPING	♦ Monthly Monitoring*; *or* ♦ Line Tightness Testing (every 3 years); *or* ♦ No Requirements (if the system has the characteristics described on page 11)
SPILL & OVERFILL PROTECTION *(see pages 12-15)*	
ALL TANKS	♦ Catchment Basins; AND ♦ Automatic Shutoff Devices *or* Overfill Alarms *or* Ball Float Valves
CORROSION PROTECTION *(see pages 16-17 and 20-21)*	
NEW TANKS & PIPING	♦ Coated and Cathodically Protected Steel; *or* ♦ Fiberglass Reinforced Plastic (FRP); *or* ♦ Steel Tank Clad With FRP (does not apply to piping)
EXISTING TANKS & PIPING	♦ Same Options As For New Tanks & Piping; *or* ♦ Cathodically Protected Steel; *or* ♦ Tank Interior Lining; *or* ♦ Tank Interior Lining AND Cathodic Protection

* Monthly Monitoring includes: Interstitial Monitoring; Automatic Tank Gauging; Vapor Monitoring, Groundwater Monitoring; Statistical Inventory Reconciliation; and other methods approved by the regulatory authority.

**Tanks 2,000 gallons and smaller may be able to use manual tank gauging (see page 9).

How Does Financial Responsibility Work?

<table>
<tr><th>GROUP OF UST OWNERS AND OPERATORS</th><th>COMPLIANCE DEADLINE</th><th>PER OCCURRENCE COVERAGE</th><th>AGGREGATE COVERAGE</th></tr>
<tr><td>GROUP 1:
Petroleum marketers with 1,000 or more tanks
OR
Nonmarketers with net worth of $20 million or more (for nonmarketers, the "per occurrence" amount is the same as Group 4-B below)</td><td>January 1989</td><td rowspan="4">$1 million</td><td rowspan="7">$1 million if you have 100 or fewer tanks

OR

$2 million if you have more than 100 tanks</td></tr>
<tr><td>GROUP 2:
Petroleum marketers with 100-999 tanks</td><td>October 1989</td></tr>
<tr><td>GROUP 3:
Petroleum marketers with 13-99 tanks</td><td>April 1991</td></tr>
<tr><td>GROUP 4-A:
Petroleum marketers with 1-12 tanks</td><td>December 1993</td></tr>
<tr><td>GROUP 4-B:
Nonmarketers with net worth of less than $20 million</td><td>December 1993</td><td rowspan="3">$500,000 if throughput is 10,000 gallons monthly or less

OR

$1 million if throughput is more than 10,000 gallons monthly</td></tr>
<tr><td>GROUP 4-C:
Local governments (including Indian tribes not part of Group 5)</td><td>February 1994</td></tr>
<tr><td>GROUP 5:
Indian tribes owning USTs on Indian lands (USTs must be in compliance with UST technical requirements)</td><td>December 1998</td></tr>
</table>

WHAT DO NEW PETROLEUM USTs NEED?

✔ *Installed Correctly*

✔ *Leak Detection*

✔ *Spill And Overfill Protection*

✔ *Corrosion Protection*

New UST systems are those installed after December 22, 1988.

You must meet four requirements when you install a **new** UST system:

- ✔ You must certify that the tank and piping are **installed** according to industry codes. See page 7.
- ✔ Your UST must have **leak detection**. See pages 8-11.
- ✔ Your UST must have devices that provide **spill and overfill protection**. Also, you must follow correct tank filling practices. See pages 12-15.
- ✔ Your UST must have **corrosion protection**. See pages 16 and 17.

The table on page 4 displays basic UST requirements and deadlines.

The following pages provide information on these requirements...

What you must do for installation:

✔ *Installed Correctly*

- **Make sure your UST is installed correctly** by using qualified installers who follow industry codes. (See pages 31 and 34 for information on industry codes and installation practices.)

- **Certify** on a **notification form** (see page 26) that you have used a qualified installer who can assure you that your UST has been installed correctly.

Installation problems result from careless installation practices that do not follow standard industry codes and procedures. Improper installation is a significant cause of fiberglass-reinforced plastic (FRP) and steel UST failures, particularly piping failures. Installation includes excavation, tank system siting, burial depth, tank system assembly, backfilling around the tank system, and surface grading.

Many mistakes can be made during installation. For example, mishandling of the tank during installation can cause structural failure of FRP tanks or damage to steel tank coatings and cathodic protection. Improper layout of piping runs, incomplete tightening of joints, inadequate cover pad construction, and construction accidents can lead to failure of delivery piping.

You need to make sure that installers carefully follow the correct installation procedures called for by industry codes.

✔ Leak Detection

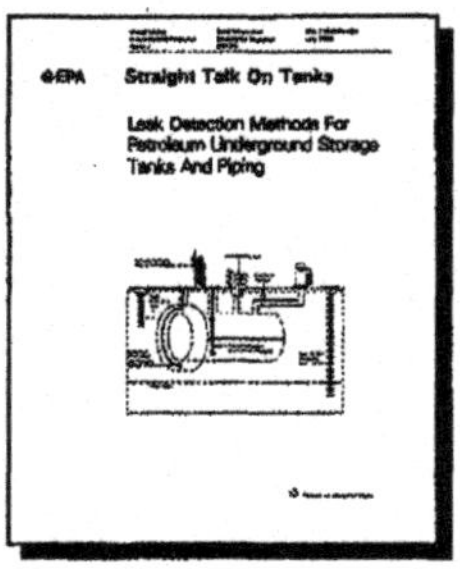

What you must do for leak detection:

You must provide your UST system with leak detection that allows you to meet three basic requirements:

- You can detect a leak from any portion of the tank or its piping that **routinely** contains petroleum;
- Your leak detection is installed, calibrated, operated, and maintained in accordance with the manufacturer's instructions; and
- Your leak detection meets the performance requirements described in the federal regulations (sections 280.43 and 280.44).

The **monthly monitoring methods** that you may use to meet the federal leak detection requirements are listed on the next page. As a ***temporary*** method, you may also use one of the two methods described on page 10. Please **note the additional leak detection requirements for piping on page 11.**

For a detailed booklet that focuses only on leak detection methods and requirements, order EPA's free booklet, **Straight Talk On Tanks,** by calling EPA's toll-free Hotline at 800 424-9346. Also, see page 32 and after for additional sources of information on leak detection.

Some Leak Detection Methods

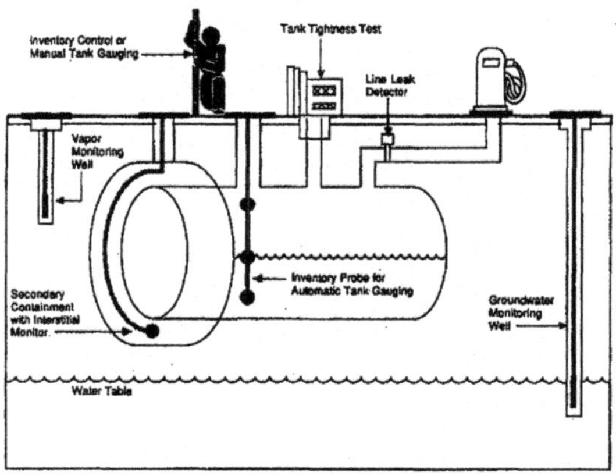

Monthly Monitoring Methods

Leak detection for USTs may consist of one or a combination of the monthly monitoring methods listed on this page.

Interstitial Monitoring: This method detects leaks in the space between the UST and a second barrier. The regulations describe general performance requirements for interstitial monitoring with double-walled USTs, USTs fitted with internal liners, and USTs using interception barriers.

Automatic Tank Gauging Systems: This method uses automated processes to monitor product level and inventory control.

Monitoring For Vapors In The Soil: This method samples vapors in the soil gas surrounding the UST. Leaked petroleum produces vapors that can be detected in the soil gas. The regulations describe several requirements for using this leak detection method. For example, this method requires using porous soils in the backfill and locating the monitoring devices in these porous soils near the UST system.

Monitoring For Liquids On The Groundwater: This method monitors the groundwater table near an UST for the presence of released free product on the water table. Monitoring wells near the UST are checked frequently to see if petroleum can be detected. The regulations describe several requirements for the use of this method. For example, this method cannot be used if the water table is more than 20 feet below the surface of the ground.

Statistical Inventory Reconciliation: In this method, a trained professional uses sophisticated computer software to conduct a statistical analysis of inventory, delivery, and dispensing data, which you must supply regularly. (EPA has a booklet available, **Introduction To Statistical Inventory Reconciliation**, that describes the use of this method. You can order this free booklet by calling EPA's toll-free Hotline at 800 424-9346 and asking for **Introduction To Statistical Inventory Reconciliation**.)

Other Methods Approved By The Regulatory Authority: If other methods can be shown to work as effectively as the methods described above for leak detection, these alternative methods can be approved by the regulatory authority.

Alternate Leak Detection Method Good For 10 Years

Instead of using one of the monthly monitoring methods noted above, you can combine inventory control with tank tightness testing, **but only for 10 years after you install a new UST.** Inventory control involves taking daily measurements of tank contents and recording deliveries and amount pumped. Based upon some daily and monthly calculations, you can discover if your tank may be leaking. Tank tightness testing usually requires taking the UST out of service while changes in level or volume over time are measured. Your UST will need a tank tightness test every 5 years. **After 10 years, you must use one of the monthly monitoring methods.**

The success of this temporary combined method depends on your performing inventory control correctly. EPA has a booklet available, **Doing Inventory Control Right**, that clearly explains how to do inventory control with simple step-by-step directions. The booklet also includes standard forms used to record inventory data. You can order this free booklet by calling EPA's toll-free Hotline at 800 424-9346 and asking for **Doing Inventory Control Right.**

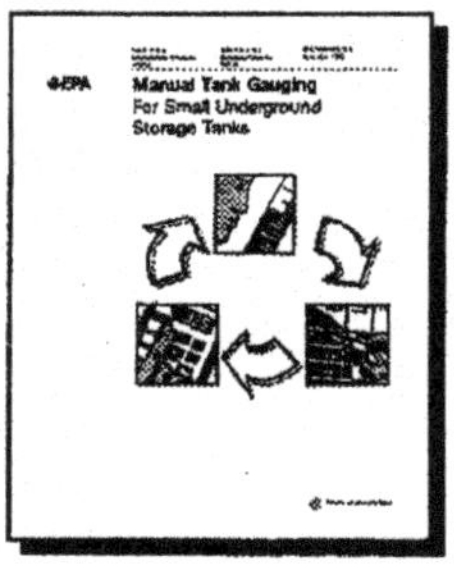

One Additional Leak Detection Method For Small Tanks

Tanks of 2,000 gallons capacity or less may be able to use **manual tank gauging** as a leak detection method, either by itself or in combination with tank tightness testing. This method involves keeping the tank undisturbed for at least 36 hours each week, during which the tank's contents are measured, twice at the beginning and twice at the end of the test period. Manual tank gauging can be used as the sole method of leak detection for the life of the tank ***only*** for tanks up to 1,000 gallons. Tanks between 1,001 and 2,000 gallons can use this method only in combination with tank tightness testing. This combined method, however, can be used only during the first 10 years following tank installation.

EPA has a booklet available, **Manual Tank Gauging: For Small Underground Storage Tanks**, that clearly explains how to do manual tank gauging with simple step-by-step directions. The booklet also includes standard forms used to record inventory data. You can order this free booklet by calling EPA's toll-free Hotline at 800 424-9346 and asking for **Manual Tank Gauging**.

Additional Leak Detection For Piping

Leak detection for piping is particularly important, because most leaks come from an UST's piping.

Pressurized piping must meet the following requirements:

- The piping must have devices that automatically shut off or restrict flow or have an alarm that indicates a leak.
- You must either conduct an **annual** tightness test of the piping or use one of the following monthly methods noted above for tanks: interstitial monitoring, vapor monitoring, groundwater monitoring, statistical inventory reconciliation, or other approved monthly methods.

If your UST has **suction** piping, your leak detection requirements will depend on which type of suction piping you have.

- One type of suction piping does **not** require leak detection if it has the following characteristics:
  - Below-grade piping operating at less than atmospheric pressure is sloped so that the piping's contents will drain back into the storage tank if the suction is released.
  - Only one check valve is included in each suction line and is located directly below the suction pump.
- Suction piping that does **not** exactly match the characteristics noted above must have leak detection, either monthly monitoring (using one of the monthly methods noted above for use on pressurized piping) or tightness testing of the piping every 3 years.

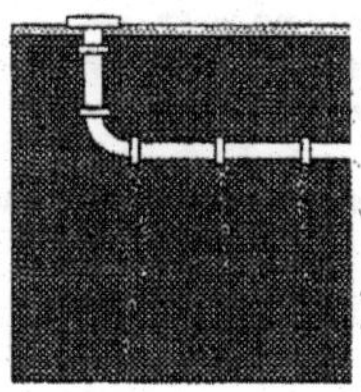

✔ *Spill Protection*

What you must do for spill protection:

- **Your USTs must have catchment basins** to contain spills. Catchment basins are described below and on the next page. New USTs must have catchment basins when they are installed.
- You and your fuel deliverer must **follow industry standards for correct filling practices.**

Many releases at UST sites come from spills. Spills often occur at the fill pipe when the delivery truck's hose is disconnected. Although these spills are usually small, repeated small releases can cause big environmental problems.

You and your fuel deliverer should watch "Keeping It Clean," a video that shows how deliveries can be made safely with no spills (see page 34 for ordering information).

Human error causes most spills. These mistakes can be avoided by following standard tank filling practices. For example, you must make sure there is room in the UST for the delivery, and the delivery driver must watch the delivery at all times. If you and the delivery driver follow standard practices, nearly all spills can be prevented. For this reason, **federal UST regulations require that you follow standard filling practices.**

If an UST never receives more than 25 gallons at a time, the UST does not have to meet the spill protection requirements. Many small used oil tanks fall in this category.

What Are Catchment Basins?

Catchment basins are also called "spill containment manholes" or "spill buckets." Basically, a catchment basin is a bucket sealed around the fill pipe (see illustration below).

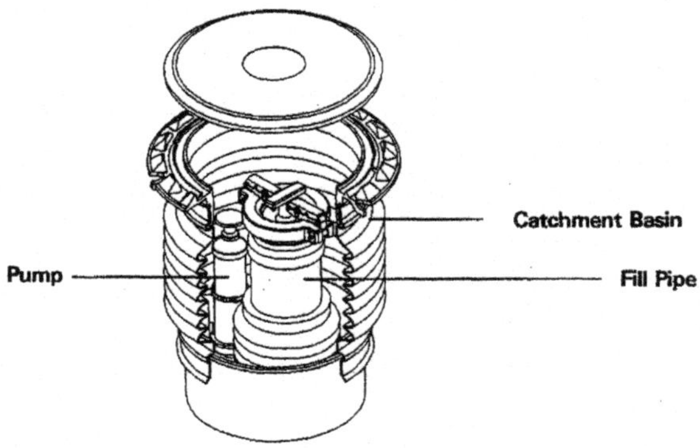

To protect against spills, the basin should be large enough to contain what may spill when the delivery hose is uncoupled from the fill pipe. Basins range in size from those capable of holding only a few gallons to those that are much larger--the larger the catchment basin, the more spill protection it provides.

Your equipment supplier can help you choose the size and type of catchment basin that meets your needs.

You need a way to remove liquid from catchment basins. Manufacturers equip catchment basins with either pumps or drains to remove liquid. The illustration on the previous page shows a catchment basin with a pump; the illustration below shows a catchment basin with a drain.

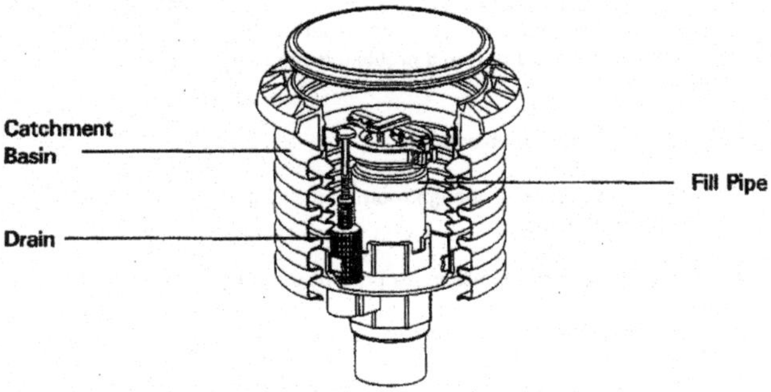

You should try to keep water out of catchment basins. Some catchment basins can collect enough water and sediment, along with spilled product, to make draining this mixture into the tank unwise. If this happens, you may pump out the catchment basin and dispose of the liquid properly. If the liquid contains fuel or chemicals, it could be considered a hazardous waste. Contact your state agency responsible for hazardous waste for information on testing and handling requirements.

Having the surrounding surface slope away from the top of catchment basins helps keep water out of them.

✔ Overfill Protection

What you must do for overfill protection:

- **Your UST must have overfill protection.** New USTs must have overfill protection devices when they are installed. The three main types of overfill protection devices (automatic shutoff devices, overfill alarms, and ball float valves) are described below and on the next page.

- You and your fuel deliverer must **follow industry standards for correct filling practices.**

Overfills usually release much larger volumes than spills. When a tank is overfilled, large volumes can be released at the fill pipe and through loose fittings on the top of the tank or a loose vent pipe. The tightness of these fittings normally would not be a problem if the tank were not filled beyond its capacity.

You can solve overfill problems by:

- Making sure there is enough room in the tank for the delivery BEFORE the delivery is made;

- Watching the entire delivery to prevent overfilling or spilling; and

- **Using overfill protection devices.**

If an UST never receives more than 25 gallons at a time, the UST does not have to meet the overfill protection requirements. Many small used oil tanks fall in this category.

NOTE: If you have "pumped delivery" where fuel is delivered under pressure, you must make sure your overfill protection device works compatibly with pumped deliveries. Also, remember that overfill protection devices are effective only when combined with careful filling practices (as noted on page 12).

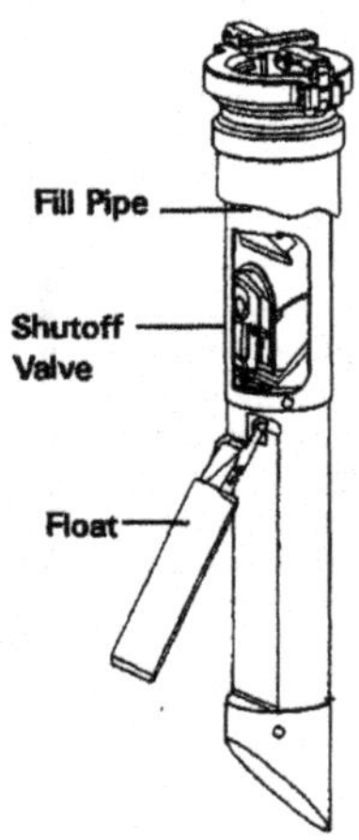

What Are *Automatic Shutoff Devices*?

An automatic shutoff device installed in an UST's fill pipe can slow down and then stop the delivery when the product has reached a certain level in the tank. This device--sometimes simply called a "fill pipe device"--has one or two valves that are operated by a float mechanism (the illustration on the left shows one kind of automatic shutoff device).

Some automatic shutoff devices work in two stages. The first stage drastically reduces the flow of product to alert the driver that the tank is nearly full. The driver can then close the delivery valve and still have room in the tank for the product left in the delivery hose.

If the driver does not pay attention and the liquid level rises higher, the valve closes completely and no more liquid can be delivered into the tank, leaving the driver with a delivery hose full of product.

To work properly, all overfill devices must be installed carefully at the correct distance below the tank top specified by the manufacturer.

What Are *Overfill Alarms*?

Overfill alarms use probes installed in the tank (see illustration on the right) to activate an alarm when the tank is either 90 percent full or within 1 minute of being overfilled. Either way, the alarm should provide enough time for the driver to close the truck's shutoff valve before an overfill happens. Alarms must be located where the driver can see or hear them easily. (Overfill alarms are often a part of automatic tank gauging systems.)

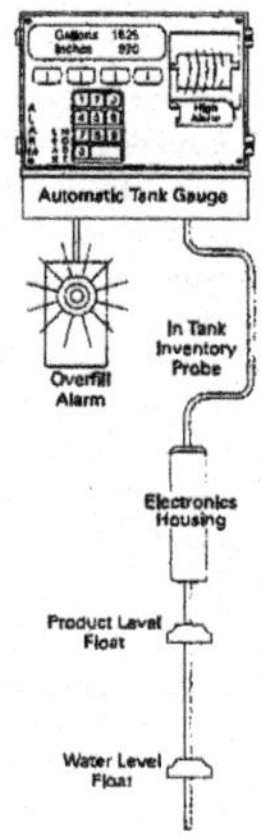

Overfill alarms work only if they alert the driver at the right time and the driver responds quickly. Remember to put the alarm on an electrical circuit that is active all the time so that the alarm will always work. Many deliveries are made at night when the facility is closed. You don't want to turn off your alarm when you turn off the office lights.

What Are *Ball Float Valves*?

Ball float valves (see illustration on the right) are placed at the bottom of the vent line several inches below the top of the UST. The ball floats on the product and rises with product level during delivery until it restricts vapor flowing out the vent line BEFORE the tank is full. If all tank fittings are tight, the ball float valve can create enough back pressure to restrict product flow into the tank--which can notify the driver to close the truck's shutoff valve. However, if the UST has loose fittings, sufficient back pressure may not develop and will result in an overfill. *Note: Manufacturers do not recommend using ball float valves with suction piping, pressurized delivery, or coaxial Stage I vapor recovery.*

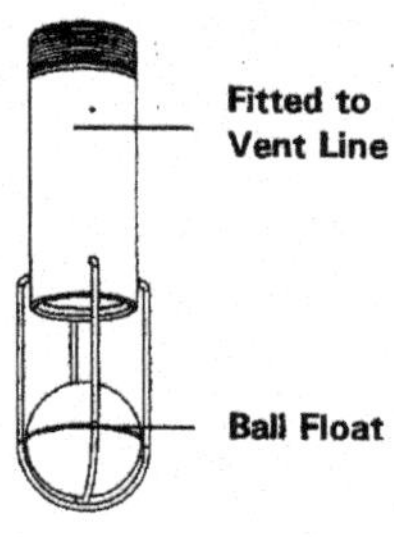

✔ Corrosion Protection

Corrosion results when bare metal and soil and moisture conditions combine to produce an underground electric current that destroys hard metal. Over time, corrosion creates holes and leaks develop.

What you must do for corrosion protection:

Your new USTs need to match one of the following performance standards:

- **Tank and piping completely made of noncorrodible material, such as fiberglass.** Corrosion protection is also provided if tank and piping are completely isolated from contact with the surrounding soil by being enclosed in or "jacketed" in noncorrodible material.

- **Tank and piping made of steel having a corrosion-resistant coating AND having cathodic protection** (such as an sti-P_3® tank with appropriate piping). A corrosion-resistant coating electrically isolates the coated metal from the surrounding environment to help protect against corrosion. *Asphaltic coating does NOT qualify as a corrosion-resistant coating.* Methods of cathodic protection are briefly explained on the next page.

- **Tank made of steel clad with a thick layer of noncorrodible material** (such as an ACT-100® tank). This option does not apply to piping. *Galvanized steel is NOT a noncorrodible material.*

Federal rules require corrosion protection for USTs because unprotected steel USTs corrode and release product through corrosion holes.

What Are Cathodic Protection Methods?

For more information on corrosion and how USTs can be protected from it, contact NACE International or other professionals in this field (see page 32).

- ***Sacrificial Anode System:*** Sacrificial anodes can be attached to the UST for corrosion protection (as illustrated below). Sacrificial anodes are pieces of metal more electrically active than the steel UST. Because these anodes are more active, the corrosive current will exit from them rather than the UST. Thus, the UST is protected while the attached anode is "sacrificed."

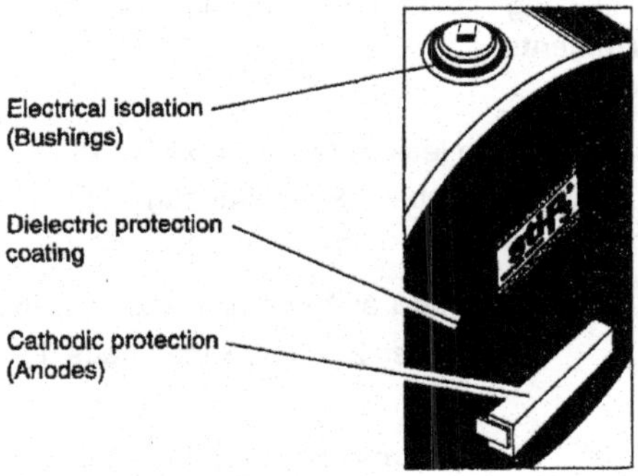

- ***Impressed Current System:*** An impressed current system uses a rectifier to convert alternating current to direct current (see illustration below). This current is sent through an insulated wire to the anodes, which are special metal bars buried in the soil near the UST. The current then flows through the soil to the UST system, and returns to the rectifier through an insulated wire attached to the UST. The UST system is protected because the current going to the UST system overcomes the corrosion-causing current normally flowing away from it.

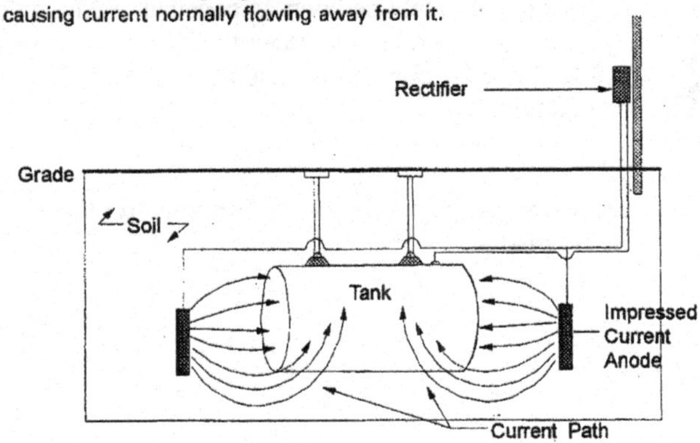

WHAT DO EXISTING PETROLEUM USTs NEED?

✔ ***Leak Detection***

✔ ***Spill And Overfill Protection***

✔ ***Corrosion Protection***

Existing UST systems are those installed before December 22, 1988.

Your existing USTs must meet the following requirements:

- ✔ Have leak detection no later than December 1993 (see page 19).
- ✔ Have spill and overfill protection by December 22, 1998 (see page 19).
- ✔ Have corrosion protection by December 22, 1998 (see pages 20-21).

The table on page 4 displays basic UST requirements and deadlines.

Although the regulatory deadline for spill, overfill, and corrosion protection is in 1998, you should make these improvements as soon as possible. Early action reduces your costs and reduces the chance that you will be liable for damages caused by releases from unimproved USTs.

The following pages provide information on these requirements...

All USTs must now have leak detection. USTs installed before December 22, 1988 had compliance deadlines that varied with the age of the USTs. By December 22, 1993, all of these "older" USTs had to be in compliance with leak detection requirements.

✔ *Leak Detection*

The leak detection requirements for existing USTs are similar to those for new tanks and piping, so please refer to pages 8-11.

- You can use any of the **monthly monitoring methods** listed for new USTs on page 9.
- You can use the combined method described on page 10 that combines inventory control and tank tightness testing. However, you can use this combined method only for 10 years after upgrading the UST with spill, overfill, and corrosion protection. After 10 years, you must use one of the monthly monitoring methods noted above. USTs that have not been upgraded can combine inventory control and ***annual*** tank tightness testing, but only until December 1998.
- Tanks of 2,000 gallons capacity or less may use manual tank gauging (see page 10). However, USTs between 1,001 and 2,000 gallons using the combination of manual tank gauging and tank tightness testing can only do so for 10 years after upgrading with spill, overfill, and corrosion protection. After 10 years, you must use one of the monthly monitoring methods noted above. USTs that have not been upgraded can combine manual tank gauging and ***annual*** tank tightness testing, but only until December 1998.

Other types of leak detection methods may be used if they have been approved by the regulatory authority.

All existing USTs must be upgraded with spill and overfill protection by December 1998. Otherwise, these USTs must be replaced with new USTs or permanently closed. EPA has a booklet available, **Don't Wait Until 1998**, that explains the requirements to upgrade, replace, or close USTs by 1998. You can order this free booklet by calling EPA's toll-free Hotline at 800 424-9346 and asking for **Don't Wait Until 1998.**

✔ *Spill And Overfill Protection*

The requirements for spill and overfill are exactly the same for existing and new USTs, so please refer to pages 12-15 for information on spill and overfill requirements. However, you should already be following industry standards for tank filling procedures that will prevent spills and overfills (see pages 12 and 14).

✔ *Corrosion Protection*

All existing USTs must have corrosion protection by December 1998. Otherwise, these USTs must be replaced with new USTs or permanently closed. EPA has a booklet available, **Don't Wait Until 1998,** that explains the requirements to upgrade, replace, or close USTs by 1998. You can order this free booklet by calling EPA's toll-free Hotline at 800 424-9346 and asking for **Don't Wait Until 1998.**

Your USTs already meet the requirements for corrosion protection if they match one of the performance standards for new USTs described on page 16:

- **Tank and piping completely made of noncorrodible material, such as fiberglass.**
- **Tank and piping made of steel having a corrosion-resistant coating AND having cathodic protection.** (Methods of cathodic protection are briefly explained on page 17.)
- **Tank made of steel clad with a thick layer of noncorrodible material** (such as an ACT-100® tank). This option does not apply to piping.

How Can You Add Corrosion Protection To Existing Steel USTs?

It is not practical to add coatings or claddings to existing steel USTs that have no corrosion protection. Instead, you must choose **ONE** of the following **THREE methods** to add corrosion protection to existing steel tanks:

Only tanks proven to be structurally sound can have cathodic protection

1. Add cathodic protection. If you are adding ONLY cathodic protection, you must do the following:

- **First, assess tank integrity.** Satisfy ONE of the following methods to make sure that the tank is structurally sound:
  - If the tank is LESS THAN 10 YEARS OLD, you can use results from one of the monthly leak detection methods to show the UST is not leaking (interstitial monitoring, automatic tank gauging, vapor monitoring, groundwater monitoring, statistical inventory reconciliation, or other approved methods).
  - If the tank is LESS THAN 10 YEARS OLD, you can use results from two tank tightness tests to show the UST is not leaking. The first test takes place before you install cathodic protection, and the second test takes place between 3 and 6 months after installation.

- If the tank is 10 YEARS OLD OR MORE, it can be internally inspected, tested, and assessed to make sure that the tank is structurally sound and free of corrosion holes (see page 31 for industry codes).

- You can assess the tank for corrosion holes by a method that the regulatory authority determines is no less protective than those above.

Page 17 describes cathodic protection methods. Impressed current systems work best with existing USTs.

■ **Second, install cathodic protection.** Regulations require a qualified corrosion expert to design cathodic protection systems installed at the UST site. The system must be tested by a qualified cathodic protection tester within 6 months of installation and at least every 3 years thereafter. You will need to keep the results of the last two tests to prove that the cathodic protection is working. In addition, you must inspect an impressed current system every 60 days to verify that the system is operating. Keep results of your last three inspections to prove that the impressed current system is operating properly.

Only tanks proven to be structurally sound can be lined.

2. **Add interior lining to the tank.** The interior of a tank can be lined with a thick layer of noncorrodible material (see page 31 for industry codes). Tanks using ***only*** an interior lining for corrosion protection must pass an internal reinspection in 10 years and every 5 years after that to make sure that the lining is sound. Keep records of the inspection results.

3. **Combine cathodic protection and interior lining.** You can add both cathodic protection and interior lining. The advantages for you of this combined method are simple: your USTs receive more protection; and you are not required to have the interior lining periodically inspected (which saves you the cost of these inspections). You will still need to have the cathodic protection system periodically tested and inspected and to keep records (as explained above).

And what about piping?

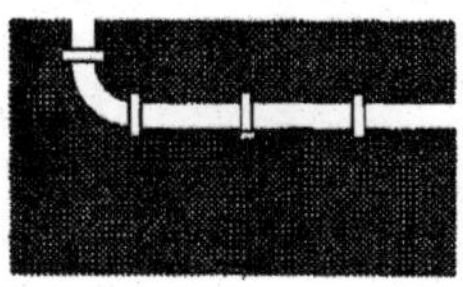

Existing steel piping must have cathodic protection. Note that cathodic protection needs to be designed by a qualified corrosion expert, tested and inspected periodically, and records kept (as explained above).

Piping entirely made of (or enclosed in) noncorrodible material, such as fiberglass, does not need cathodic protection.

WHAT DO YOU DO ABOUT UST RELEASES?

Warning signals indicate that your UST may be leaking and creating problems for the environment and your business. You can minimize these problems by paying careful attention to early warning signals and reacting to them quickly before major problems develop.

You should suspect a leak when you discover the following warning signals:

- **Unusual operating conditions** (such as erratic behavior of the dispensing pump). Check first to see if this problem results from equipment failure that can be repaired.
- **Results from leak detection monitoring and testing that indicate a leak.** What at first appears to be a leak may be the result of faulty equipment that is part of your UST system or its leak detection. Double check this equipment carefully for failures.

You need to call your regulatory authority and report suspected leaks. Then find out quickly if these ***suspected leaks*** are ***actual leaks*** using the following investigative steps:

- Conduct tightness testing of the entire UST system.
- Check the site for additional information on the presence and source of contamination.

If these system tests and site checks confirm a leaking UST, follow the actions for responding to **confirmed** leaks described on the next page.

You must also respond quickly to any evidence of leaked petroleum that appears at or near your site. For example, neighbors might tell you they smell petroleum vapors in their basements or taste petroleum in their drinking water. If evidence of this type is discovered, you must report this discovery **immediately** to the regulatory authority and take the investigative steps and follow-up actions noted above.

Your response to **confirmed** releases comes in two stages: **short-term** and **long-term**.

Short-Term Actions

- Take immediate action to stop and contain the release.
- Report the release to the regulatory authority within 24 hours. However, petroleum spills and overfills of less than 25 gallons do not have to be reported if you immediately contain and clean up these releases.
- Make sure the release poses no immediate hazard to human health and safety by removing explosive vapors and fire hazards. Your fire department should be able to help or advise you with this task. You must also make sure you handle contaminated soil properly so that it poses no hazard (for example, from vapors or direct contact).
- Remove petroleum from the UST system to prevent further release into the environment.
- Find out how far the petroleum has moved and begin to recover the leaked petroleum (such as product floating on the water table). Report your progress and any information you have collected to the regulatory authority no later than 20 days after confirming a release.
- Investigate to determine if the release has damaged or might damage the environment. This investigation must determine the extent of contamination both in soils and groundwater. You must report to the regulatory authority what you have learned from an investigation of your site according to the schedule established by the regulatory authority. At the same time, you must also submit a report explaining how you plan to clean up the site. Additional site studies may be required.

> ***Contact your regulatory authority for specific time frames, necessary steps, and guidance in doing site assessments and Corrective Action Plans.***

Long-Term Actions

Based on the information you have provided, the regulatory authority will decide if you must take further action at your site. You may need to take two more actions:

- Develop and submit a Corrective Action Plan that shows how you will meet requirements established for your site by the regulatory authority.
- Make sure you implement the steps approved by the regulatory authority for your site.

Can Leaking Tanks Be Repaired?

You can repair a leaking tank if the person who does the repair carefully follows standard industry codes that establish the correct way to conduct repairs. (See page 31 for industry codes and standards.)

Within 30 days of the repair, you must prove that the tank repair has worked by doing one of the following:

- Have the tank inspected internally or tightness tested following standard industry codes; or
- Use one of the monthly leak detection monitoring methods (see page 9); or
- Use other methods approved by the regulatory authority.

Within 6 months of repair, USTs with cathodic protection must be tested to show that the cathodic protection is working properly.

You must keep records for each repair as long as you keep the UST in service.

Can Leaking Piping Be Repaired?

Damaged **metal** piping **cannot** be repaired and must be replaced. Loose fittings can be tightened, and in some cases that may solve the problem.

Piping made of fiberglass-reinforced plastic can be repaired, but **only** in accordance with the manufacturer's instructions or national codes of practice. Within 30 days of the repair, piping must be tested in the same ways noted above for testing tank repairs (except for internal inspection).

HOW DO YOU CLOSE USTs?

You can close your UST **temporarily** or **permanently**.

Closing Temporarily

You may temporarily close your UST for up to 12 months by following these requirements for **temporary closure**:

- Continue to monitor for leaks by maintaining the UST's leak detection. (If your UST is empty, however, you do not need to maintain leak detection.) Also, continue to monitor and maintain any corrosion protection systems. If a release is discovered, respond as you would for a release from an active UST, as described on page 23.

- If the UST remains temporarily closed for more than 3 months, you must leave vent lines open, but cap and secure all other lines, pumps, manways, and ancillary equipment.

After 12 months of **temporary** closure, you must **permanently** close your UST. Your UST, however, can remain "temporarily" closed indefinitely if it meets the requirements for new or upgraded USTs and the requirements above for temporarily closed USTs.

Closing Permanently

If your UST does not meet the requirements for corrosion protection and it remains closed for more than 12 months, you must close it permanently. If you decide for any reason to close your UST permanently, follow these requirements for **permanent closure**:

- You must notify the regulatory authority 30 days before you close your UST.

- You must determine if contamination from your UST is present in the surrounding environment. If there is contamination, you will have to take the corrective actions described on page 23.

- You can either remove the UST from the ground or leave it in the ground. In both cases, the tank must be emptied and cleaned by removing all liquids, dangerous vapor levels, and accumulated sludge. These potentially very hazardous actions need to be carried out carefully by following standard safety practices. (See pages 31 and 34 for sources of information on good closure practices.) If you leave the UST in the ground, you must also fill it with a harmless, chemically inactive solid, like sand. The regulatory authority can help you decide how best to close your UST so that it meets local requirements for closure.

WARNING: People are killed or injured every year while closing or removing tanks. Use safe removal practices (see page 31 for a safe closure standard). Only trained professionals should close or remove USTs.

WHAT DO YOU NEED TO REPORT?

You need to report to the regulatory authority on the following occasions:

- When you install an UST, you have to fill out a **notification form** available from your state. This form provides information about your UST, including a certification of correct installation. (You should have already used this form to identify your existing USTs. If you haven't done that yet, be sure you do so now.)
- You must report suspected releases to the regulatory authority (see page 22). If a release is confirmed, you must also report follow-up actions you plan or have taken to correct the damage caused by your UST (see page 23).
- You must notify the regulatory authority 30 days before you permanently close your UST (see page 25).

You should check with your regulatory authority about the particular reporting requirements in your area, including any additional or more stringent requirements than those noted above.

If you are not sure who your regulatory authority is, see the list of state UST contacts starting on page 35.

WHAT RECORDS MUST YOU KEEP?

You will have to keep records that can be provided to an inspector during an on-site visit that prove your facility meets certain requirements. These records must be kept long enough to show your facility's recent compliance status in four major areas:

- You will have to keep records of **leak detection** performance and maintenance:
  - The last year's monitoring results, and the most recent tightness test.
  - Copies of performance claims provided by leak detection manufacturers.
  - Records of recent maintenance, repair, and calibration of on-site leak detection equipment.
- You will have to keep records showing the required inspections and tests of your **corrosion protection** system.
- You must keep records showing that a **repaired or upgraded UST system** was properly repaired or upgraded.
- For at least 3 years after **closing an UST**, you must keep records of the site assessment results required for permanent closure. (These results show what impact your UST has had on the surrounding area.)
- You must keep records that document your financial responsibility, as explained in EPA's booklet, **Dollars And Sense** (to order this booklet, see page 33).

You should check with your regulatory authority about the particular recordkeeping requirements in your area. Generally, you should follow this useful rule of thumb for recordkeeping: When in doubt, keep it.

FOR HAZARDOUS SUBSTANCE USTs ONLY

Several hundred substances were designated as "hazardous" in Section 101(14) of the Comprehensive Environmental Response, Compensation, and Liability Act of 1980 (CERCLA).

PARTIAL LIST OF HAZARDOUS SUBSTANCES

HAZARDOUS SUBSTANCE	CASRN*
Acenephthene	83329
Acenephthylene	208968
Acetaldehyde	75070
Acetaldehyde, chloro-	107200
Acetaldehyde, tricloro-	75876
...	...
Chromic sulfate	1010153
Choromium	8
CHROMIUM AND COMPOUNDS	7440473
Chromous chloride	--
Chrysene	1004905
Cobaltous bromide	5
Cobaltous formate	218019
Cobaltous sulfamate	7789437
Copper	544183
Copper cyanide	1401741
	5
	7440508
	544923
...	...
Famphur	52857
Ferric ammonium citrate	1185576
Ferric ammonium oxalate	2944674
Ferric chloride	7705080
Ferric dextran	9004664
Ferric fluoride	7783508
Ferric nitrate	1042148
Ferric sulfate	4
	1002822
	5
...	...
Kelthane	115322
Kepone	143500
Lasiocarpine	303344
LEAD AND COMPOUNDS	--
Lead	7439921
Lead acetate	301042
Lead arsenate	7784409
Lead chloride	7758954
Lead fluoborate	1381496
Lead iodide	5
	1010163
	0
...	...
Zinc nitrate	7779886
Zinc phenosulfonate	127882
Zinc phosphide	1314847
Zinc silicoflouride	1687171
Zinc sulfate	9
Zirconium nitrate	7733020
Zirconium sulfate	1374889
Zirconium tetrachloride	9
	1464461
	2
	1002611
	6

* Chemical Abstracts Service Registry Number

The UST regulations apply to the same hazardous substances identified by CERCLA, except for those listed as hazardous **wastes.** These hazardous wastes are already regulated under Subtitle C of the Resource Conservation and Recovery Act and are **not** covered by the UST regulations. (See 40 **CFR** Parts 260-270 for the hazardous waste regulations.) Information on the CERCLA hazardous substances is available from EPA through the RCRA/CERCLA Hotline at 800 424-9346.

What Requirements Apply To New Hazardous Substance USTs?

New hazardous substance USTs are those installed after December 22, 1988. These USTs have to meet the same requirements described earlier for new petroleum USTs concerning correct installation, spill, overfill and corrosion protection, corrective action, and closure. (See pages 7 and 12-17.)

In addition, new hazardous substance USTs must have secondary containment and interstitial monitoring for leak detection, as described below.

Secondary Containment

All new hazardous substance USTs must have "secondary containment." A single-walled tank is the first or "primary" containment. Using only primary containment, a leak can escape into the environment. But by enclosing an UST within a second wall, leaks can be contained and detected quickly before harming the environment.

There are several ways to construct secondary containment:

- Placing one tank inside another tank or one pipe inside another pipe (making them double-walled systems).
- Placing the UST system inside a concrete vault.
- Lining the excavation zone around the UST system with a liner that cannot be penetrated by the hazardous substance.

Interstitial Monitoring

The hazardous substance UST must have a leak detection system that can indicate the presence of a leak in the confined space between the first and the second wall. Several devices are available to monitor this confined "interstitial" space. ("Interstitial" simply means "between the walls.") The UST regulations describe these various methods and the requirements for their proper use.

You can apply for an exception, called a **variance**, from the requirement for secondary containment and interstitial monitoring. To obtain a variance you must demonstrate to the regulatory authority that your alternative leak detection method will work effectively by providing detailed studies of your site, proposed leak detection method, and available methods for corrective action.

What About Existing Hazardous Substance USTs?

Existing UST systems are those installed before December 22, 1988. In addition to immediately starting tank filling procedures that prevent spills and overfills, you will need to meet the following requirements for existing USTs.

Leak Detection

Deadlines for compliance with leak detection requirements were phased in according to the age of the UST. **By December 22, 1993, all existing USTs were required to have leak detection.** Pressurized piping must meet the requirements for **new** pressurized piping (described on page 11).

You can meet the leak detection requirements in one of the following three ways:

- Until December 1998, you can use any of the leak detection methods described on page 19 ***but only if*** the method you choose can effectively detect releases of the hazardous substance stored in the UST.
- After December 22, 1998, your UST must meet the same requirements for secondary containment and interstitial monitoring that apply to new hazardous substance USTs.
- After December 22, 1988, a **variance** can be granted if you meet the same requirements described above for receiving a variance for a new hazardous substance UST.

Spill, Overfill, and Corrosion Protection

By December 22, 1998, you must improve your USTs:

- By using devices that **prevent spills and overfills.**
- By adding **corrosion protection** to steel tanks and piping.

Although the regulatory deadline is in 1998, you should make these improvements as soon as you can to reduce the chance that you will be liable for damages caused by your unimproved UST.

What If You Have A Hazardous Substance Release?

You must follow the same short-term and long-term actions described earlier for petroleum releases on page 23, with two exceptions.

First, you must immediately report hazardous substance spills or overfills that meet or exceed their "reportable quantities" to the National Response Center at 800 424-8802 or 202 267-2675.

Second, you must also report hazardous substance spills or overfills that meet or exceed their "reportable quantities" to the regulatory authority within 24 hours. However, if these spills or overfills are smaller than their "reportable quantities" and are immediately contained and cleaned up, they do not need to be reported. You can get information on the "reportable quantities" by calling the EPA RCRA/CERCLA Hotline at 800 424-9346.

Industry Codes And Standards

Installation

API Recommended Practice 1615 (1987), "Installation of Underground Petroleum Storage Systems"

PEI RP100-94 (1994), "Recommended Practices for Installation of Underground Liquid Storage Systems"

Tank Filling Practices

NFPA 385 (1985), "Standard for Tank Vehicles for Flammable and Combustible Liquids"

Closure

API Recommended Practice 1604 (1987), "Removal and Disposal of Used Underground Petroleum Storage Tanks"

Assessing Tank Integrity, Repairing Tanks, and Interior Lining Of Tank

API Recommended Practice 1631 (1992), "Interior Lining of Underground Storage Tanks"

NLPA Standard 631 (1991), "Entry, Cleaning, Interior Inspection, Repair, and Lining of Underground Storage Tanks"

ASTM ES 40-94 (1994), "Emergency Standard Practice for Alternative Procedures for the Assessment of Buried Steel Tanks Prior to the Addition of Cathodic Protection"

Corrosion Protection

API Recommended Practice 1632 (1987), "Cathodic Protection of Underground Petroleum Storage Tanks and Piping Systems"

NACE RP-0169-92 (1992), "Standard Recommended Practice: Control of Corrosion on Underground or Submerged Metallic Piping Systems"

NACE RP-0285-85 (1985), "Standard Recommended Practice: Control of External Corrosion on Metallic Buried, Partially Buried, or Submerged Liquid Storage Systems"

STI R892-91 (1991), "Recommended Practice for Corrosion Protection of Underground Piping Networks Associated with Liquid Storage and Dispensing Systems"

General (Repair, Spill and Overfill, Installation, Compatibility)

API Recommended Practice 1621 (1993), "Bulk Liquid Stock Control at Retail Outlets"

API Recommended Practice 1626 (1985), "Storing and Handling Ethanol and Gasoline-Ethanol Blends at Distribution Terminals and Service Stations"

API Recommended Practice 1627 (1986), "Storage and Handling of Gasoline-Methanol/Cosolvent Blends at Distribution Terminals and Service Stations"

API Recommended Practice 1635 (1987), "Management of Underground Petroleum Storage Systems at Marketing and Distribution Facilities"

NFPA 30 (1993), "Flammable and Combustible Liquids Code"

NFPA 30A (1993), "Automotive and Marine Service Station Code"

*Organizations are identified on the next page.

Organizations To Contact For UST Information

API -- American Petroleum Institute
1220 L Street, N.W.
Washington, DC 20005
(202) 682-8000

Fiberglass Petroleum Tank and
Pipe Institute
9801 Westheimer; Suite 606
Houston, TX 77042-3951
(713) 465-3310

NACE International (formerly the National
Association of Corrosion Engineers)
Box 218340
Houston, TX 77218-8340
(713) 492-0535

NFPA -- National Fire Protection
Association
1 Batterymarch Park
Box 9101
Quincy, MA 02269-9101
(617) 770-3000

NLPA -- National Leak Prevention
Association
Box 1643
Boise, ID 83701
(208) 336-6941

PEI -- Petroleum Equipment Institute
Box 2380
Tulsa, OK 74101-2380
(918) 494-9696

Steel Tank Institute
570 Oakwood Road
Lake Zurich, IL 60047
(708) 438-TANK [8265]

Publications And Videos About USTs

PUBLICATIONS

TITLE	AVAILABLE FREE FROM
Normas Y Procedimientos Para T.S.A. Spanish translation of **Musts For USTs: A Summary Of The Federal Regulations For Underground Storage Tank Systems.** (About 40 pages.)	You can call EPA's toll-free RCRA/Superfund Hotline at 800 424-9346 and order free copies. Just identify the titles you want. Or you can write and ask for titles by addressing your requests to: NCEPI Box 42419 Cincinnati, OH 45242 Or you can fax your order to NCEPI at 513 891-6685.

Straight Talk On Tanks: Leak Detection Methods For Petroleum Underground Storage Tanks
Booklet explains federal regulatory requirements for leak detection and briefly describes allowable leak detection methods. (About 30 pages.)

Doing Inventory Control Right: For Underground Storage Tanks
Booklet describes how owners and operators of USTs can use inventory control and periodic tightness testing to meet federal leak detection requirements. Contains reporting forms. (About 16 pages.)

Manual Tank Gauging: For Small Underground Storage Tanks
Booklet provides simple, step-by-step directions for conducting manual tank gauging for tanks 2,000 gallons or smaller. Contains reporting forms. (About 12 pages.)

Don't Wait Until 1998: Spill, Overfill, And Corrosion Protection For Underground Storage Tanks
Information to help owners and operators of USTs meet the 1998 deadline for compliance with requirements to upgrade, replace, or close USTs installed before December 1988. (About 16 pages.)

Introduction To Statistical Inventory Reconciliation: For Underground Storage Tanks
Booklet describes the use of Statistical Inventory Reconciliation (SIR) to meet federal leak detection requirements. (About 12 pages.)

Dollars And Sense: Financial Responsibility Requirements For Underground Storage Tanks
Booklet clearly summarizes the "financial responsibility" required of UST owners and operators by federal UST regulations. (About 16 pages.)

An Overview Of Underground Storage Tank Remediation Options
Fact sheets provide information about technologies that can be used to remediate petroleum contamination in soil and groundwater. (About 26 pages.)

Controlling UST Cleanup Costs
Fact sheet series on the cleanup process includes: *Hiring a Contractor, Negotiating the Contract, Interpreting the Bill, Managing the Process, and Understanding Contractor Code Words*. (About 10 pages.)

Federal Register Reprints
Not simple summaries, these reprints are extensive records of the rulemaking process including technical information, explanatory preambles, and the rules as they appear in the Code of Federal Regulations. Reprints dated 9/23/88; 10/26/88; 11/9/89; 5/2/90; and 2/18/93. Over 300 pages.

Publications And Videos About USTs

VIDEOS

TITLE/COST	AVAILABLE FROM
Doing It Right Illustrates proper installation of underground tanks and piping for installation crews. Part 1: Tanks (24 minutes); Part 2: Piping (16 minutes). Cost: $25 **Doing It Right II: Installing Required UST Equipment** Illustrates installation of spill and overfill equipment, observation wells, and piping leak detection (23 minutes). Cost: $60 **Doing It Right** and **Doing it Right II** Set Cost: $75 **Keeping It Clean: Making Safe And Spill-Free Motor Fuel Deliveries** Making pollution-free deliveries to USTs. Includes Stage 1 vapor recovery, overfill prevention and spill containment. For fuel tanker drivers and UST owner/operators (25 minutes). Cost: $60 **Petroleum Leaks Underground** How liquids and vapors move in the subsurface and why early response to leaked petroleum is so important. Part 1: How Liquids Move (14 minutes); Part 2: How Vapors Move (15 minutes). Cost: $75 **Straight Talk On Leak Detection** Overview of the leak detection methods available for complying with federal regulations. Part 1: Straight Talk From Tank Owners (owners address the problems of UST compliance [5 minutes]); Part 2: Straight Talk On Leak Detection (30 minutes). Cost: $40	**Environmental Media Center** Box 30212 Bethesda, MD 20814 301 654-7141 800 522-0362 Visa and MasterCard accepted
Tank Closure Without Tears: An Inspector's Safety Guide Focuses on explosive vapors and safe tank removal (30 minutes). Video and Booklet Cost: $35; Booklet: $5 **What Do We Have Here?: An Inspector's Guide To Site Assessment At Tank Closure** Inspecting sites for contamination where tanks have been removed. Part 1: Site Assessment Overview (30 minutes); Part 2: Field Testing Instruments At A Glance (14 minutes); Part 3: Soil And Water Sampling At A Glance (7 minutes). Video and Booklet Cost: $45; Booklet: $5 **Searching For The Honest Tank: A Guide To UST Facility Compliance Inspection** Covers major steps of UST inspections from protocols and equipment to enforcement and followup; from cathodic protection to leak detection. Directed at inspectors, yet also helpful to owners and operators (30 minutes). Video and Booklet Cost: $40; Booklet: $5	**New England Interstate Environmental Training Center** ATTN:VIDEOS 2 Fort Road South Portland, ME 04106 207 767-2539

State Contacts For UST Information

[LUST contacts focus on leaking underground storage tanks.]

Alabama Dept. of **Env. Management**
Groundwater Section/
Water Division
Box 301463
Montgomery, AL
36130-1463
UST phone:
334 271-7986
LUST phone:
334 271-7834

Alaska Dept. of **Env. Conservation**
410 Willoughby Ave.
Juneau, AK
99801-1795
phone: 907 465-5203

Arizona Dept. of Env. Quality
3033 N. Central Ave.
Phoenix, AZ 85012
phone: 602 207-4324

Arkansas Dept. of Pollution Control & Ecology
Regulated Storage
Tank Division
8001 National Dr.
Little Rock, AR
72209
phone: 501 570-2801

California State Water Resources Control Board
Div. of Clean Water
Program
Box 944212
Sacramento, CA
94244-2120
phone: 916 227-4313

[UST contact]
Colorado State Oil Inspection Office
1515 Arapahoe Street
Suite 525, Tower 3
Denver, CO
80202-2117
phone: 303 620-4300

[LUST contact]
Colorado Dept. of Health
Hazardous Materials &
Waste Mgt. Div.
UST Program
4300 Cherry Creek Dr. S.
Denver, CO 80220
phone: 303 692-3453

Connecticut Dept. of Env. Protection
UST Program
State Office Bldg.
79 Elm Street
Hartford, CT 06106
phone: 203 424-3374

DC Env. Regulatory Administration
Pesticides, Hazardous
Waste & UST Div.
2100 Martin Luther King
Ave. S.E., Suite 203
Washington, D.C. 20020
phone: 202 645-6080

Delaware Dept. of **Natural Resources & Env. Control**
UST Branch
715 Grantham Lane
New Castle, DE 19720
phone: 302 323-4588

Florida Dept. of Env. Regulation
Tank Section
Twin Towers Office
Bldg., Rm. 403
2600 Blair Stone Road
Tallahassee, FL
32399-2400
phone: 904 488-3935

Georgia Dept. of Natural Resources
UST Mgt. Program
4244 Intl. Parkway,
Suite 100
Atlanta, GA 30354
phone: 404 362-2687

Hawaii Dept. of Health
Solid and Hazardous
Waste Branch
919 Ala Moana Blvd.,
Suite 212
Honolulu, HI 96814
phone: 808 586-4226

Idaho Dept. of Health & Welfare
Div. of Env. Quality
1410 North Hilton
Boise, ID 83706
phone: 208 334-0542

[UST contact]
Illinois Office of State Fire Marshall
Div. of Petroleum &
Chem. Safety
1035 Stevenson Dr.
Springfield, IL 62703
phone: 217 785-5878

[LUST contact]
Illinois EPA
Div. of Remediation Mgt.
LUST Section
Box 19276
Springfield, IL
62794-9276
phone: 217 782-6760

Indiana Dept. of **Env. Mgt.**
Office of Env. Response
Box 7015
Indianapolis, IN 46206
phone: 317 233-6418

Iowa Dept. of Natural Resources
UST Section
Wallace State Office
Bldg.
900 East Grand
Des Moines, IA 50319
phone: 515 281-8135

Kansas Dept. of Health & Env.
Bureau of Env.
Remediation
Storage Tank Section
Forbes Field, Bldg. 740
Topeka, KS 66620
phone: 913 296-1678

Kentucky Div. of Waste Mgt.
UST Branch
14 Reilly Road
Frankfort, KY 40601
phone: 502 564-6716

Louisiana Dept. of **Env. Quality**
UST Division
Box 82178
Baton Rouge, LA
70810
phone: 504 765-0243

Maine Dept. of Env. Protection
State House, Station 17
Hospital St., Ray Bldg.
Augusta, ME 04333
phone: 207 287-2651

Maryland Dept. of **Env.**
Waste Mgt. Adm.
Oil Control Program
2500 Broening Highway
Baltimore, MD 21224
phone: 410 631-3442

[UST contact]
Massachusetts Dept. of Public Safety
UST Program
1010 Commonwealth
Ave.
Boston, MA 02215
phone: 617-351-6000

[LUST contact]
Massachusetts Dept. of Env. Protection
Bureau of Waste Site
Cleanup
One Winter St.
Boston, MA 02108
phone: 617 556-1044

State Contacts For UST Information

Michigan Dept. of Natural Resources
UST Division
Box 30157
Lansing, MI 48909
UST phone:
517 373-8168
MUSTFA (state fund):
517 373-6247

Minnesota Pollution Control Agency
Tanks and Spills Section
520 Lafayette Road North
St. Paul, MN
55155-3898
phone: 612 297-8609

Mississippi Dept. of Env. Quality
Bureau of Pollution Control, UST Section
Box 10385
Jackson, MS
39289-0385
phone: 601 961-5171

Missouri Dept. of Natural Resources
Water Pollution Control Program (UST)
Environmental Services Program (LUST)
Box 176
Jefferson City, MO
65102-0176
UST phone:
314 751-7428
LUST phone:
314 526-3384

Montana Dept. of Health & Env. Sciences
Solid & Hazardous Waste Bureau
UST Program
Cogswell Bldg.
Helena, MT 59620
phone: 406 444-5970

[UST contact]
Nebraska State Fire Marshal
Flammable Liquid Storage
246 South 14th St.
Lincoln, NE 68508
phone: 402 471-9465

[LUST contact]
Nebraska Dept. of Env. Control
LUST/ER Section
Box 98922
Lincoln, NE
68509-8922
phone: 402 471-4230

Nevada Dept. of Conserv. & Natural Resources
Div. of Env. Protection
Capitol Complex
333 W. Nye Lane
Carson City, NV 89710
phone: 702 687-5872

New Hampshire Dept. of Env. Services
Oil Compliance Section
Groundwater Protection Bureau
Box 95
Concord, NH 03301
phone: 603 271-3644

New Jersey Dept. of Env. Protection
Responsible Party Site Remediation
401 E. State (CN-028)
Trenton, NJ
08625-0028
phone: 609 984-3156

New Mexico Env. Dept.
UST Bureau
Box 26110
Santa Fe, NM
87502
phone: 505 827-0188

New York Dept. of Env. Conservation
Bulk Storage Section
50 Wolf Rd., Rm. 340
Albany, NY
12233-3750
phone: 518 457-4351

North Carolina Pollution Control Branch
Div. of Env. Mgt.
Dept. of Env. Health & Natural Resources
441 N. Harrington St.
Raleigh, NC
27603-1323
phone: 919 733-8486

North Dakota Div. of Waste Mgt.
Dept. of Health and Consolidated Lab.
Box 5520
Bismarck, ND
58502-5520
phone: 701 328-5166

Ohio Dept. of Commerce
Bureau of UST Regulations
8895 East Main St.
Box 687
Reynoldsburg, OH 43068
phone: 614 752-7938

Oklahoma Corporation Commission
Fuel Storage Division
Box 52000-2000
Oklahoma City, OK
73152-2000
UST phone:
405 521-3107
LUST phone:
405 521-6575

[UST contact]
Oregon Dept. of Env. Quality
UST Compliance Section
811 S.W. Sixth Ave.,
7th Floor
Portland, OR 97204
phone: 503 229-5774

[LUST contact]
Oregon Dept. of Env. Quality
UST Cleanup Program
811 S.W. Sixth Ave.,
9th Floor
Portland, OR 97204
phone: 503 229-6642

Pennsylvania Dept. of Env. Resources
Storage Tank Program
Box 8762
Harrisburg, PA
17101-2301
phone: 717 772-5599

Rhode Island Dept. of Env. Mgt.
UST Section
291 Promenade St.
Providence, RI 02908
phone: 401 277-2234

South Carolina Dept. of Health and Env.Control
Groundwater Protection Division
2600 Bull St.
Columbia, SC 29201
UST phone:
803 734-5335
LUST phone:
803 734-5331

South Dakota Dept. of Env. & Nat. Resources
UST Program
523 East Capitol
Pierre, SD 57501
phone: 605 773-3296

Tennessee Dept. of Env. & Conservation
Div. of USTs
4th Floor, L&C Tower
401 Church St.
Nashville, TN
37243-1541
phone: 615 532-0945

Texas Natural Resources Conservation Comm.
Petroleum Storage Tank Div.
Box 13087
Austin, TX 78711-3087
phone: 512 239-2000

Utah Dept. of Env. Quality
Div. of Env. Response and Remediation
UST Branch
168 North 1950 West,
1st Floor
Salt Lake City, UT 84116
phone: 801 536-4100

Vermont Dept. of Natural Resources
UST Program
103 South Main St.,
West Bldg.
Waterbury, VT 05676
phone: 802 244-8702

[UST contact]
Virginia Dept. Env. Quality
Response Program
4951 Cox Road
Glen Allen, VA 23060
phone: 804 527-5189

State Contacts For UST Information

[LUST contact]
Virginia Dept. Env. Quality
Groundwater Mgt. Program
4951 Cox Road
Glen Allen, VA 23060
phone: 804 527-5189

Washington Dept. of Ecology
Box 47655
Olympia, WA 98504-7655
UST phone: 206 407-7211
LUST phone: 206 407-7218

West Virginia Div. of Env. Protection
Office of Waste Mgt.
UST Section
1356 Hansford St.
Charleston, WV 25301
phone: 304 558-6371

[UST contact]
Wisconsin Dept. of Industry, Labor & Human Relations
Division of Safety and Buildings
Box 7969
Madison, WI 53707
phone: 608 267-7605

[LUST contact]
Wisconsin Dept. of Natural Resources
Bureau of Solid and Hazardous Waste Mgt.
Box 7921
Madison, WI 53707-7921
phone: 608 267-7560

Wyoming Dept. of Env. Quality
Water Quality Div.
Herschler Bldg., 4th Floor
122 West 25th St.
Cheyenne, WY 82002
phone: 307 777-7096

U.S. TERRITORIES

American Samoa Env. Protection Agency
Office of the Governor
American Samoa Government
ATTN: UST Program
Pago Pago, American Samoa 96799
phone: 684 633-2304

Commonwealth of Northern Mariana Islands Div. of Env. Quality
Box 1304
Dr. Torres Hospital
Saipan, MP 96950
phone: 607 234-6984

Guam Env. Protection Agency
IT&E
Harmon Plaza, Complex Unit D-107
130 Rojas St.
Harmon, Guam 96911
phone: 671 646-8863

Puerto Rico Env. Quality Board
Box 11488
Commonwealth of Puerto Rico
Santurce, Puerto Rico 00910
phone: 809 767-8109

Virgin Islands Div. of Env. Protection
Dept. of Planning and Natural Resources
Nisky Center 231
45A Estate Nisky
Charlotte Amalie
St. Thomas, Virgin Islands 00802
phone: 809 774-3320

■ APPENDIX E.5 Introduction to Statistical Inventory Reconciliation

United States Environmental Protection Agency | Solid Waste And Emergency Response 5403W | EPA 510-B-95-009 September 1995

EPA

Introduction To Statistical Inventory Reconciliation

For Underground Storage Tanks

Contents

Why You Should Read This Booklet

Federal and state laws require underground storage tank systems (USTs) to have leak detection. One of the available leak detection methods is Statistical Inventory Reconciliation (SIR). In this method, a trained professional uses sophisticated computer software to conduct a statistical analysis of inventory, delivery, and dispensing data.

SIR can allow the owner or operator of an UST facility to meet leak detection requirements without an extensive outlay of capital, using only the equipment that most facilities have readily at hand—a tank stick and a tank chart used for inventory control. The SIR analysis itself is usually provided as a service by vendors who charge a monthly fee based on the number of tanks.

This booklet provides basic information on the method—what it is, how it works, factors that impact data quality—to assist you in determining if SIR is appropriate to your needs.

If you need information on federal leak detection requirements and the various methods of leak detection available to you, see **Straight Talk on Tanks.** For a free copy, call the U.S. Environmental Protection Agency's (EPA's) toll-free Hotline at 800 424-9346 and ask for publication EPA 510-K-95-003.

How Does SIR Work?

On the face of it, SIR looks very similar to old-fashioned inventory control—the owner or operator, using simple equipment, tracks tank volumes, deliveries, and sales. However, the similarity ends there. Simple inventory control is relatively imprecise. Depending on your system throughput, you could be losing hundreds of gallons every month without realizing anything is wrong!

To many people, SIR may seem like magic. But it's based on sound mathematical principles.

By contrast, SIR analysis can be very sensitive and accurate. A SIR vendor can take the same inventory data and analyze them for releases so small that many would go unnoticed with inventory control. By using a month's worth of good tank data, it is possible for SIR methods to detect a release of just over 1½ pints per hour (that's about 145 gallons per month) from a tank or its product lines 95 times out of a hundred.

The mechanics of how SIR works are beyond the scope of this booklet. SIR vendors actually use a variety of statistical tools to evaluate inventory data, and no two vendors' methods are exactly alike—the information they collect and the results they provide can vary. Still, for fundamental release detection purposes, there are only three possible bottom-line responses for any SIR test: *PASS*, *FAIL*, or *INCONCLUSIVE*. These bottom-line responses are described below and on the following pages.

A 'FAIL' does not necessarily mean your system is leaking, but you must still notify your local UST agency.

PASS—According to the analyzed data, the UST system tests tight.

FAIL—Analyzed data indicate a loss of product from the system or an influx of groundwater. However, a *FAIL* does not *necessarily* indicate that your system is leaking. A *FAIL* may indicate miscalibrated dispensers, inaccurately metered deliveries, or stolen product. There is also a chance that a *FAIL* is a false alarm. **If you receive a *FAIL*, you must first notify your local UST regulatory agency.** Then, you should explore possible reasons for the *FAIL* (see page 7). Keep your local UST agency informed as to your findings.

INCONCLUSIVE—Analyzed data cannot make the call. There is a chance that the information provided to the SIR vendor is so bad that it is not possible to make a determination. This often can be traced back to poor tank sticking or bookkeeping practices (for example, a new hire who has received inadequate training). Whatever the reason, an *INCONCLUSIVE* result means, in effect, that you have failed to perform leak detection on the UST in question for that month. You are in violation of federal leak detection requirements. Contact your state UST program office to find out local policy on how *INCONCLUSIVE* results are handled. See page 7 for additional information.

Necessary Equipment

One of the major attractions of SIR for UST owners and operators is that it does not require a large, up-front investment of capital—the primary cost is subscribing to the SIR vendor's services. The equipment needed to use the method is usually already found on-site at most UST facilities.

Gauge Stick Or Other Gauges

A gauge stick, made of wood or other non-sparking material, is used to measure the depth of liquid in the UST. Typically, such sticks are marked or notched in 1/8-inch increments starting with the bottom of the stick. It is important that the stick be in good condition. Sticks that have worn ends, cut-off ends, worn-off numbers, or worn-off varnish coatings are not acceptable and should be replaced.

Good sticking practices are essential to good SIR analyses.

Other forms of gauges can also be used if they are available and in good operating condition. Automatic tank gauges, for instance, can simplify measuring tank volumes. (Keep in mind, of course, that some automatic tank gauging systems can serve as acceptable monthly tank leak detection methods by themselves.)

Whatever form of gauge you choose to use, you must follow the SIR vendor's instructions carefully to gather useful data. For instance, many providers of SIR services require that the tank measurements are made to the nearest 1/8-inch. If you fail to follow the vendor's instructions, you may end up with inconclusive test results.

Pastes For Finding Fuel Or Water

If you use a gauge stick, you can improve the quality of your readings if you use a fuel-sensitive paste smeared over about six inches of the stick where you expect the fuel level to be. The paste changes color where it comes into contact with the fuel.

Similarly, you can use a water-sensitive paste on the end of the stick to monitor for the presence of water in the bottom of the tank. While water in the tank can come with your deliveries or as a result of condensation of moisture inside the tank, it can also come from groundwater leaking in through holes or through loose fittings high in your tank.

Tank Chart

The strapping chart used to convert stick measurements into gallons must be the right one for the tank. The chart should have stick measurements listed to 1/8 of an inch to minimize math errors that occur when using charts marked off to the nearest inch. SIR vendors can quickly determine if the chart is inappropriate to your tank, and will often generate a proper one for your tank.

Calibrated Dispensing Meters

A poorly calibrated totalizer can produce bad data that may be mistaken for some types of releases. While many SIR vendors can identify this pattern as a possible cause of a *FAIL*, it is wise to avoid the problem entirely. Keep your dispensers in good operating condition and have them periodically recalibrated as recommended by your equipment manufacturer and as required by state and local weights and measures agencies.

Forms

The SIR vendor typically provides forms on which daily stick readings, sales, and deliveries are recorded. These forms often resemble the inventory sheets usually maintained at UST facilities. In some instances, SIR vendors may allow submission of the data on a facility's own inventory sheets. Some vendors may also permit submission of data in electronic format, such as computer spreadsheets.

SIR Reporting And Recordkeeping

What You Should Provide To The Vendor

Although SIR vendors may ask for a variety of information, some of the more common elements include:

- Tank size (capacity, diameter, and length).
- Tank type, material of construction, and manufacturer.
- Product type.
- Date each stick measurement was taken.
- Daily opening stick measurement and volume.
- Daily closing stick measurement and volume.
- Daily sales volume.
- Gross deliveries over the course of the month.
- Thirty days of observations.

What The Vendor Should Provide To You

Vendors supply different levels of service to their clients. You will need to consult with individual vendors to find the collection of features you desire. However, there is a core of reporting elements that should be common to all SIR analyses (see sample on page 6). These include:

- Clear and timely reporting of results in terms of *PASS*, *FAIL*, or *INCONCLUSIVE*.
- Complete and annotated copies of inventory records used in the analysis, showing such problems as errors in delivery records or bad measurements tossed out by the test.
- Suggestions as to the likely cause of any test failure or inconclusive result.
- Instructions on follow-up actions to be taken in the event of a *FAIL* or *INCONCLUSIVE* (for example: "Notify your local UST agency of a failed test result within 24 hours").

Also, in the case of quantitative testing methods, the form should report the calculated leak rate in gallons per hour and the leak threshold at which a leak would be declared based on the data provided for each tank. The minimum detectable leak rate (MDL) for your data may also be provided by some vendors. (See page 9.)

Your SIR vendor may also supply you with other useful information and services beyond the basics itemized above. SIR vendors may further provide:

- Off-site storage of leak detection records.
- Potential reasons for a *FAIL* other than a release of product:
  - Apparent product theft
  - Missed product delivery entry
  - Suspected totalizer miscalibration
- Potential reasons and possible solutions for any *INCONCLUSIVE* results.
- Possible location of leak within the system.
- Assessment of tank sticking practices.
- Special tank-specific strapping charts for those tanks needing them (such as tilted tanks and odd-sized tanks).

What You Should Keep On File

The minimal recordkeeping requirements for facilities using SIR are the same as for other release detection methods:

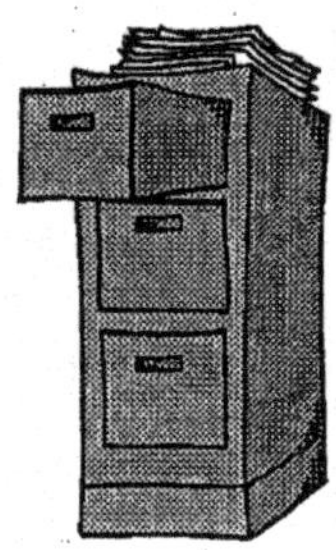

- All written performance claims pertaining to the SIR method used and the manner in which those claims were justified or tested by the vendor (such as a third-party evaluation of the method) must be kept on file for five years from the date you started using the method at the facility.
- The monthly SIR reports, along with the results of any other sampling, testing, or monitoring, must be kept for at least one year.

Sample Cover Sheet Of A Quantitative SIR Report

Monthly Statistical Inventory Reconciliation Report

Sammy's SIR Services
12739 N. South St.
Nowhere, USA

For the Period of: October 1-31, 1995

Facility: Joe's Gas & Grill
123 E. West Street
Anytown, XX

Simple, concise description of system status.

October 1995

Tank ID	Product	System Status	Measured Leak Rate	Threshold	MDL
001	Premium Unleaded	PASS	0.037	0.09	0.18
002[1]	Regular Unleaded	FAIL!	0.735	0.13	0.26
003[2]	Diesel	INCONCLUSIVE!	0.120	0.17	0.34

(All rates are in gallons per hour)

[1] Tank number 002 has an unexplained loss of product. You must notify your local underground storage tank agency of the failed test within 24 hours and take appropriate steps to confirm or refute.

[2] Tank number 003 could not be analyzed to the necessary performance levels. The tank has failed to meet federal leak detection requirements for the period in question. Improved tank sticking methods may help. See report detail for more information.

What to do if problems are discovered.

Detailed copies of inventory records included.

Copies of your analyzed inventory records accompany this report.

Keep this report on file for at least one year.

Questions? Call xxx-xxx-xxxx.

Number to call for help.

- Records of equipment calibration and maintenance must be kept for at least one year. Any schedules of required calibration and maintenance provided by the SIR vendor must be kept for five years from the date you began using the method at the facility.

You should check with your local UST agency to determine if there are any additional recordkeeping requirements.

What To Do When You Get A '*FAIL*'

When your UST system fails a SIR monthly analysis, you must report the incident to your local UST program agency within 24 hours or whatever time period your local agency requires.

At the same time, you need to begin to investigate the cause of the failed test. Within seven days, you must determine the cause of the *FAIL* and report back to your local agency. Your SIR vendor may, on the basis of the test results, be able to provide you with areas to examine, such as a miscalibrated totalizer. You must have any defective equipment repaired or replaced immediately.

If the *FAIL* cannot be linked to equipment problems, you must have the system tightness tested or the site checked for evidence of a release (such as sampling in the excavation zone). You must report the results to the local agency. If a release is confirmed, the agency will provide instructions for any necessary cleanup action.

What To Do When You Get An '*INCONCLUSIVE*'

An *INCONCLUSIVE* means you have failed to meet leak detection requirements. However, the steps you must take upon getting an *INCONCLUSIVE* depend on the requirements of your local UST agency. In some instances, you may be required to perform a system tightness test to be sure the UST is not leaking. In others, you may be given an additional month to come into compliance. Be sure to know what is required locally. A list of UST agency phone numbers can be found in the back of this booklet.

An 'INCONCLUSIVE' means that you effectively have no leak detection for that month.

An *INCONCLUSIVE* should in no way be taken as demonstrating the failings of a given vendor's method—it is inherent to *all* methods. Even if vendors use terms other than "inconclusive," they represent the same condition.

In all cases, you will want to double check your operating procedures to see what caused the *INCONCLUSIVE* and prevent its recurrence. Your SIR vendor will provide assistance in locating the problem and offer suggestions to improve your data collection.

Answers To Frequently Asked Questions

"Can SIR be used on manifolded tanks?"

SIR methods can be used on tank systems that have multiple tanks linked together by siphon bars. This generally requires that each tank in the manifolded system be individually stuck for inventory measurements. As with single tank systems, no product deliveries or sales should be made during the time the sticking and totalizer readings are taking place.

Check with your local UST agency to determine if it permits use of SIR on manifolded systems or has additional requirements.

"Can SIR be used as an annual tightness test?"

For facilities that are still using inventory control with tightness testing as their means of leak detection, it is possible to use SIR in place of more traditional tightness tests such as an overfill test. The performance requirements for a tightness test are more stringent than for monthly monitoring methods, however, so be sure to check that your SIR vendor can meet those requirements. Tanks must be tested for releases of 0.1 gph with a probability of detection (P_D) of 95% and a probability of false alarm (P_{FA}) of 5%. To act as a replacement for piping tightness testing, the requirements are even more rigorous—the SIR method must be able to detect releases of 0.08 gph with a P_D of 95% and a P_{FA} of 5%. To find releases of this magnitude, SIR vendors often need several months of good data.

Be sure to contact your local UST agency to see if it allows use of SIR as an annual test. Also, remember that inventory control with tightness testing can only be used for a limited time. You may want to consider moving now to an approved method of monthly monitoring, such as automatic tank gauge systems, monitoring wells, or monthly SIR analyses.

"Why did a SIR vendor fail my tank for a leak under 0.2 gph?"

> ***There is no such thing as an "acceptable" leak.* Any *leak will cost you in the long run and should be fixed.***

First of all, it is a misconception that *any* leakage into the environment is acceptable. Even small leaks over long periods of time can result in extensive contamination that can cost you substantial time and money for soil and groundwater clean up.

Secondly, the performance standard by which leak detection methods (including SIR) are measured says that leaks of 0.2 gph must be detected in 95 out of 100 times. Further, false alarms should not happen more than five times in a hundred. What this means is that the SIR vendor looks at the estimated leak rate determined for a tank—say 0.15 gph—and asks the question "What is the likelihood that the *true* leak rate is actually 0.2 gph?" On the basis of a statistical analysis of the data you provide the vendor, the SIR vendor can make the call as to whether your system tests tight or not.

Typically, a *FAIL* will be called for apparent releases of around 0.1 gph. See the question on page 10 on 'estimated leak,' 'threshold,' and 'MDL' for additional information.

"What is the difference between 'qualitative' and 'quantitative' SIR methods?"

Although there are many methods that are employed by vendors performing SIR analyses, they break down into two major classifications: *qualitative* and *quantitative*.

NOTE: EPA neither certifies nor approves any leak detection vendor or method. Method evaluations are generally done by third-party testing organizations.

Qualitative methods do not provide estimated leak rates. When a vendor's qualitative method is evaluated to demonstrate its capability of meeting the EPA performance standard, it simply reports results in terms of *PASS*, *FAIL*, or *INCONCLUSIVE*. These results are compared with the evaluator's knowledge of which tanks are leaking in a test set of tank records.

Quantitative methods also categorize results in terms of *PASS*, *FAIL*, or *INCONCLUSIVE*, but they go further by actually providing a numerical estimate of the leak rate, typically in gallons per hour. In evaluating the performance of the method, the evaluator compares the method's estimates with the actual leak rates imposed on the test set of tank records.

"What is this 'estimated leak rate,' 'threshold,' and 'MDL' stuff all about?"

These are rather technical statistical terms often used by quantitative SIR vendors to provide their clients with more detailed information on their analyses. They provide insight beyond the simple *PASS*, *FAIL*, and *INCONCLUSIVE*, including just how bad a leak appears to be (estimated leak rate) and how good the data are that you have been providing to the vendor for analysis (MDL).

SIR vendors offer a diverse number of supplemental reporting options. Look over what each has to offer and choose the one that best meets your needs.

The ***estimated leak rate*** is the number a quantitative SIR method comes up with for the amount of product your tank appears to be losing. The number is usually expressed in gallons per hour since the EPA regulations use those units.

This estimated leak rate is rarely, if ever, zero. All tanks, whether leaking or tight, will generally show a leak rate. The question is, is this leak rate significant? This is where the threshold comes in.

The ***threshold*** is basically an action level leak rate. That is, if the estimated leak rate exceeds the threshold leak rate, the SIR vendor declares a *FAIL*. It is important to note that the threshold is *not* a fixed number, such as 0.1 gph. Instead, it is typically the value associated with a fixed percentage set to the probability of false alarms (that is, declaring a leak on a system that is actually tight) the SIR vendor is willing to accept. EPA's regulations allow no more than 5% of analyses to turn out to be false alarms. However, many SIR vendors consider one failure in twenty analyses to be too high and set their thresholds to a 1% probability of false alarm.

Finally, the ***MDL*** is the ***Minimum Detectable Leak***. The MDL is the smallest leak rate the vendor can determine for the data provided with a P_D of 95% or better. The MDL is tied to the threshold and is usually twice the threshold leak rate. The MDL must be less than or equal to the EPA performance standard rate of 0.2 gph at a P_D of 95% and a P_{FA} of 5% in order to make a *PASS/FAIL* call. If the MDL exceeds the performance standard, your system cannot be given a *PASS*—an *INCONCLUSIVE* is the best you can get.

Fortunately, most vendors who provide this level of detail often provide a "plain English" translation as well.

"Can SIR be used as a monthly test of my piping, too?"

NOTE: State programs may require a different form of testing for lines. Check with your local UST agency.

Yes. SIR is a test of the entire UST system. Losses are reported regardless of their origins. So, whether you are losing product as a result of a tank leak, a line leak, miscalibrated equipment, or theft, a *FAIL* will result if the estimated leak rate exceeds the threshold for calling a leak. Remember, though, that if you are using pressurized lines, you will also need to have an automatic flow restrictor, shutoff device, or continuous alarm in place to fully meet piping leak detection requirements.

"How much does SIR cost?"

Unlike most other methods, SIR has no installation costs and equipment costs are minimal—a well-calibrated dispensing meter and a good stick are about all you need. While vendor costs will vary, monthly monitoring for a facility with three USTs costs about $800 to $1200 per year. SIR used as an annual tightness test costs about $200 to $600. (These figures are based on estimates in 1995.)

"There are so many vendors. How do I choose?"

Whether you have decided to invest in SIR services or other leak detection methods, the basic steps are similar:

- Request information from the vendors you are interested in. Compare their services, option packages, and prices to see which vendors best meet your needs. Ask for references and check them.
- Contact your local UST agency to see if it has a certification program for leak detection vendors. Consult the agency's list of certified service providers.
- Contact the Better Business Bureau to see if there have been any complaints lodged against the vendor.

Publications And Videos About USTs

PUBLICATIONS

TITLE	AVAILABLE FREE FROM
Musts For USTs: A Summary Of The Federal Regulations For Underground Storage Tank Systems Booklet clearly summarizes federal UST requirements for installation, release detection, spill, overfill, and corrosion protection, corrective action, closure, reporting and recordkeeping. (About 40 pages.)	You can call EPA's toll-free RCRA/Superfund Hotline at 800 424-9346 and order free copies. Just identify the titles you want. Or you can write and ask for titles by addressing your requests to: NCEPI Box 42419 Cincinnati, OH 45242 Or you can fax your order to NCEPI at 513 891-6685.
Normas Y Procedimientos Para T.S.A. Spanish translation of **Musts For USTs.** (About 40 pages.)	
Straight Talk On Tanks: Leak Detection Methods For Petroleum Underground Storage Tanks Booklet explains federal regulatory requirements for leak detection and briefly describes allowable leak detection methods. (About 30 pages.)	
Doing Inventory Control Right: For Underground Storage Tanks Booklet describes how owners and operators of USTs can use inventory control and periodic tightness testing to meet federal leak detection requirements. Contains reporting forms. (About 16 pages.)	
Manual Tank Gauging: For Small Underground Storage Tanks Booklet provides simple, step-by-step directions for conducting manual tank gauging for tanks 2,000 gallons or smaller. Contains reporting forms. (About 12 pages.)	
Don't Wait Until 1998: Spill, Overfill, And Corrosion Protection For Underground Storage Tanks Information to help owners and operators of USTs meet the 1998 deadline for compliance with requirements to upgrade, replace, or close USTs installed before December 1988. (About 16 pages.)	
Dollars And Sense: Financial Responsibility Requirements For Underground Storage Tanks Booklet clearly summarizes the "financial responsibility" required of UST owners and operators by federal UST regulations. (About 16 pages.)	
An Overview Of Underground Storage Tank Remediation Options Fact sheets provide information about technologies that can be used to remediate petroleum contamination in soil and groundwater. (About 26 pages.)	
Controlling UST Cleanup Costs Fact sheet series on the cleanup process includes: *Hiring a Contractor, Negotiating the Contract, Interpreting the Bill, Managing the Process, and Understanding Contractor Code Words.* (About 10 pages.)	
Federal Register Reprints Not simple summaries, these reprints are extensive records of the rulemaking process including technical information, explanatory preambles, and the rules as they appear in the Code of Federal Regulations. Reprints dated 9/23/88; 10/26/88; 11/9/89; 5/2/90; and 2/18/93. Over 300 pages.	

Publications And Videos About USTs

PUBLICATIONS

TITLE — **AVAILABLE FROM**

Doing It Right
Illustrates proper installation of underground tanks and piping for installation crews. Part 1: Tanks (24 minutes); Part 2: Piping (16 minutes). Cost: $25

Doing It Right II: Installing Required UST Equipment
Illustrates installation of spill and overfill equipment, observation wells, and piping leak detection (23 minutes). Cost: $60

Doing It Right and **Doing It Right II** Set Cost: $75

Keeping It Clean: Making Safe And Spill-Free Motor Fuel Deliveries
Making pollution-free deliveries to USTs. Includes Stage 1 vapor recovery, overfill prevention and spill containment. For fuel tanker drivers and UST owner/operators (25 minutes). Cost: $60

Petroleum Leaks Underground
How liquids and vapors move in the subsurface and why early response to leaked petroleum is so important. Part 1: How Liquids Move (14 minutes);
Part 2: How Vapors Move (15 minutes). Cost: $75

Straight Talk On Leak Detection
Overview of the leak detection methods available for complying with federal regulations. Part 1: Straight Talk From Tank Owners (owners address the problems of UST compliance [5 minutes]); Part 2: Straight Talk On Leak Detection (30 minutes). Cost: $40

Available from:
Environmental Media Center
Box 30212
Bethesda, MD 20814
301 654-7141
800 522-0362
Visa and MasterCard accepted

Tank Closure Without Tears: An Inspector's Safety Guide
Focuses on explosive vapors and safe tank removal (30 minutes). Video and Booklet Cost: $35; Booklet: $5

What Do We Have Here?: An Inspector's Guide To Site Assessment At Tank Closure
Inspecting sites for contamination where tanks have been removed.
Part 1: Site Assessment Overview (30 minutes); Part 2: Field Testing Instruments At A Glance (14 minutes); Part 3: Soil And Water Sampling At A Glance (7 minutes). Video and Booklet Cost: $45; Booklet: $5

Searching For The Honest Tank: A Guide To UST Facility Compliance Inspection
Covers major steps of UST inspections from protocols and equipment to enforcement and followup; from cathodic protection to leak detection. Directed at inspectors, yet also helpful to owners and operators (30 minutes). Video and Booklet Cost: $40; Booklet: $5

Available from:
New England Interstate
Environmental Training Center
ATTN:VIDEOS
2 Fort Road
South Portland, ME 04106
207 767-2539

☆U.S. GOVERNMENT PRINTING OFFICE: 1995- 624-581 - 82353

State UST Program Office Numbers

Alabama Dept. of Env. Management
334 271-7986

Alaska Dept. of Env. Conservation
907 465-5203

Arizona Dept. of Env. Quality
602 207-4324

Arkansas Dept. of Pollution Control & Ecology
501 570-2801

California State Water Resources Control Board
916 227-4313

Colorado State Oil Inspection Office
303 620-4300

Connecticut Dept. of Env. Protection
203 424-3374

DC Env. Regulatory Administration
202 645-6080

Delaware Dept. of Natural Resources & Env. Control
302 323-4588

Florida Dept. of Env. Regulation
904 488-3935

Georgia Dept. of Natural Resources
404 362-2687

Hawaii Dept. of Health
808 586-4226

Idaho Dept. of Health & Welfare
208 334-0542

Illinois Office of State Fire Marshall
217 785-5878

Indiana Dept. of Env. Mgt.
317 233-6418

Iowa Dept. of Natural Resources
515 281-8135

Kansas Dept. of Health & Env.
913 296-1678

Kentucky Div. of Waste Mgt.
502 564-6716

Louisiana Dept. of Env. Quality
504 765-0243

Maine Dept. of Env. Protection
207 287-2651

Maryland Dept. of Env.
410 631-3442

Massachusetts Dept. of Public Safety
617 351-6000

Michigan Dept. of Natural Resources
517 373-8168

Minnesota Pollution Control Agency
612 297-8609

Mississippi Dept. of Env. Quality
601 961-5171

Missouri Dept. of Natural Resources
314 751-7428

Montana Dept. of Health & Env. Sciences
406 444-5970

Nebraska State Fire Marshal
402 471-9465

Nevada Dept. of Conserv. &Natural Resources
702 687-5872

New Hampshire Dept. of Env. Services
603 271-3644

New Jersey Dept. of Env. Protection
609 984-3156

New Mexico Env. Dept.
505 827-0188

New York Dept. of Env. Conservation
518 457-4351

North Carolina Pollution Control Branch
919 733-8486

North Dakota Div. of Waste Mgt.
701 328-5166

Ohio Dept. of Commerce
614 752-7938

Oklahoma Corporation Commission
405 521-3107

Oregon Dept. of Env. Quality
503 229-5774

Oregon Dept. of Env. Quality
503 229-6642

Pennsylvania Dept. of Env. Resources
717 772-5599

Rhode Island Dept. of Env. Mgt.
401 277-2234

South Carolina Dept. of Health and Env.Control
803 734-5335

South Dakota Dept. of Env. & Nat. Resources
605 773-3296

Tennessee Dept. of Env. & Conservation
615 532-0945

Texas Natural Resources Conservation Comm.
512 239-2000

Utah Dept. of Env. Quality
801 536-4100

Vermont Dept. of Natural Resources
802 244-8702

Virginia Dept. Env. Quality
804 527-5189

Virginia Dept. Env. Quality
804 527-5189

Washington Dept. of Ecology
206 407-7211

West Virginia Div. of Env. Protection
304 558-6371

Wisconsin Dept. of Industry, Labor & Human Relations
608 267-7605

Wisconsin Dept. of Natural Resources
608 267-7560

Wyoming Dept. of Env. Quality
307 777-7096

U.S. TERRITORIES

American Samoa Env. Protection Agency
684 633-2304

Commonwealth of Northern Mariana Islands Div. of Env. Quality
607 234-6984

Guam Env. Protection Agency
671 646-8863

Puerto Rico Env. Quality Board
809 767-8109

Virgin Islands Div. of Env. Protection
809 774-3320

■ APPENDIX E.6 Leak Detection Methods for Petroleum Underground Storage Tanks and Piping

United States Environmental Protection Agency
Solid Waste And Emergency Response 5403W
EPA 510-K-95-003
July 1995

EPA

Straight Talk On Tanks

Leak Detection Methods For Petroleum Underground Storage Tanks And Piping

Contents

Do You Have Questions About Leak Detection?

As an owner or operator of underground storage tanks (USTs) storing petroleum:

- Do you understand the basic leak detection requirements for USTs?
- Are you confused about choosing the most appropriate leak detection method for your UST?

If your USTs do not meet the leak detection requirements described in this booklet, you can be cited for violations and fined.

These are important questions, because your UST and its underground piping must have leak detection ***NOW***.

This booklet begins with an overview of the regulatory requirements for leak detection. Each following section focuses on one leak detection method or the special requirements for piping.

In this booklet, you will find answers to many basic questions about how leak detection methods work and which methods are most appropriate for your UST site.

Why is leak detection so important?

As of April 1995, over 287,000 confirmed UST leaks had been reported. At sites without leak detection, leaks were discovered late, after contamination had spread, requiring difficult and costly cleanups.

By contrast, if you have effective leak detection, you can respond quickly to signs of leaks. You can minimize the extent of environmental damage and the threat to human health and safety. Early action on your part also protects you from the high costs that can result from cleaning up extensive leaks and responding to third-party liability claims.

If you need an overview of all the federal requirements for USTs, please refer to **Musts For USTs**, a booklet developed by the U.S. Environmental Protection Agency (EPA). You can order a free copy of this booklet by calling EPA's toll-free Hotline at 800 424-9346 and asking for **Musts For USTs**.

An Overview Of Leak Detection Requirements

All new USTs (those installed after December 1988) must have leak detection when they are installed.

USTs installed before December 1988 (called "existing USTs") had compliance deadlines for leak detection phased in over 5 years. **By December 1993, *all "existing USTs"* had to have leak detection.**

EPA has identified the following methods that owners and operators may use to meet the federal leak detection requirements:

- Secondary Containment With Interstitial Monitoring
- Automatic Tank Gauging Systems
- Vapor Monitoring
- Groundwater Monitoring
- Statistical Inventory Reconciliation
- Other Methods Approved By The Regulatory Authority

The leak detection methods noted above are all ***monthly monitoring methods*** and eventually everyone must use at least one of them. However, as a ***temporary*** method (until 1998 or for 10 years after installing or upgrading an UST), you can combine tank tightness testing with inventory control (or with manual tank gauging if you have a small tank), as explained on page 4.

State or local regulations may differ from the federal requirements, so find out which requirements apply to your UST. See the list of state UST contacts starting on page 28.

Brief descriptions of leak detection methods appear on the next two pages. More complete descriptions appear in the following sections.

Leak detection requirements for piping differ somewhat from those for tanks. Leak detection methods for piping include secondary containment with interstitial monitoring, vapor monitoring, groundwater monitoring, statistical inventory reconciliation, and tightness testing. Pressurized piping must also have an automatic line leak detector. See pages 22-25 for descriptions of the requirements for suction and pressurized piping.

- **Secondary Containment With Interstitial Monitoring** (see pages 6-7)

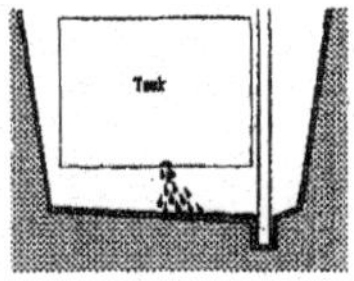

Secondary containment consists of using a barrier, an outer wall, a vault, or a liner around the UST or piping. Leaked product from the inner tank or piping is directed towards an "interstitial" monitor located between the inner tank or piping and the outer barrier. Interstitial monitoring methods range from a simple dipstick to a continuous, automated vapor or liquid sensor permanently installed in the system.

- **Automatic Tank Gauging Systems** (see pages 8-9)

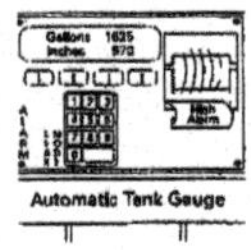

A probe permanently installed in the tank is wired to a monitor to provide information on product level and temperature. These systems automatically calculate the changes in product volume that can indicate a leaking tank.

- **Vapor Monitoring** (see pages 10-11)

Vapor monitoring measures product "fumes" in the soil around the UST to check for a leak. This method requires installation of carefully placed monitoring wells. Vapor monitoring can be performed manually on a periodic basis or continuously using permanently installed equipment.

- **Groundwater Monitoring** (see pages 12-13)

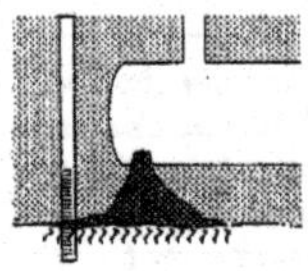

Groundwater monitoring senses the presence of liquid product floating on the groundwater. This method requires installation of monitoring wells at strategic locations in the ground near the tank and along the piping runs. To discover if leaked product has reached groundwater, these wells can be checked periodically by hand or continuously with permanently installed equipment. This method cannot be used at sites where groundwater is more than 20 feet below the surface.

- **Statistical Inventory Reconciliation** (see pages 14-15)

In this method, a trained professional uses sophisticated computer software to conduct a statistical analysis of inventory, delivery, and dispensing data, which you must supply regularly.

- **Other Methods Approved By The Regulatory Authority**

Regulatory authorities can approve any technology that meets a performance standard of detecting a leak of 0.2 gallons per hour with a probability of detection of at least 95 percent and a probability of false alarm of no more than 5 percent.

- **Tank Tightness Testing With Inventory Control** (see pages 16-19)

  This method **combines** periodic tank tightness testing with monthly inventory control. Tightness tests are sophisticated tests performed by trained professionals. You must also use monthly inventory control. Inventory control involves taking daily measurements of tank contents and recording deliveries and amount pumped. Note: this combined method can be used only during the first 10 years following new tank installation or upgrade of your existing UST. After that, you must use monthly monitoring methods. (This combined method can also be used temporarily, until December 1998, with tanks that have not been upgraded with spill, overfill, and corrosion protection.)

- **Manual Tank Gauging** (see pages 20-21)

  Manual tank gauging can be used only for tanks of 2,000 gallons or less capacity. This method requires keeping the tank undisturbed for at least 36 hours each week, during which the contents of the tank are measured twice at the beginning and twice at the end of the test period. At the end of the month, you average the weekly tests and compare the results to the standards shown on page 21 to see if your tank may be leaking. This method can be used by itself only for tanks up to 1,000 gallons. Tanks between 1,001 and 2,000 gallons can use this method only in combination with periodic tank tightness testing. This combined method, however, can be used only during the first 10 years after tank installation or upgrade. (The combined method can also be used temporarily, until December 1998, with tanks that have not been upgraded with spill, overfill, and corrosion protection.)

Make sure the vendor of the leak detection method you use has provided you with evidence that your leak detection meets regulatory requirements for performance standards.

Look For The "Proof" Of A Third-Party Evaluation

An evaluation performed by a third party (someone who is independent of the manufacturer or vendor of the leak detection system) shows that a leak detection system can work as designed. The evaluation follows required evaluation procedures, and often takes place in a laboratory. EPA and third parties have developed evaluation procedures for all leak detection systems.

Although an evaluation and its resulting documentation are technical, you should be familiar with the evaluation's "results" form and, when provided, its "description" form. You should obtain these forms from the leak detection vendor and keep them on file. They contain a signed certification that the system performed as described, as well as documenting any limitations of the system. This information is important to your compliance with the UST requirements. For example, if a tank tightness test was evaluated and certified only for tests taking 2 hours or more, then your UST must be tested for at least 2 hours or it would fail to meet the leak detection requirements. EPA also advises that a leak detection system should not be used on UST systems more than 1.5 times the capacity of the UST used in the evaluation.

Required "Probabilities" For Some Leak Detection

The regulations require not only that leak detection methods be able to detect certain leak rates, but that they also give the correct answer consistently. In general, methods must detect the specified leak rate with a probability of detection of at least 95 percent and a probability of false alarm of no more than 5 percent. Simply stated, this means that, of 100 tests of USTs leaking at the specified rate, at least 95 of them must be correctly detected. It also means that, of 100 tests of non-leaking USTs, no more than 5 can be incorrectly called leaking. This is what is meant by the "probabilities" noted in this booklet.

Which leak detection method is best for you?

Some states maintain lists of those leak detection systems that have been evaluated by third parties. Check with your state to see which systems are allowed in your state. See list of state contacts starting on page 28.

There is no one leak detection system that is best for all sites, nor is there a particular type of leak detection that is consistently the least expensive.

Each leak detection method has unique characteristics. For example, vapor detection devices work rapidly and most effectively in porous soils, while liquid detectors are only appropriate for areas with a high water table.

Identifying the best leak detection choice for your UST depends on a number of factors including cost (both initial installation cost and long-term operation and maintenance cost), facility configuration (such as complexity of piping runs and manifolded tanks), groundwater depth, soil type, seasonal rainfall and temperature ranges, availability of experienced installers, and other variables.

You should look around extensively for experienced, professional vendors and installers of leak detection. Ask questions that help you find the most reliable, cost-effective leak detection for your type of facility. Some possible information sources are: references from fellow UST owners, oil marketers, equipment suppliers, trade journals, trade associations, state and local trade associations (especially those for petroleum marketers and UST owners), and state and local regulatory authorities. (See the list of state UST contacts starting on page 28.) Your state may also have an assistance fund that may be able to help you pay for your UST's leak detection.

For additional information about federal UST requirements, contact EPA's RCRA/Superfund Hotline at 800 424-9346.

Secondary Containment With Interstitial Monitoring

Will you be in compliance?

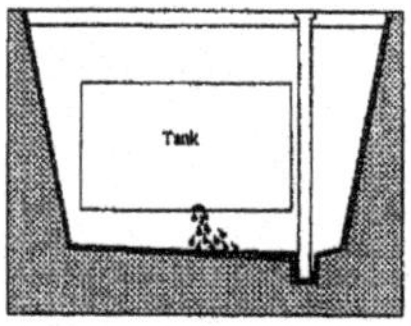

When installed and operated according to the manufacturer's specifications, secondary containment with interstitial monitoring meets the federal leak detection requirements for new and existing USTs. Operation of the monitoring device at least once each month fulfills the requirements for the life of the tank. Secondary containment with interstitial monitoring can also be used to detect leaks from piping (see the section on leak detection for piping starting on page 22).

How does the leak detection method work?

Secondary containment

- Secondary containment provides a barrier between the tank and the environment. The barrier holds the leak between the tank and the barrier so that the leak is detected. The barrier is shaped so that a leak will be directed towards the interstitial monitor.
- Barriers include:
  - Double-walled or "jacketed" tanks, in which an outer wall partially or completely surrounds the primary tank;
  - Concrete vaults, with or without lining;
  - Internally fitted liners ("bladders"); and
  - Leakproof excavation liners that partially or completely surround the tank.
- Clay and other earthen materials cannot be used as barriers.

Interstitial monitors

- Monitors are used to check the area between the tank and the barrier for leaks and alert the operator if a leak is suspected.
- Some monitors indicate the physical presence of the leaked product, either liquid or gaseous. Other monitors check for a change in condition that indicates a hole in the tank, such as a loss of vacuum or a change in the level of a monitoring liquid between the walls of a double-walled tank.
- Monitors can be as simple as a dipstick used at the lowest point of the containment to see if liquid product has leaked and pooled there. Monitors can also be sophisticated automated systems that continuously check for leaks.

What are the regulatory requirements?

- The barrier must be immediately around or beneath the tank.
- The interstitial monitor must be checked at least once every 30 days.
- A double-walled system must be able to detect a release through the inner wall.
- An excavation liner must:
  - Direct a leak towards the monitor;
  - Not allow the specific product being stored to pass through it any faster than 10^{-6} cm/sec;
  - Be compatible with the product stored in the tank;
  - Not interfere with the UST's cathodic protection;
  - Not be disabled by moisture;
  - Always be above the groundwater and the 25-year flood plain; and
  - Have clearly marked and secured monitoring wells, if they are used.

Find out if there are state or local requirements on the use of secondary containment that differ from those described here.

Will it work at your site?

- In areas with high groundwater or a lot of rainfall, it may be necessary to select a secondary containment system that completely surrounds the tank to prevent moisture from interfering with the monitor.

Anything else you should consider?

- This method works effectively only if the barrier and the interstitial monitor are installed correctly. Therefore, trained and experienced installers are necessary.

Automatic Tank Gauging Systems

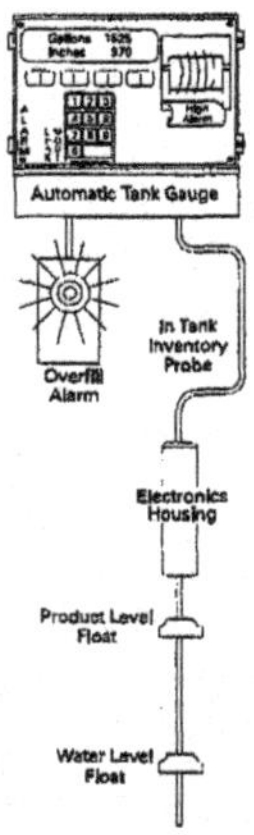

Will you be in compliance?

When installed and operated according to the manufacturer's specifications, automatic tank gauging systems (ATGS) meet the federal leak detection requirements for new and existing USTs. A test performed each month fulfills the requirements for the life of the tank. (For additional requirements for piping, see the section on leak detection for piping starting on page 22.)

How does the leak detection method work?

- The product level and temperature in a tank are measured continuously and automatically analyzed and recorded by a computer.
- In the "inventory mode," the ATGS replaces the use of the gauge stick to measure product level and perform inventory control. This mode records the activities of an in-service tank, including deliveries.
- In the "test mode," the tank is taken out of service and the product level and temperature are measured for at least one hour. Some systems, known as "continuous ATGS," do not require the tank to be taken out of service to perform a test. This is because these systems can gather and analyze data during many short periods when no product is being added to or taken from the tank.

What are the regulatory requirements?

- The ATGS must be able to detect a leak no larger than 0.2 gallons per hour with certain probabilities of detection and of false alarm. Some ATGS can also detect a leak of 0.1 gallons per hour with the required probabilities.

Will it work at your site?

- ATGS have been used primarily on tanks containing gasoline or diesel, with a capacity of up to 15,000 gallons. If considering using an ATGS for larger tanks or products other than gasoline or diesel, discuss its applicability with the manufacturer's representative.
- Water around a tank may hide a leak by temporarily preventing the product from leaving the tank. To detect a leak in this situation, the ATGS should be capable of detecting water in the bottom of a tank.

Anything else you should consider?

- The ATGS probe is permanently installed through an opening (not the fill pipe) on the top of the tank. Each tank at a site must be equipped with a separate probe.
- The ATGS probe is connected to a monitor that displays ongoing product level information and the results of the monthly test. Printers can be connected to the monitor to record this information.
- ATGS are often equipped with alarms for high and low product level, high water level, and theft.
- ATGS can be linked with computers at other locations, from which the system can be programmed or read.
- For ATGS that are not of the "continuous" type, no product should be delivered to the tank or withdrawn from it for at least 6 hours before the monthly test or during the test (which generally takes 1 to 6 hours).
- An ATGS can be programmed to perform a test more often than once per month (a recommended practice).

Find out if there are state or local requirements on the use of ATGS that differ from those described here.

Vapor Monitoring

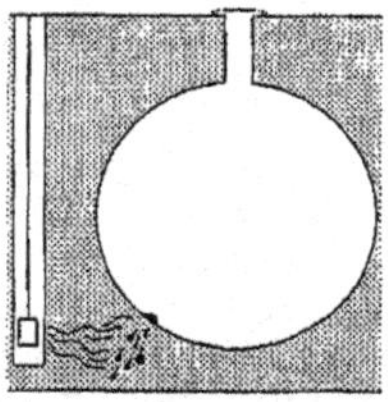

Will you be in compliance?

When installed and operated according to the manufacturer's instructions, vapor monitoring meets the federal leak detection requirements for new and existing USTs. Operation of a vapor monitoring system at least once each month fulfills the requirements for the life of the tank. Vapor monitoring can also be installed to detect leaks from piping (see the section on leak detection for piping starting on page 22).

How does the leak detection method work?

- Vapor monitoring senses or measures "fumes" from leaked product in the soil around the tank to determine if the tank is leaking.

- Fully automated vapor monitoring systems have permanently installed equipment to continuously or periodically gather and analyze vapor samples and respond to a release with a visual or audible alarm.

- Manually operated vapor monitoring systems range from equipment that immediately analyzes a gathered vapor sample to devices that gather a sample that must be sent to a laboratory for analysis. Monitoring results from manual systems are generally less accurate than those from automated systems. Manual systems must be used at least once a month to monitor a site.

- All vapor monitoring devices should be periodically calibrated according to the manufacturer's instructions to ensure that they are properly responding.

- Before installation, a site assessment is necessary to determine the soil type, groundwater depth and flow direction, and the general geology of the site. This can only be done by a trained professional.

- The number of wells and their placement is very important. Only an experienced contractor can properly design and construct an effective monitoring well system. Vapor monitoring requires the installation of monitoring wells within the tank backfill. A minimum of two wells is recommended for a single tank excavation. Three or more wells are recommended for an excavation with two or more tanks. Some state and local agencies have developed regulations for monitoring well placement.

What are the regulatory requirements?

- The UST backfill must be sand, gravel or another material that will allow the vapors to easily move to the monitor.
- The backfill should be clean enough that previous contamination does not interfere with the detection of a current leak.
- The substance stored in the UST must vaporize easily so that the vapor monitor can detect a release.
- High ground water, excessive rain, or other sources of moisture must not interfere with the operation of vapor monitoring for more than 30 consecutive days.
- Monitoring wells must be secured and clearly marked.

Find out if there are state or local requirements on the use of vapor monitoring that differ from those described here.

Will it work at your site?

- Before installing a vapor monitoring system, a site assessment must be done to determine whether vapor monitoring is appropriate at the site. A site assessment usually includes at least a determination of the groundwater level, background contamination, stored product type, and soil type. This assessment can only be done by a trained professional.

Groundwater Monitoring

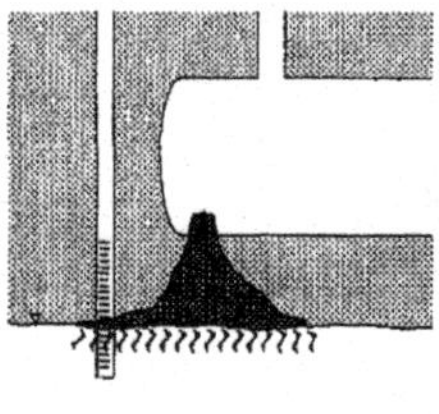

Will you be in compliance?

When installed and operated according to the manufacturer's instructions, a groundwater monitoring system meets the federal leak detection requirements for new and existing USTs. Operation of a groundwater monitoring system at least once each month fulfills the requirements for the life of a tank. Groundwater monitoring can also be used to detect leaks in piping (see the section on leak detection for piping starting on page 22).

How does the leak detection method work?

- Groundwater monitoring involves the use of permanent monitoring wells placed close to the UST. The wells are checked at least monthly for the presence of product that has leaked from the UST and is floating on the groundwater surface.
- The two main components of a groundwater monitoring system are the monitoring well (typically a well of 2-4 inches in diameter) and the monitoring device.
- Detection devices may be permanently installed in the well for automatic, continuous measurements for leaked product.
- Detection devices are also available in manual form. Manual devices range from a bailer (used to collect a liquid sample for visual inspection) to a device that can be inserted into the well to electronically indicate the presence of leaked product. Manual devices must be operated at least once a month.
- Before installation, a site assessment is necessary to determine the soil type, groundwater depth and flow direction, and the general geology of the site. This assessment can only be done by a trained professional.
- The number of wells and their placement is very important. Only an experienced contractor can properly design and construct an effective monitoring well system. A minimum of two wells is recommended for a single tank excavation. Three or more wells are recommended for an excavation with two or more tanks. Some state and local agencies have developed regulations for monitoring well placement.

> ***NOTE: Groundwater monitoring cannot be used at sites where groundwater is more than 20 feet below the surface.***

What are the regulatory requirements?

- Groundwater monitoring can only be used if the stored substance does not easily mix with water and floats on top of water.

- If groundwater monitoring is to be the sole method of leak detection, the groundwater must not be more than 20 feet below the surface, and the soil between the well and the UST must be sand, gravel or other coarse materials.

- Monitoring wells must be properly designed and sealed to keep then from becoming contaminated from outside sources. The wells must also be clearly marked and secured.

- Wells should be placed in the UST backfill so that they can detect a leak as quickly as possible.

- Product detection devices must be able to detect one-eighth inch or less of leaked product on top of the groundwater.

Find out if there are state or local requirements on the use of groundwater monitoring that differ from those described here.

Will it work at your site?

- In general, groundwater monitoring works best at UST sites where:
  - Monitoring wells are installed in the tank backfill; and
  - There are no previous releases of product that would falsely indicate a current release.

- A professionally conducted site assessment is critical for determining these site-specific conditions.

Statistical Inventory Reconciliation

Will you be in compliance?

Statistical inventory reconciliation (SIR), when performed according to the vendor's specifications, meets federal leak detection requirements for new and existing USTs as follows. SIR with a 0.2 gallon per hour leak detection capability meets the federal requirements for monthly monitoring for the life of the tank and piping. SIR with a 0.1 gallon per hour leak detection capability meets the federal requirements as an equivalent to tank tightness testing. SIR can, if it has the capability of detecting even smaller leaks, meet the federal requirements for line tightness testing as well. (For additional requirements for piping, see the section on leak detection for piping starting on page 22.)

How does the leak detection method work?

- SIR analyzes inventory, delivery, and dispensing data collected over a period of time to determine whether or not a tank system is leaking.

- Each operating day, the product level is measured using a gauge stick or other tank level monitor. You also keep complete records of all withdrawals from the UST and all deliveries to the UST. After data have been collected for the period of time required by the SIR vendor, you provide the data to the SIR vendor.

- The SIR vendor uses sophisticated computer software to conduct a statistical analysis of the data to determine whether or not your UST may be leaking. The SIR vendor provides you with a test report of the analysis.

> ***You can order a booklet,*** **Introduction To Statistical Reconciliation,** ***that describes the use of this method. To order this free booklet, call EPA's toll-free Hotline at 800 424-9346***

What are the regulatory requirements?

- To be allowable as monthly monitoring, a SIR method must be able to detect a leak at least as small as 0.2 gallons per hour and meet the federal regulatory requirements regarding probabilities of detection and of false alarm. Data must be submitted at least monthly.

- To be allowable as an equivalent to tank tightness testing, a SIR method must be able to detect a leak at least as small 0.1 gallons per hour and meet the federal regulatory requirements regarding probabilities of detection and of false alarm.

- The individual SIR method must have been evaluated with a test procedure to certify that it can detect leaks at the required level and with the appropriate probabilities of detection and of false alarm.

- If the test report is not conclusive, you must take the steps necessary to find out conclusively whether your tank is leaking. Because SIR requires multiple days of data, you will probably have to use another method.

- You must keep on file both the test reports and the documentation that the SIR method used is certified as valid for your UST system.

Will it work at your site?

- Generally, few product or site restrictions apply to the use of SIR.

- SIR has been used primarily on tanks no more than 18,000 gallons in capacity. If you are considering using a SIR method for larger tanks, discuss its applicability with the vendor.

- Water around a tank may hide a hole in the tank or distort the data to be analyzed by temporarily preventing a leak. To detect a leak in this situation, you must check for water at least once a month.

Find out if there are state or local requirements on the use of statistical inventory reconciliation that differ from those described here.

Anything else you should consider?

- Data, including product level measurements, dispensing data, and delivery data, should all be carefully collected according to the SIR vendor's specifications. Poor data collection produces inconclusive results and noncompliance.

- The SIR vendor will generally provide forms for recording data, a calibrated chart converting liquid level to volume, and detailed instructions on conducting measurements.

- SIR should not be confused with other release detection methods that also rely on periodic reconciliation of inventory, withdrawal, and delivery data. Unlike manual tank gauging or inventory control, SIR uses a sophisticated statistical analysis of data to detect releases. This analysis can only be done by competent, trained practitioners.

Tank Tightness Testing With Inventory Control

Will you be in compliance?

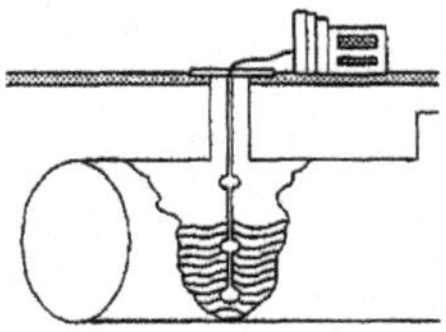

When performed according to the manufacturer's specifications, periodic tank tightness testing combined with monthly inventory control can temporarily (as described below) meet the federal leak detection requirements for new and existing USTs.

These two leak detection methods must be used together, because neither method alone meets the federal requirements for leak detection for tanks. Tightness testing is also an option for underground piping, as described in the section on leak detection for piping starting on page 22.

Because they must be used together, both tank tightness testing and inventory control are discussed in this section. Tank tightness testing is discussed first, followed by inventory control.

Tank Tightness Testing

How does the leak detection method work?

Tightness tests include a wide variety of methods. Other terms used for these methods include "precision," "volumetric," and "nonvolumetric" testing.

- Many tightness test methods are "volumetric" methods in which the change in product level in a tank over several hours is measured very precisely (in milliliters or thousandths of an inch).
- Other methods use acoustics or tracer chemicals to determine the presence of a hole in the tank. With such methods, all of the factors in the following bullets may not apply.
- For most methods, changes in product temperature also must be measured very precisely (thousandths of a degree) at the same time as level measurements, because temperature changes cause volume changes that interfere with finding a leak.
- For most methods, a net decrease in product volume (subtracting out volume changes caused by temperature) over the time of the test indicates a leak.
- The testing equipment is temporarily installed in the tank, usually through the fill pipe.

- The tank must be taken out of service for the test, generally for several hours, depending on the method.
- Many test methods require that the product in the tank be a certain level before testing, which often requires adding product from another tank on-site or purchasing additional product.
- Some tightness test methods require all of the measurements and calculations to be made by hand by the tester. Other tightness test methods are highly automated. After the tester sets up the equipment, a computer controls the measurements and analysis.
- A few methods measure properties of the product that are independent of temperature, such as the mass of the product, and so do not need to measure product temperature.
- Some automatic tank gauging systems are capable of meeting the regulatory requirements for tank tightness testing and can be considered as an equivalent method.

What are the regulatory requirements?

Find out if there are state or local requirements on the use of these methods that differ from those described here.

- The tightness test method must be able to detect a leak at least as small as 0.1 gallon per hour with certain probabilities of detection and of false alarm.
- Tightness tests must be performed periodically. New tanks must be tightness tested every 5 years for 10 years following installation. Upgraded existing tanks must be tightness tested every 5 years for 10 years following upgrade. ("Upgraded" tanks have spill, overfill, and corrosion protection.) Existing tanks that have not been upgraded must be tightness tested every year until 1998 (subsequently these tanks must be upgraded, replaced, or closed).
- After the applicable time period noted above, you must have a monitoring method that can be performed at least once per month. See the other sections of this booklet for allowable monthly monitoring options.

Will it work at your site?

- Tank tightness testing has been used primarily on tanks no more than 15,000 gallons in capacity containing gasoline and diesel. If you are considering using tightness testing for larger tanks or products other than gasoline or diesel, discuss the method's applicability with the manufacturer's representative.

Anything else you should consider?

- For most methods, the test is performed by a testing company. You just observe the test.
- Manifolded tanks generally should be disconnected and tested separately.
- Depending on the method, up to 4 tanks can be tested at one time. Generally, an automated system is necessary to test 3 or 4 tanks at a time.
- Procedure and personnel, not equipment, are usually the most important factors in a successful tightness test. Therefore, well-trained and experienced testers are very important. Some states and local authorities have tester certification programs.

Inventory Control

How does the leak detection method work?

Inventory control requires daily measurements of tank contents and math calculations that let you compare your "stick" inventory (what you've measured) to your "book" inventory (what your recordkeeping indicates you should have). Some people call this process "inventory reconciliation." If the difference between your "stick" and "book" inventory is too large, your tank may be leaking.

EPA has a booklet, **Doing Inventory Control Right**, that fully explains how to do inventory control. The booklet also contains standard reporting forms. You can order this free booklet by calling EPA's toll-free Hotline at 800 424-9346 and asking for **Doing Inventory Control Right**.

- UST inventories are determined daily by using a gauge stick and the data is recorded on a form. The level on the gauge stick is converted to a volume of product in the tank using a calibration chart, which is often furnished by the UST manufacturer.
- The amounts of product delivered to and withdrawn from the UST each day are also recorded. At least once each month, the gauge stick data and the sales and delivery data are reconciled and the month's overage or shortage is determined. If the overage or shortage is greater than or equal to 1.0 percent of the tank's flow-through volume plus 130 gallons of product, the UST may be leaking.

What are the regulatory requirements?

- Inventory control must be used in conjunction with periodic tank tightness tests.
- The gauge stick should be long enough to reach the bottom of the tank and marked so that the product level can be determined to the nearest one-eighth of an inch.
- A monthly measurement must be taken to identify any water at the bottom of the tank.
- Deliveries must be made through a drop tube that extends to within one foot of the tank bottom.
- Product dispensers must be calibrated to the local weights and measures standards.

Will it work at your site?

If your tank is not level, inventory control may need to be modified. You will need to get a corrected tank chart.

Anything else you should consider?

- Inventory control is a practical, commonly used management tool that does not require closing down the tank operation for long periods.
- You can perform inventory control yourself.
- The accuracy of tank gauging can be greatly increased by spreading product-finding paste on the gauge stick before taking measurements (or by using in-tank product level monitoring devices).

Manual Tank Gauging

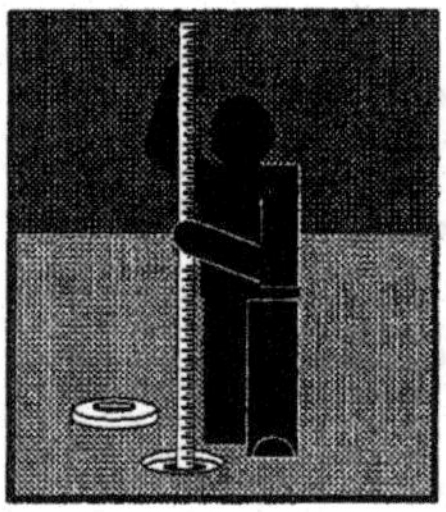

Will you be in compliance?

NOTE: Manual tank gauging can only be used for smaller tanks. Tanks 1,000 gallons or less can use this method alone, but tanks from 1,001-2,000 gallons can only use manual tank gauging when it is combined with tank tightness testing. Manual tank gauging cannot be used for tanks over 2,000 gallons. When performed according to recommended practices, manual tank gauging meets the federal leak detection requirements for USTs with a capacity of 1,000 gallons or less for the life of the tank. (For additional requirements for piping, see the section on leak detection for piping starting on page 22.)

How does the leak detection method work?

EPA has a booklet, **Manual Tank Gauging**, that fully explains how to do manual tank gauging correctly. The booklet also contains standard reporting forms. You can order this free booklet by calling EPA's toll-free Hotline at 800 424-9346 and asking for **Manual Tank Gauging**.

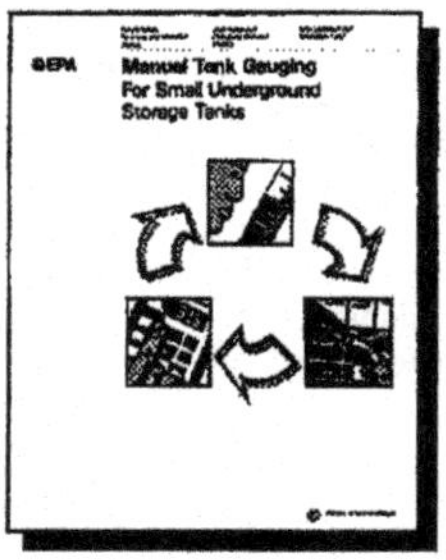

- Four measurements of the tank's contents must be taken weekly, two at the beginning and two at the end of **at least a 36-hour period** during which nothing is added to or removed from the tank. See the table on the next page.

- The average of the two consecutive ending measurements are subtracted from the average of the two beginning measurements to indicate the change in product volume.

- Every week, the calculated change in tank volume is compared to the standards shown in the table on the next page. If the calculated change exceeds the weekly standard, the UST may be leaking. Also, monthly averages of the four weekly test results must be compared to the monthly standard in the same way. See the table on the next page.

What are the regulatory requirements?

- Liquid level measurements must be taken with a gauge stick that is marked to measure the liquid to the nearest one-eighth of an inch.

- Manual tank gauging may be used as the sole method of leak detection for tanks with a capacity of 1,000 gallons or less for the life of the tank. Tanks between 551 and 1,000 gallons have two testing standards based on their diameter (see table). These tanks may use a combination of manual tank gauging and periodic tank tightness for the life of the tank (see table for testing standards).

Table of Test Standards for Manual Tank Gauging

Tank Size	Minimum Duration Of Test	Weekly Standard (1 test)	Monthly Standard (4-test average)
up to 550 gallons	36 hours	10 gallons	5 gallons
551-1,000 gallons (when tank diameter is 64")	44 hours	9 gallons	4 gallons
551-1,000 gallons (when tank diameter is 48")	58 hours	12 gallons	6 gallons
551-1,000 gallons (also requires periodic tank tightness testing)	36 hours	13 gallons	7 gallons
1,001-2,000 gallons (also requires periodic tank tightness testing)	36 hours	26 gallons	13 gallons

- For tanks with a capacity of 1,001-2,000 gallons, manual tank gauging must be combined with periodic tightness testing. New tanks must be tightness tested every 5 years for 10 years following installation. Upgraded existing tanks must be tightness tested every 5 years for 10 years following upgrade. ("Upgraded" tanks have spill, overfill, and corrosion protection.) Existing tanks that have not been upgraded must be tightness tested every year until 1998. See the earlier section on tank tightness testing for details on this method.

- Unless the tank is 1,000 gallons or less, this combined method will meet the federal requirements only ***temporarily*** (as explained above). You must eventually have another monitoring method that can be performed at least once a month. See the other sections of this booklet for allowable monthly monitoring options.

- Tanks greater than 2,000 gallons in capacity may not use this method of leak detection to meet these regulatory requirements.

Find out if there are state or local requirements on the use of manual tank gauging that differ from those described here.

Will it work at your site?

- Manual tank gauging is inexpensive and can be an effective leak detection method when used as described above with tanks of the appropriate size.

Anything else you should consider?

- You can perform manual tank gauging yourself. Correct gauging, recording, and interpretation are the most important factors for successful tank gauging. The accuracy of tank gauging can be greatly increased by spreading product-finding paste on the gauge stick before taking measurements.

Leak Detection For Underground Piping

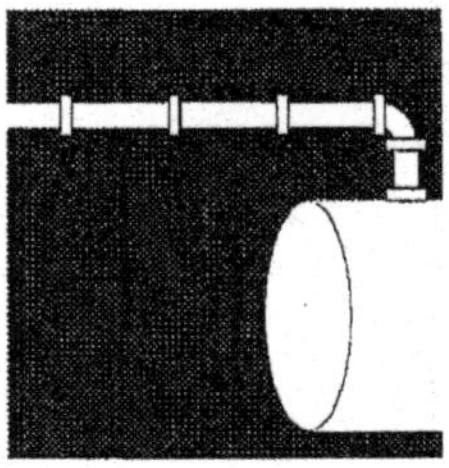

Will you be in compliance?

When installed and operated according to the manufacturer's specifications, the leak detection methods discussed here meet the federal regulatory requirements for the life of new and existing underground piping systems. Your UST may have ***suction*** or ***pressurized*** piping, which are discussed below.

What are the regulatory requirements for* suction *piping?

- No leak detection is required if the suction piping has (1) enough slope so that the product in the pipe can drain back into the tank when suction is released and (2) has only one check valve, which is as close as possible beneath the pump in the dispensing unit. If a suction line is to be considered exempt based on these design elements, there must be some way to check that the line was actually installed according to these plans.

- If a suction line does not meet all of the design criteria noted above, one of the following leak detection methods must be used:

  - A line tightness test at least every 3 years; or

  - Monthly interstitial monitoring; or

  - Monthly vapor monitoring; or

  - Monthly groundwater monitoring; or

  - Monthly statistical inventory reconciliation.

The line tightness test must be able to detect a leak at least as small as 0.1 gallon per hour with certain probabilities of detection and of false alarm.

Interstitial monitoring, vapor monitoring, groundwater monitoring, and statistical inventory reconciliation have the same regulatory requirements for piping as they do for tanks. See the earlier sections of this booklet on those methods.

What are the regulatory requirements for* pressurized *piping?

Each pressurized piping run must have one leak detection method from each set below:

> ***Find out if there are state or local requirements on the use of leak detection methods for piping that differ from those described here.***

An Automatic Line Leak Detector:

- Automatic flow restrictor; or
- Automatic flow shutoff; or
- Continuous alarm system.

And One Other Method:

- Monthly interstitial monitoring; or
- Monthly vapor monitoring; or
- Monthly groundwater monitoring; or
- Monthly statistical inventory reconciliation; or
- Annual tightness test.

- The automatic line leak detector (LLD) must be designed to detect a leak at least as small as 3 gallons per hour at a line pressure of 10 pounds per square inch within 1 hour by shutting off the product flow, restricting the product flow, or triggering an audible or visual alarm.
- The line tightness test must be able to detect a leak at least as small as 0.1 gallon per hour when the line pressure is 1.5 times its normal operating pressure. The test must be conducted each year. If the test is performed at pressures lower than 1.5 times operating pressure, the leak rate to be detected must be correspondingly lower.
- Automatic LLDs and line tightness tests must also be able to meet the federal regulatory requirements regarding probabilities of detection and false alarm.
- Interstitial monitoring, vapor monitoring, groundwater monitoring, and statistical inventory reconciliation have the same regulatory requirements for piping as they do for tanks. See the earlier sections of this booklet on those methods.

How do the leak detection methods work?

Automatic line leak detectors (LLDs)

- Flow restrictors and flow shutoffs can monitor the pressure within the line in a variety of ways: whether the pressure decreases over time; how long it takes for a line to reach operating pressure; and combinations of increases and decreases in pressure.

- If a suspected leak is detected, a *flow restrictor* keeps the product flow through the line well below the usual flow rate. If a suspected leak is detected, a *flow shutoff* completely cuts off product flow in the line or shuts down the pump.

- A *continuous alarm system* constantly monitors line conditions and immediately triggers an audible or visual alarm if a leak is suspected. Automated internal, vapor, or interstitial line monitoring systems can also be set up to operate continuously and sound an alarm, flash a signal on the console, or even ring a telephone in a manager's office when a leak is suspected.

- Both automatic flow restrictors and shutoffs are permanently installed directly into the pipe or the pump housing.

- Vapor and interstitial monitoring systems can be combined with automatic shutoff systems so that whenever the monitor detects a suspected release the piping system is shut down. This would qualify as a continuous alarm system. Such a setup would meet the monthly monitoring requirement as well as the LLD requirement.

Line tightness testing

- Tracer methods do not measure pressure or flow rates of the product. Instead they use a tracer chemical to determine if there is a hole in the line. With tracer methods, all of the factors below may not apply.

- The line is taken out of service and pressurized, usually above the normal operating pressure. A drop in pressure over time, usually an hour or more, suggests a possible leak.

- Suction lines are not pressurized very much during a tightness test (about 7 to 15 pounds per square inch).

- Most line tightness tests are performed by a testing company. You just observe the test.

- Some *tank* tightness test methods can be performed to include a tightness test of the connected piping.

- For most line tightness tests, no permanent equipment is installed.

- In the event of trapped vapor pockets, it may not be possible to conduct a valid line tightness test. There is no way to tell definitely before the test begins if this will be a problem, but long complicated piping runs with many risers and dead ends are more likely to have vapor pockets.

- Some permanently installed electronic systems (such as some ATGS) can meet the requirements of a line tightness test.

Secondary containment with interstitial monitoring

- A barrier is placed between the piping and the environment. Double-walled piping or a leakproof liner in the piping trench can be used.
- A monitor is placed between the piping and the barrier to sense a leak if it occurs. Monitors range from a simple stick that can be put in a sump to see if a liquid is present, to continuous automated systems, such as those that monitor for the presence of liquid product or vapors.
- Proper installation of secondary containment is the most important and the most difficult aspect of this leak detection method. Trained and experienced installers are necessary.
- See the section on secondary containment for additional information. Secondary containment for piping is similar to that for tanks.

Vapor or groundwater monitoring

- Vapor monitoring detects product that leaks into the soil and evaporates.
- Groundwater monitoring checks for leaked product floating on the groundwater near the piping.
- A site assessment must be used to determine monitoring well placement and spacing.
- UST systems using vapor or groundwater monitoring for the tanks are well suited to use the same monitoring method for the piping.
- See the earlier sections on vapor and groundwater monitoring. Use of these methods with piping is similar to that for tanks.

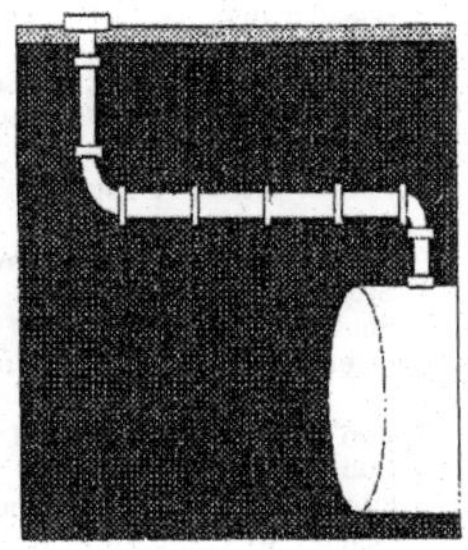

Publications And Videos About USTs

PUBLICATIONS

TITLE	AVAILABLE FREE FROM

Musts For USTs: A Summary Of The Federal Regulations For Underground Storage Tank Systems
Booklet clearly summarizes federal UST requirements for installation, release detection, spill, overfill, and corrosion protection, corrective action, closure, reporting and recordkeeping. (About 40 pages.)

Normas Y Procedimientos Para T.S.A.
Spanish translation of **Musts For USTs**. (About 40 pages.)

Tank Issues: Site Characterization For External Leak Monitoring
Booklet discusses natural and man-made site characteristics that need to be considered in the design of a vapor or groundwater monitoring system. (About 5 pages.)

Tank Issues: Design And Placement Of Floating Liquid Monitoring Wells
Booklet discusses site conditions that affect the performance of liquid product monitors and the construction and placement of monitoring wells. (About 5 pages.)

Doing Inventory Control Right: For Underground Storage Tanks
Booklet describes how owners and operators of USTs can use inventory control and periodic tightness testing to meet federal leak detection requirements. Contains reporting forms. (About 16 pages.)

Manual Tank Gauging: For Small Underground Storage Tanks
Booklet provides simple, step-by-step directions for conducting manual tank gauging for tanks 2,000 gallons or smaller. Contains reporting forms. (About 12 pages.)

Introduction To Statistical Inventory Reconciliation: For Underground Storage Tanks
Booklet describes the use of Statistical Inventory Reconciliation (SIR) to meet federal leak detection requirements. (About 12 pages.)

Don't Wait Until 1998: Spill, Overfill, And Corrosion Protection For Underground Storage Tanks
Information to help owners and operators of USTs meet the 1998 deadline for compliance with requirements to upgrade, replace, or close USTs installed before December 1988. (About 16 pages.)

Dollars And Sense: Financial Responsibility Requirements For Underground Storage Tanks
Booklet clearly summarizes the "financial responsibility" required of UST owners and operators by federal UST regulations. (About 16 pages.)

An Overview Of Underground Storage Tank Remediation Options
Fact sheets provide information about technologies that can be used to remediate petroleum contamination in soil and groundwater. (About 26 pages.)

Controlling UST Cleanup Costs
Fact sheet series on the cleanup process includes: *Hiring a Contractor, Negotiating the Contract, Interpreting the Bill, Managing the Process, and Understanding Contractor Code Words*. (About 10 pages.)

You can call EPA's toll-free RCRA/Superfund Hotline at 800 424-9346 and order free copies. Just identify the titles you want. Or you can write and ask for titles by addressing your requests to:

NCEPI
Box 42419
Cincinnati, OH 45242

Or you can fax your order to NCEPI at 513 891-6685.

Publications And Videos About USTs

VIDEOS

TITLE/COST	AVAILABLE FROM
Doing It Right Illustrates proper installation of underground tanks and piping for installation crews. Part 1: Tanks (24 minutes); Part 2: Piping (16 minutes). Cost: $25 **Doing It Right II: Installing Required UST Equipment** Illustrates installation of spill and overfill equipment, observation wells, and piping leak detection (23 minutes). Cost: $60 **Doing It Right and Doing it Right II** Set Cost: $75 **Keeping It Clean: Making Safe And Spill-Free Motor Fuel Deliveries** Making pollution-free deliveries to USTs. Includes Stage 1 vapor recovery, overfill prevention and spill containment. For fuel tanker drivers and UST owner/operators (25 minutes). Cost: $60 **Petroleum Leaks Underground** How liquids and vapors move in the subsurface and why early response to leaked petroleum is so important. Part 1: How Liquids Move (14 minutes); Part 2: How Vapors Move (15 minutes). Cost: $75 **Straight Talk On Leak Detection** Overview of the leak detection methods available for complying with federal regulations. Part 1: Straight Talk From Tank Owners (owners address the problems of UST compliance [5 minutes]); Part 2: Straight Talk On Leak Detection (30 minutes). Cost: $40	**Environmental Media Center** Box 30212 Bethesda, MD 20814 301 654-7141 800 522-0362 Visa and MasterCard accepted
Tank Closure Without Tears: An Inspector's Safety Guide Focuses on explosive vapors and safe tank removal (30 minutes). Video and Booklet Cost: $35; Booklet: $5 **What Do We Have Here?: An Inspector's Guide To Site Assessment At Tank Closure** Inspecting sites for contamination where tanks have been removed. Part 1: Site Assessment Overview (30 minutes); Part 2: Field Testing Instruments At A Glance (14 minutes); Part 3: Soil And Water Sampling At A Glance (7 minutes). Video and Booklet Cost: $45; Booklet: $5 **Searching For The Honest Tank: A Guide To UST Facility Compliance Inspection** Covers major steps of UST inspections from protocols and equipment to enforcement and followup; from cathodic protection to leak detection. Directed at inspectors, yet also helpful to owners and operators (30 minutes). Video and Booklet Cost: $40; Booklet: $5	**New England Interstate Environmental Training Center** ATTN:VIDEOS 2 Fort Road South Portland, ME 04106 207 767-2539

State Contacts For UST Information

[LUST contacts focus on leaking underground storage tanks.]

Alabama Dept. of Env. Management
Groundwater Section/ Water Division
Box 301463
Montgomery, AL 36130-1463
UST phone: 334 271-7986
LUST phone: 334 271-7834

Alaska Dept. of Env. Conservation
410 Willoughby Ave.
Juneau, AK 99801-1795
phone: 907 465-5203

Arizona Dept. of Env. Quality
3033 N. Central Ave.
Phoenix, AZ 85012
phone: 602 207-4324

Arkansas Dept. of Pollution Control & Ecology
Regulated Storage Tank Division
8001 National Dr.
Little Rock, AR 72209
phone: 501 570-2801

California State Water Resources Control Board
Div. of Clean Water Program
Box 944212
Sacramento, CA 94244-2120
phone: 916 227-4313

[UST contact]
Colorado State Oil Inspection Office
1515 Arapahoe Street
Suite 525, Tower 3
Denver, CO 80202-2117
phone: 303 620-4300

[LUST contact]
Colorado Dept. of Health
Hazardous Materials & Waste Mgt. Div.
UST Program
4300 Cherry Creek Dr. S.
Denver, CO 80220
phone: 303 692-3453

Connecticut Dept. of Env. Protection
UST Program
State Office Bldg.
79 Elm Street
Hartford, CT 06106
phone: 203 424-3374

DC Env. Regulatory Administration
Pesticides, Hazardous Waste & UST Div.
2100 Martin Luther King Ave. S.E., Suite 203
Washington, D.C. 20020
phone: 202 645-6080

Delaware Dept. of Natural Resources & Env. Control
UST Branch
715 Grantham Lane
New Castle, DE 19720
phone: 302 323-4588

Florida Dept. of Env. Regulation
Tank Section
Twin Towers Office Bldg., Rm. 403
2600 Blair Stone Road
Tallahassee, FL 32399-2400
phone: 904 488-3935

Georgia Dept. of Natural Resources
UST Mgt. Program
4244 Intl. Parkway, Suite 100
Atlanta, GA 30354
phone: 404 362-2687

Hawaii Dept. of Health
Solid and Hazardous Waste Branch
919 Ala Moana Blvd., Suite 212
Honolulu, HI 96814
phone: 808 586-4226

Idaho Dept. of Health & Welfare
Div. of Env. Quality
1410 North Hilton
Boise, ID 83706
phone: 208 334-0542

[UST contact]
Illinois Office of State Fire Marshall
Div. of Pet. & Chem. Safety
1035 Stevenson Dr.
Springfield, IL 62703
phone: 217 785-5878

[LUST contact]
Illinois EPA
Div. of Remediation Mgt.
LUST Section
Box 19276
Springfield, IL 62794-9276
phone: 217 782-6760

Indiana Dept. of Env. Mgt.
Office of Env. Response
Box 7015
Indianapolis, IN 46206
phone: 317 233-6418

Iowa Dept. of Natural Resources
UST Section
Wallace State Office Bldg.
900 East Grand
Des Moines, IA 50319
phone: 515 281-8135

Kansas Dept. of Health & Env.
Bureau of Env. Remediation
Storage Tank Section
Forbes Field, Bldg. 740
Topeka, KS 66620
phone: 913 296-1678

Kentucky Div. of Waste Mgt.
UST Branch
14 Reilly Road
Frankfort, KY 40601
phone: 502 564-6716

Louisiana Dept. of Env. Quality
UST Division
Box 82178
Baton Rouge, LA 70810
phone: 504 765-0243

Maine Dept. of Env. Protection
State House, Station 17
Hospital St., Ray Bldg.
Augusta, ME 04333
phone: 207 287-2651

Maryland Dept. of Env.
Waste Mgt. Adm.
Oil Control Program
2500 Broening Highway
Baltimore, MD 21224
phone: 410 631-3442

[UST contact]
Massachusetts Dept. of Public Safety
UST Program
1010 Commonwealth Ave.
Boston, MA 02215
phone: 617-351-6000

[LUST contact]
Massachusetts Dept. of Env. Protection
Bureau of Waste Site Cleanup
One Winter St.
Boston, MA 02108
phone: 617 556-1044

Michigan Dept. of Natural Resources
UST Division
Box 30157
Lansing, MI 48909
UST phone: 517 373-8168
MUSTFA (state fund): 517 373-6247

Minnesota Pollution Control Agency
Tanks and Spills Section
520 Lafayette Road North
St. Paul, MN 55155-3898
phone: 612 297-8609

Mississippi Dept. of Env. Quality
Bureau of Pollution Control, UST Section
Box 10385
Jackson, MS 39289-0385
phone: 601 961-5171

Missouri Dept. of Natural Resources
Water Pollution Control Program (UST)
Environmental Services Program (LUST)
Box 176
Jefferson City, MO 65102-0176
UST phone: 314 751-7428
LUST phone: 314 526-3384

Montana Dept. of Health & Env. Sciences
Solid & Hazardous Waste Bureau
UST Program
Cogswell Bldg.
Helena, MT 59620
phone: 406 444-5970

[UST contact]
Nebraska State Fire Marshal
Flammable Liquid Storage
246 South 14th St.
Lincoln, NE 68508
phone: 402 471-9465

[LUST contact]
Nebraska Dept. of Env. Control
LUST/ER Section
Box 98922
Lincoln, NE 68509-8922
phone: 402 471-4230

*U.S. Government Printing Office: 1995 — 398-029

State Contacts For UST Information

Nevada Dept. of Conserv. &Natural Resources
Div. of Env. Protection
Capitol Complex
333 W. Nye Lane
Carson City, NV 89710
phone: 702 687-5872

New Hampshire Dept. of Env. Services
Oil Compliance Section
Groundwater Protection Bureau
Box 95
Concord, NH 03301
phone: 603 271-3644

New Jersey Dept. of Env. Protection
Responsible Party Site Remediation
401 E. State (CN-028)
Trenton, NJ
08625-0028
phone: 609 984-3156

New Mexico Env. Dept.
UST Bureau
Box 26110
Santa Fe, NM
87502
phone: 505 827-0188

New York Dept. of Env. Conservation
Bulk Storage Section
50 Wolf Rd., Rm. 340
Albany, NY
12233-3750
phone: 518 457-4351

North Carolina Pollution Control Branch
Div. of Env. Mgt.
Dept. of Env. Health & Natural Resources
441 N. Harrington St.
Raleigh, NC
27603-1323
phone: 919 733-8486

North Dakota Div. of Waste Mgt.
Dept. of Health and Consolidated Lab.
Box 5520
Bismarck, ND
58502-5520
phone: 701 328-5166

Ohio Dept. of Commerce
Bureau of UST Regulations
8895 East Main St.
Box 687
Reynoldsburg, OH 43068
phone: 614 752-7938

Oklahoma Corporation Commission
Fuel Storage Division
Box 52000-2000
Oklahoma City, OK 73152-2000
UST phone:
405 521-3107
LUST phone:
405 521-6575

[UST contact]
Oregon Dept. of Env. Quality
UST Compliance Section
811 S.W. Sixth Ave.,
7th Floor
Portland, OR 97204
phone: 503 229-5774

[LUST contact]
Oregon Dept. of Env. Quality
UST Cleanup Program
811 S.W. Sixth Ave.,
9th Floor
Portland, OR 97204
phone: 503 229-6642

Pennsylvania Dept. of Env. Resources
Storage Tank Program
Box 8762
Harrisburg, PA
17101-2301
phone: 717 772-5599

Rhode Island Dept. of Env. Mgt.
UST Section
291 Promenade St.
Providence, RI 02908
phone: 401 277-2234

South Carolina Dept. of Health and Env.Control
Groundwater Protection Division
2600 Bull St.
Columbia, SC 29201
UST phone:
803 734-5335
LUST phone:
803 734-5331

South Dakota Dept. of Env. & Nat. Resources
UST Program
523 East Capitol
Pierre, SD 57501
phone: 605 773-3296

Tennessee Dept. of Env. & Conservation
Div. of USTs
4th Floor, L&C Tower
401 Church St.
Nashville, TN
37243-1541
phone: 615 532-0945

Texas Natural Resources Conservation Comm.
Petroleum Storage Tank Div.
Box 13087
Austin, TX 78711-3087
phone: 512 239-2000

Utah Dept. of Env. Quality
Div. of Env. Response and Remediation
UST Branch
168 North 1950 West,
1st Floor
Salt Lake City, UT 84116
phone: 801 536-4100

Vermont Dept. of Natural Resources
UST Program
103 South Main St.,
West Bldg.
Waterbury, VT 05676
phone: 802 244-8702

[UST contact]
Virginia Dept. Env. Quality
Response Program
4951 Cox Road
Glen Allen, VA 23060
phone: 804 527-5189

[LUST contact]
Virginia Dept. Env. Quality
Groundwater Mgt. Program
4951 Cox Road
Glen Allen, VA 23060
phone: 804 527-5189

Washington Dept. of Ecology
Box 47655
Olympia, WA
98504-7655
UST phone:
206 407-7211
LUST phone:
206 407-7218

West Virginia Div. of Env. Protection
Office of Waste Mgt.
UST Section
1356 Hansford St.
Charleston, WV 25301
phone: 304 558-6371

[UST contact]
Wisconsin Dept. of Industry, Labor & Human Relations
Division of Safety and Buildings
Box 7969
Madison, WI 53707
phone: 608 267-7605

[LUST contact]
Wisconsin Dept. of Natural Resources
Bureau of Solid and Hazardous Waste Mgt.
Box 7921
Madison, WI
53707-7921
phone: 608 267-7560

Wyoming Dept. of Env. Quality
Water Quality Div.
Herschler Bldg.,
4th Floor
122 West 25th St.
Cheyenne, WY 82002
phone: 307 777-7096

U.S. TERRITORIES

American Samoa Env. Protection Agency
Office of the Governor
American Samoa Government
ATTN: UST Program
Pago Pago, American Samoa 96799
phone: 684 633-2304

Commonwealth of Northern Mariana Islands Div. of Env. Quality
Box 1304
Dr. Torres Hospital
Saipan, MP 96950
phone: 607 234-6984

Guam Env. Protection Agency
IT&E
Harmon Plaza,
Complex Unit D-107
130 Rojas St.
Harmon, Guam 96911
phone: 671 646-8863

Puerto Rico Env. Quality Board
Box 11488
Commonwealth of Puerto Rico
Santurce, Puerto Rico
00910
phone: 809 767-8109

Virgin Islands Div. of Env. Protection
Dept. of Planning and Natural Resources
Nisky Center 231
45A Estate Nisky
Charlotte Amalie
St. Thomas, Virgin Islands 00802
phone: 809 774-3320

■ APPENDIX E.7 Financial Responsibility Requirements for Underground Storage Tanks

United States Environmental Protection Agency | Solid Waste And Emergency Response 5403W | EPA 510-K-95-004 July 1995

EPA

Dollars And Sense

Financial Responsibility Requirements For Underground Storage Tanks

CONTENTS

Why are there financial responsibility requirements?

When Congress amended Subtitle I of the Resource Conservation and Recovery Act in 1986, it directed the U.S. Environmental Protection Agency (EPA) to develop financial responsibility regulations for owners and operators of underground storage tanks (USTs) storing petroleum.

This booklet briefly describes federal financial responsibility requirements for underground storage tanks (USTs) storing petroleum.

EPA estimates that there are about 1.2 million federally regulated petroleum USTs buried at over 500,000 sites nationwide. These sites are owned by marketers who sell gasoline to the public (such as service stations and convenience stores) and nonmarketers who use USTs solely for their own needs (such as fleet service operators and local governments).

Many of these USTs have released or will release petroleum into the environment through spills, overfills, or failures in the tank and piping system. EPA estimates that the total number of confirmed releases could reach 400,000 in the next several years. After this peak, EPA expects fewer releases as owners of USTs comply with federal and state requirements for new USTs and upgrade older USTs with spill, overfill, and corrosion protection.

Cleaning up these leaks can be costly. Congress wanted owners and operators of USTs to demonstrate that they have the financial resources--through insurance or other means explained below--to pay for the costs of corrective action and third-party liability that can result from leaking USTs.

The financial responsibility requirements are designed to make sure that someone can pay the costs of cleaning up leaks and compensating third-parties for bodily injury and property damage caused by leaking USTs.

If you need an overview of all the federal requirements for USTs, please refer to EPA's free booklet, **Musts For USTs**. Ordering information for **Musts For USTs** and several other publications and videos about USTs appears on pages 13-14.

Why should your USTs meet these requirements?

It is particularly important that someone be prepared to pay cleanup costs so that cleanup activities can begin as quickly as possible. Without fast action at contaminated sites, contamination can spread and increase significantly the chance of damaging the environment and human health.

Complying with the financial responsibility requirements also protects ***you***, as an owner or operator of USTs. If your UST leaks, you may be faced with high cleanup costs or with lawsuits brought by third parties. You will be able to pay these costs if you have met the financial responsibility requirements.

The financial responsibility requirements are found in the Code of Federal Regulations, 40 CFR Part 280.

Also, if you cannot demonstrate financial responsibility by the compliance deadline that applies to your UST facility, you can be cited for violations and fined.

Who needs to demonstrate financial responsibility?

Either the owner or the operator of the UST must demonstrate financial responsibility, if the owner and operator are different individuals or firms. It is the responsibility of the owner and operator to decide which one will demonstrate financial responsibility.

Federal and state governments and their agencies that own USTs are not required to demonstrate financial responsibility. Local governments, however, must comply with the financial responsibility requirements.

What kinds of USTs need to meet these requirements?

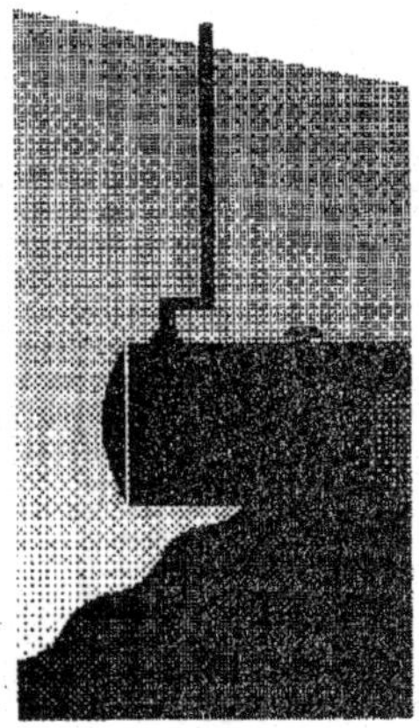

An underground storage tank system (UST) is a tank and any underground piping connected to the tank that has at least 10 percent of its combined volume underground. The federal UST regulations for financial responsibility apply only to underground tanks and piping storing petroleum.

The financial responsibility requirements apply to **all USTs holding petroleum**, with the following exceptions. These tanks are **NOT** covered by the financial responsibility requirements:

- Farm and residential tanks of 1,100 gallons or less capacity holding motor fuel used for noncommercial purposes;
- Tanks storing heating oil used on the premises where it is stored;
- Tanks on or above the floor of underground areas, such as basements or tunnels;
- Septic tanks and systems for collecting storm water and wastewater;
- Flow-through process tanks;
- Tanks of 110 gallons or less capacity, and tanks holding a minimal concentration of petroleum; and
- Emergency spill and overfill tanks.

Other storage sites not covered by the federal financial responsibility requirements (such as surface impoundments and field-constructed tanks) are identified in the **Code of Federal Regulations, 40 CFR Part 280.**

How much financial responsibility coverage is required? When do you need it?

Both the amount of financial responsibility coverage you need and the date by which you need it are determined by the type of business you operate, the amount of throughput of your tank, and the number of tanks you have. On the next page you will find a table that displays five groups of UST owners and operators, compliance deadlines for each group, and required coverage amounts.

Basically, your financial responsibility amount falls into two types: **per occurrence** and **annual aggregate**.

- **Per occurrence.** "Per occurrence" means the amount of money that must be available to pay the costs from one occurrence of a leaking UST. If your tank is used in petroleum production, refining or marketing (such as service stations and truck stops), you must be able to demonstrate that you have $1 million of per occurrence coverage. The per occurrence amount may be less if your tanks are located at a facility **NOT** engaged in petroleum production, refining or marketing. In this case, if your facility has a monthly throughput of 10,000 gallons or less, you need $500,000 of per occurrence coverage.

The table on page 5 displays compliance deadlines and coverage amounts for all categories of UST owners and operators.

- **Annual aggregate.** You must also have coverage for an annual aggregate amount. The annual aggregate amount is the total amount of financial responsibility that you must have to cover all leaks that might occur in one year. The amount of aggregate coverage that you need depends on the number of tanks that you own or operate: $1 million annual aggregate for 100 or fewer tanks; $2 million annual aggregate for more than 100 tanks.

Financial Responsibility Requirements

<table>
<tr><th>Group Of UST Owners And Operators</th><th>Compliance Deadline</th><th>Per Occurrence Coverage</th><th>Aggregate Coverage</th></tr>
<tr><td>GROUP 1:
Petroleum marketers with 1,000 or more tanks
OR
Nonmarketers with net worth of $20 million or more (for nonmarketers, the "per occurrence" amount is the same as Group 4-B below)</td><td>January 1989</td><td rowspan="4">$1 million</td><td rowspan="7">$1 million
if you have 100 or fewer tanks
OR
$2 million
if you have more than 100 tanks</td></tr>
<tr><td>GROUP 2:
Petroleum marketers with 100-999 tanks</td><td>October 1989</td></tr>
<tr><td>GROUP 3:
Petroleum marketers with 13-99 tanks</td><td>April 1991</td></tr>
<tr><td>GROUP 4-A:
Petroleum marketers with 1-12 tanks</td><td>December 1993</td></tr>
<tr><td>GROUP 4-B:
Nonmarketers with net worth of less than $20 million</td><td>December 1993</td><td rowspan="3">$500,000
if throughput is 10,000 gallons monthly or less
OR
$1 million
if throughput is more than 10,000 gallons monthly</td></tr>
<tr><td>GROUP 4-C:
Local governments (including Indian tribes not part of Group 5)</td><td>February 1994</td></tr>
<tr><td>GROUP 5:
Indian tribes owning USTs on Indian lands (USTs must be in compliance with UST technical requirements)</td><td>December 1998</td></tr>
</table>

Dollars And Sense 5

How can you demonstrate financial responsibility?

There are several options for demonstrating financial responsibility (each option is described fully in the **Code of Federal Regulations, 40 CFR Part 280**):

- **Use state financial assurance funds.** Your state may pay for some cleanup and third-party liability costs (see page 7).

- **Obtain insurance coverage.** Insurance may be available from a private insurer or a risk retention group (see page 7).

- **Obtain a guarantee.** You may secure a guarantee for the coverage amount from another firm with whom you have a substantial business relationship. The provider of the guarantee has to pass one of the financial tests described on page 9.

- **Obtain a surety bond.** A surety bond is a guarantee by a surety company that it will meet your financial responsibility obligations.

- **Obtain a letter of credit.** A letter of credit is a contract involving you, an issuer (usually a bank), and a third party (such as the implementing agency) that obligates the issuer to help you demonstrate your financial responsibility.

- **Pass a financial test.** If your firm has a tangible net worth of at least $10 million, you can prove your financial responsibility by passing one of the two financial tests described on page 9.

- **Set up a trust fund.** You may set up a fully-funded trust fund administered by a third party to demonstrate financial responsibility.

- **Use other state methods.** You may also use any additional methods of coverage approved by your state.

You may also use a combination of methods to demonstrate financial responsibility. The methods you choose must cover all the costs for which you are responsible (both corrective action and third-party liability) and add up to the amount of coverage you are required to demonstrate.

Local governments have four additional compliance methods tailored to their special characteristics: a bond rating test, a financial test, a guarantee, and a dedicated fund (all fully described in 40 CFR Part 280).

What about state financial assurance funds?

Most states have established programs that can help pay for cleanup and third-party liability costs resulting from leaking petroleum USTs. Owners and operators of USTs may use these state financial assurance funds to demonstrate financial responsibility.

If you don't know how to reach your state agency, see the list of state UST/LUST contacts starting on page 15.

Many state funds, however, pay only part of the cleanup costs or require that you pay a deductible amount. Some state funds do not pay for third-party liability costs. State assurance funds often have eligibility requirements, such as proof that the UST facility is in compliance with requirements for leak detection and recordkeeping.

You should contact your state agency to determine if the state has a fund that you may use to demonstrate financial responsibility. Find out what the state will pay for and what additional amount of financial responsibility, if any, you must obtain. See the list of state UST/LUST contacts starting on page 15. (LUST means "leaking underground storage tanks.")

What about insurance coverage?

To make sure your insurance policy meets the financial responsibility requirements, have your insurer fill out one of the two following forms. If your UST insurance coverage is an expansion of a policy you already have, your insurer should give you an "Endorsement" document that completes and reproduces the model form shown on page 10. If you get a new policy, your insurer should give you a "Certificate of Insurance" document that completes and reproduces the model form shown on page 11. **Keep the completed form in your records, as explained on page 8.**

If you belong to a trade association, it may be able to provide you with information about insurers and risk retention groups that cover USTs.

You should be aware that insurers may require you to meet certain conditions for coverage. For example, your insurer may require you to test your tank for tightness or make improvements in your tank system, such as adding corrosion protection.

You may also be able to get insurance coverage through a risk retention group. A risk retention group is an insurance company formed by businesses or individuals with similar risks to provide insurance coverage for those risks.

What happens if your coverage is cancelled?

Your financial responsibility method must specify that the provider may cancel your coverage only after sending you advance written notice. Insurance, risk retention group, and state fund coverage can be cancelled only 60 days after you receive the cancellation notice. Guarantees, surety bonds, and letters of credit can be cancelled only 120 days after you receive the cancellation notice. You must find replacement coverage within 60 days after you receive the notice of cancellation. If you cannot get replacement coverage, you must notify your implementing agency.

What records must you keep or report to others?

You must keep at your UST site or your place of business the following records of financial responsibility coverage:

- A current "Certification of Financial Responsibility" (see page 12 for a model of this form that you must complete and reproduce); and

- Any additional documentation that shows your financial responsibility method is valid and provides details on the method's coverage, such as signed copies of official letters, policies, and state fund agreements.

Keep these records until your UST site is properly closed.

Keeping clear records makes good business sense. If asked by UST inspectors or regulatory officials to demonstrate your financial responsibility, you can use these records to demonstrate quickly your compliance with the financial responsibility requirements.

Many states require you to file copies of your financial responsibility records with the state agency, so check with your state to see what you need to do. You do not need to report financial responsibility records to the federal EPA unless you have been requested to do so.

Two Financial Tests

Test 1

- Your firm must have a tangible net worth of at least $10 million; and
- Your firm must have a tangible net worth of a least 10 times the amount of aggregate coverage that you are required to demonstrate plus any other liability coverage for which your firm is using the test to demonstrate financial responsibility to EPA; and
- Your firm must file the firm's annual financial statements with the Securities and Exchange Commission (SEC), or annually report the firm's tangible net worth to Dun and Bradstreet and receive a rating of 4A or 5A. Utilities may file financial statements with the Energy Information Administration, or the Rural Electrification Administration instead of the SEC; and
- Your firm must have audited financial statements that do not include an adverse auditor's opinion or disclaimer of opinion.

Test 2

- Your firm must have a tangible net worth of at least $10 million; and
- Your firm must have a tangible net worth of at least 6 times the amount of aggregate coverage that you are required to demonstrate; and
- Have U.S. assets that are at least 90 percent of total assets or at least 6 times the required aggregate amount; and
- Have net working capital at least 6 times the required aggregate amount, or a bond rating AAA, AA, A, or BBB from Standard and Poor's, or Aaa, Aa, A, or Baa from Moody's; and
- Your firm must have audited financial statements that do not include an adverse auditor's opinion or disclaimer of opinion.

Model Of "Endorsement"

Name: [name of each covered location]

Address: [address of each covered location]

Policy Number:

Period of Coverage [current policy period]:

Name of [Insurer or Risk Retention Group]:

Address of [Insurer or Risk Retention Group]:

Name of Insured:

Address of Insured:

1. This endorsement certifies that the policy to which the endorsement is attached provides liability insurance covering the following underground storage tank(s):

[List the number of tanks at each facility and the name(s) and address(es) of the facility(ies) where the tanks are located. If more than one instrument is used to assure different tanks at any one facility, for each tank covered by this instrument, list the tank identification number provided in the notification submitted pursuant to 40 CFR 280.22, or the corresponding state requirement, and the name and address of the facility.]

for [insert: "taking corrective action" and/or "compensating third parties for bodily injury and property damage caused by" either "sudden accidental releases" or "nonsudden accidental releases" or "accidental release"; in accordance with and subject to the limits of liablity, exclusions, conditions, and other terms of the policy; if coverage is different for different tanks or locations, indicate the type of coverage applicable to each tank or location] arising from operating the underground storage tank(s) identified above.

The limits of liability are [insert the dollar amount of the "per occurrence" and "annual aggregate" limits of the Insurer's or Group's liability; if the amount of coverage is different for different types of coverage or for different underground storage tanks or locations, indicate the amount of coverage for each type of coverage and/or for each underground storage tank or location], exclusive of legal defense costs, which are subject to a separate limit under the policy. This coverage is provided under [policy number]. The effective date of said policy is [date].

2. The insurance afforded with respect to such occurrences is subject to all of the terms and conditions of the policy; provided, however, that any provisions inconsistent with subsections (a) through (e) of this Paragraph 2 are hereby amended to conform with subsection (a) through (e):

a. Bankruptcy or insolvency of the insured shall not relieve the ["Insurer" or "Group"] of its obligations under the policy to which this endorsement is attached.

b. The ["Insurer" or "Group"] is liable for the payment of amounts within any deductible applicable to the policy to the provider of corrective action or a damaged third party, with a right of reimbursement by the insured for any such payment made by the ["Insurer" or "Group"]. This provision does not apply with respect to that amount of any deductible for which coverage is demonstrated under another mechanism or combination of mechanisms as specified in 40 CFR 280.95-280.102.

c. Whenever requested by [a Director of an Implementing Agency], the ["Insurer" or "Group"] agrees to furnish to [the Director] a signed duplicate original of the policy and all endorsements.

d. Cancellation or any other termination of the insurance by the ["Insurer" or "Group"], except for non-payment of premium or misrepresentation by the insured, will be effective only upon written notice and only after the expiration of 60 days after a copy of such written notice is received by the insured. Cancellation for non-payment of premium or misrepresentation by the insured will be effective only upon written notice and only after expiration of a minimum of 10 days after a copy of such written notice is received by the insured.

[Insert for claims-made policies:

e. The insurance covers claims otherwise covered by the policy that are reported to the ["Insurer" or "Group"] within six months of the effective date of the cancellation or non-renewal of the policy except where the new or renewed policy has the some retroactive date or a retroactive date earlier than that of the prior policy, and which arise out of any covered occurrence that commenced after the policy retroactive date, if applicable, and prior to such policy renewal or termination date. Claims reported during such extended reporting period are subject to the terms, conditions, limits, including limits of liability, and exclusions of the policy.]

I hereby certify that the wording of this instrument is identical to the wording in 40 CFR 280.97(b)(1) and that the ["Insurer" or "Group"] is ["licensed to transact the business of insurance or eligible to provide insurance as an excess or surplus lines insurer in one or more states".]

[Signature of authorized representative of Insurer or Risk Retention Group]

[Printed name of person signing]

[Title of person signing], Authorized Representative of [name of Insurer of Risk Retention Group]

[Address of Representative]

Model Of "Certificate Of Insurance"

Name: [name of each covered location]

Address: [address of each covered location]

Policy Number:

Period of Coverage [current policy period]:

Name of [Insurer or Risk Retention Group]:

Address of [Insurer or Risk Retention Group]:

Name of Insured:

Address of Insured:

1. [Name of Insurer or Risk Retention Group], [the "Insurer" or "Group"], as identified above, hereby certifies that it has issued liability insurance covering the following underground storage tank(s):

[List the number of tanks at each facility and the name(s) and address(es) of the facility(ies) where the tanks are located. If more than one instrument is used to assure different tanks at any one facility, for each tank covered by this instrument, list the tank identification number provided in the notification submitted pursuant to 40 CFR 280.22, or the corresponding state requirement, and the name and address of the facility.]

for [insert: "taking corrective action" and/or "compensating third parties for bodily injury and property damage caused by" either "sudden accidental releases" or "nonsudden accidental releases" or "accidental releases"; in accordance with and subject to the limits of liability, exclusions, conditions, and other terms of the policy; if coverage is different for different tanks or locations, indicate the type of coverage applicable to each tank or location] arising from operating the underground storage tank(s) identified above.

The limits of liability are [insert the dollar amount of the "per occurrence" and "annual aggregate" limits of the Insurer's or Group's liability; if the amount of coverage is different for different types of coverage or for different underground storage tanks or locations, indicate the amount of coverage for each type of coverage and/or for each underground storage tank or location], exclusive of legal defense costs, which are subject to a separate limit under the policy. This coverage is provided under [policy number]. The effective date of said policy is [date].

2. The ["Insurer" or "Group"] further certifies the following with respect to the insurance described in Paragraph 1:

a. Bankruptcy or insolvency of the insured shall not relieve the ["Insurer" or "Group"] of its obligations under the policy to which this certificate applies.

b. The ["Insurer" or "Group"] is liable for the payment of amounts within any deductible applicable to the policy to the provider of corrective action or a damaged third party, with a right of reimbursement by the insured for any such payment made by the ["Insurer" or "Group"]. This provision does not apply with respect to that amount of any deductible for which coverage is demonstrated under another mechanism or combination of mechanisms as specified in 40 CFR 280.95-280.102.

c. Whenever requested by [a Director of an Implementing Agency], the ["Insurer" or "Group"] agrees to furnish to [the Director] a signed duplicate original of the policy and all endorsements.

d. Cancellation or any other termination of the insurance by the ["Insurer" or "Group"], except for non-payment of premium or misrepresentation of insured, will be effective only upon written notice and only after the expiration of 60 days after a copy of such written notice is received by the insured. Cancellation for non-renewal of premium or misrepresentation by the insured will be effective only upon written notice and only after expiration of a minimum of 10 days after a copy of such written notice is received by the insured.

[Insert for claims-made policies:

e. The insurance covers claims otherwise covered by the policy that are reported to the ["Insurer" or "Group"] within six months of the effective date of cancellation or non-renewal of the policy except where the new or renewed policy has the same retroactive date or a retroactive date earlier than that of the prior policy, and which arises out of any covered occurrence that commenced after the policy retroactive date, if applicable, and prior to such policy renewal or termination date. Claims reported during such extended reporting periods are subject to the terms, conditions, limits, including limits of liability, and exclusions of the policy.]

I hereby certify that the wording of this instrument is identical to the wording in 40 CFR 280.97(b)(1) and that the ["Insurer" or "Group"] is ["licensed to transact the business of insurance or eligible to provide insurance as an excess or surplus lines insurer in one or more states".]

[Signature of authorized representative of Insurer or Risk Retention Group]

[Printed name of person signing]

[Title of person signing], Authorized Representative of [name of Insurer or Risk Retention Group]

[Address of Representative]

Model Of "Certification Of Financial Responsibility"

[Owner or operator] hereby certifies that it is in compliance with the requirements of Subpart H of 40 CFR Part 280.

The financial assurance mechanism(s) used to demonstrate financial responsibility under Subpart H of 40 CFR Part 280 is (are) as follows:

[For each mechanism, list the type of mechanism, name of issuer, mechanism number (if applicable), amount of coverage, effective period of coverage and whether the mechanism covers "taking corrective action" and/or "compensating third parties for bodily injury and property damage caused by" either "sudden accidental releases" or "non-sudden accidental releases" or "accidental releases."]

[Signature of owner or operator]

[Printed name of owner or operator]

[Title]

[Date]

[Signature of witness or notary]

[Printed name of witness or notary]

[Date]

The owner or operator must update this certification whenever the financial insurance mechanism(s) used to demonstrate financial responsibility change(s).

Publications And Videos About USTs

PUBLICATIONS

TITLE | **AVAILABLE FREE FROM**

Musts For USTs: A Summary Of The Federal Regulations For Underground Storage Tank Systems
Booklet clearly summarizes federal UST requirements for installation, release detection, spill, overfill, and corrosion protection, corrective action, closure, reporting and recordkeeping. (About 40 pages.)

Normas Y Procedimientos Para T.S.A.
Spanish translation of Musts For USTs. (About 40 pages.)

Straight Talk On Tanks: Leak Detection Methods For Petroleum Underground Storage Tanks
Booklet explains federal regulatory requirements for leak detection and briefly describes allowable leak detection methods. (About 30 pages.)

Doing Inventory Control Right: For Underground Storage Tanks
Booklet describes how owners and operators of USTs can use inventory control and periodic tightness testing to meet federal leak detection requirements. Contains reporting forms. (About 16 pages.)

Manual Tank Gauging: For Small Underground Storage Tanks
Booklet provides simple, step-by-step directions for conducting manual tank gauging for tanks 2,000 gallons or smaller. Contains reporting forms. (About 12 pages.)

Don't Wait Until 1998: Spill, Overfill, And Corrosion Protection For Underground Storage Tanks
Information to help owners and operators of USTs meet the 1998 deadline for compliance with requirements to upgrade, replace, or close USTs installed before December 1988. (About 16 pages.)

An Overview Of Underground Storage Tank Remediation Options
Fact sheets provide information about technologies that can be used to remediate petroleum contamination in soil and groundwater. (About 26 pages.)

Controlling UST Cleanup Costs
Fact sheet series on the cleanup process includes: *Hiring a Contractor, Negotiating the Contract, Interpreting the Bill, Managing the Process, and Understanding Contractor Code Words*. (About 10 pages.)

Federal Register Reprints
Not simple summaries, these reprints are extensive records of the rulemaking process including technical information, explanatory preambles, and the rules as they appear in the Code of Federal Regulations. Reprints dated 9/23/88; 10/26/88; 11/9/89; 5/2/90; and 2/18/93. Over 300 pages.

AVAILABLE FREE FROM:

You can call EPA's toll-free RCRA/Superfund Hotline at 800 424-9346 and order free copies. Just identify the titles you want. Or you can write and ask for titles by addressing your requests to:

NCEPI
Box 42419
Cincinnati, OH 45242

Or you can fax your order to NCEPI at 513 891-6685.

Publications And Videos About USTs

VIDEOS

TITLE/COST — **AVAILABLE FROM**

Doing It Right
Illustrates proper installation of underground tanks and piping for installation crews. Part 1: Tanks (24 minutes); Part 2: Piping (16 minutes). Cost: $25

Doing It Right II: Installing Required UST Equipment
Illustrates installation of spill and overfill equipment, observation wells, and piping leak detection (23 minutes). Cost: $60

Doing It Right and Doing it Right II Set Cost: $75

Keeping It Clean: Making Safe And Spill-Free Motor Fuel Deliveries
Making pollution-free deliveries to USTs. Includes Stage 1 vapor recovery, overfill prevention and spill containment. For fuel tanker drivers and UST owner/operators (25 minutes). Cost: $60

Petroleum Leaks Underground
How liquids and vapors move in the subsurface and why early response to leaked petroleum is so important. Part 1: How Liquids Move (14 minutes); Part 2: How Vapors Move (15 minutes). Cost: $75

Straight Talk On Leak Detection
Overview of the leak detection methods available for complying with federal regulations. Part 1: Straight Talk From Tank Owners (owners address the problems of UST compliance [5 minutes]); Part 2: Straight Talk On Leak Detection (30 minutes). Cost: $40

Available from:
Environmental Media Center
Box 30212
Bethesda, MD 20814
301 654-7141
800 522-0362
Visa and MasterCard accepted

Tank Closure Without Tears: An Inspector's Safety Guide
Focuses on explosive vapors and safe tank removal (30 minutes). Video and Booklet Cost: $35; Booklet: $5

What Do We Have Here?: An Inspector's Guide To Site Assessment At Tank Closure
Inspecting sites for contamination where tanks have been removed.
Part 1: Site Assessment Overview (30 minutes); Part 2: Field Testing Instruments At A Glance (14 minutes); Part 3: Soil And Water Sampling At A Glance (7 minutes). Video and Booklet Cost: $45; Booklet: $5

Searching For The Honest Tank: A Guide To UST Facility Compliance Inspection
Covers major steps of UST inspections from protocols and equipment to enforcement and followup; from cathodic protection to leak detection. Directed at inspectors, yet also helpful to owners and operators (30 minutes). Video and Booklet Cost: $40; Booklet: $5

Available from:
New England Interstate Environmental Training Center
ATTN:VIDEOS
2 Fort Road
South Portland, ME 04106
207 767-2539

State UST/LUST Contacts

Alabama Dept. of Env. Management
Groundwater Section/ Water Division
Box 301463
Montgomery, AL 36130-1463
UST phone: 334 271-7986
LUST phone: 334 271-7834

Alaska Dept. of Env. Conservation
410 Willoughby Ave.
Juneau, AK 99801-1795
phone: 907 465-5203

Arizona Dept. of Env. Quality
3033 N. Central Ave.
Phoenix, AZ 85012
phone: 602 207-4324

Arkansas Dept. of Pollution Control & Ecology
Regulated Storage Tank Division
8001 National Dr.
Little Rock, AR 72209
phone: 501 570-2801

California State Water Resources Control Board
Div. of Clean Water Program
Box 944212
Sacramento, CA 94244-2120
phone: 916 227-4313

[UST contact]
Colorado State Oil Inspection Office
1515 Arapahoe Street
Suite 525, Tower 3
Denver, CO 80202-2117
phone: 303 620-4300

[LUST contact]
Colorado Dept. of Health
Hazardous Materials & Waste Mgt. Div.
UST Program
4300 Cherry Creek Dr. S.
Denver, CO 80220
phone: 303 692-3453

Connecticut Dept. of Env. Protection
UST Program
State Office Bldg.
79 Elm Street
Hartford, CT 06106
phone: 203 424-3374

DC Env. Regulatory Administration
Pesticides, Hazardous Waste & UST Div.
2100 Martin Luther King Ave. S.E., Suite 203
Washington, D.C. 20020
phone: 202 645-6080

Delaware Dept. of Natural Resources & Env. Control
UST Branch
715 Grantham Lane
New Castle, DE 19720
phone: 302 323-4588

Florida Dept. of Env. Regulation
Tank Section
Twin Towers Office Bldg., Rm. 403
2600 Blair Stone Road
Tallahassee, FL 32399-2400
phone: 904 488-3935

Georgia Dept. of Natural Resources
UST Mgt. Program
4244 Intl. Parkway, Suite 100
Atlanta, GA 30354
phone: 404 362-2687

Hawaii Dept. of Health
Solid and Hazardous Waste Branch
919 Ala Moana Blvd., Suite 212
Honolulu, HI 96814
phone: 808 586-4226

Idaho Dept. of Health & Welfare
Div. of Env. Quality
1410 North Hilton
Boise, ID 83706
phone: 208 334-0542

[UST contact]
Illinois Office of State Fire Marshall
Div. of Petroleum & Chem. Safety
1035 Stevenson Dr.
Springfield, IL 62703
phone: 217 785-5878

[LUST contact]
Illinois EPA
Div. of Remediation Mgt.
LUST Section
Box 19276
Springfield, IL 62794-9276
phone: 217 782-6760

Indiana Dept. of Env. Mgt.
Office of Env. Response
Box 7015
Indianapolis, IN 46206
phone: 317 233-6418

Iowa Dept. of Natural Resources
UST Section
Wallace State Office Bldg.
900 East Grand
Des Moines, IA 50319
phone: 515 281-8135

Kansas Dept. of Health & Env.
Bureau of Env. Remediation
Storage Tank Section
Forbes Field, Bldg. 740
Topeka, KS 66620
phone: 913 296-1678

Kentucky Div. of Waste Mgt.
UST Branch
14 Reilly Road
Frankfort, KY 40601
phone: 502 564-6716

Louisiana Dept. of Env. Quality
UST Division
Box 82178
Baton Rouge, LA 70810
phone: 504 765-0243

Maine Dept. of Env. Protection
State House, Station 17
Hospital St., Ray Bldg.
Augusta, ME 04333
phone: 207 287-2651

Maryland Dept. of Env.
Waste Mgt. Adm.
Oil Control Program
2500 Broening Highway
Baltimore, MD 21224
phone: 410 631-3442

[UST contact]
Massachusetts Dept. of Public Safety
UST Program
1010 Commonwealth Ave.
Boston, MA 02215
phone: 617-351-6000

[LUST contact]
Massachusetts Dept. of Env. Protection
Bureau of Waste Site Cleanup
One Winter St.
Boston, MA 02108
phone: 617 556-1044

Michigan Dept. of Natural Resources
UST Division
Box 30157
Lansing, MI 48909
UST phone: 517 373-8168
MUSTFA (state fund): 517 373-6247

Minnesota Pollution Control Agency
Tanks and Spills Section
520 Lafayette Road North
St. Paul, MN 55155-3898
phone: 612 297-8609

State UST/LUST Contacts

Mississippi Dept. of Env. Quality
Bureau of Pollution Control, UST Section
Box 10385
Jackson, MS
39289-0385
phone: 601 961-5171

Missouri Dept. of Natural Resources
Water Pollution Control Program (UST)
Environmental Services Program (LUST)
Box 176
Jefferson City, MO
65102-0176
UST phone:
314 751-7428
LUST phone:
314 526-3384

Montana Dept. of Health & Env. Sciences
Solid & Hazardous Waste Bureau
UST Program
Cogswell Bldg.
Helena, MT 59620
phone: 406 444-5970

[UST contact]
Nebraska State Fire Marshal
Flammable Liquid Storage
246 South 14th St.
Lincoln, NE 68508
phone: 402 471-9465

[LUST contact]
Nebraska Dept. of Env. Control
LUST/ER Section
Box 98922
Lincoln, NE
68509-8922
phone: 402 471-4230

Nevada Dept. of Conserv. &Natural Resources
Div. of Env. Protection
Capitol Complex
333 W. Nye Lane
Carson City, NV 89710
phone: 702 687-5872

New Hampshire Dept. of Env. Services
Oil Compliance Section
Groundwater Protection Bureau
Box 95
Concord, NH 03301
phone: 603 271-3644

New Jersey Dept. of Env. Protection
Responsible Party Site Remediation
401 E. State (CN-028)
Trenton, NJ
08625-0028
phone: 609 984-3156

New Mexico Env. Dept.
UST Bureau
Box 26110
Santa Fe, NM
87502
phone: 505 827-0188

New York Dept. of Env. Conservation
Bulk Storage Section
50 Wolf Rd., Rm. 340
Albany, NY
12233-3750
phone: 518 457-4351

North Carolina Pollution Control Branch
Div. of Env. Mgt.
Dept. of Env. Health & Natural Resources
441 N. Harrington St.
Raleigh, NC
27603-1323
phone: 919 733-8486

North Dakota Div. of Waste Mgt.
Dept. of Health and Consolidated Lab.
Box 5520
Bismarck, ND
58502-5520
phone: 701 328-5166

Ohio Dept. of Commerce
Bureau of UST Regulations
8895 East Main St.
Box 687
Reynoldsburg, OH 43068
phone: 614 752-7938

Oklahoma Corporation Commission
Fuel Storage Division
Box 52000-2000
Oklahoma City, OK
73152-2000
UST phone:
405 521-3107
LUST phone:
405 521-6575

[UST contact]
Oregon Dept. of Env. Quality
UST Compliance Section
811 S.W. Sixth Ave.,
7th Floor
Portland, OR 97204
phone: 503 229-5774

[LUST contact]
Oregon Dept. of Env. Quality
UST Cleanup Program
811 S.W. Sixth Ave.,
9th Floor
Portland, OR 97204
phone: 503 229-6642

Pennsylvania Dept. of Env. Resources
Storage Tank Program
Box 8762
Harrisburg, PA
17101-2301
phone: 717 772-5599

Rhode Island Dept. of Env. Mgt.
UST Section
291 Promenade St.
Providence, RI 02908
phone: 401 277-2234

South Carolina Dept. of Health and Env.Control
Groundwater Protection Division
2600 Bull St.
Columbia, SC 29201
UST phone:
803 734-5335
LUST phone:
803 734-5331

South Dakota Dept. of Env. & Nat. Resources
UST Program
523 East Capitol
Pierre, SD 57501
phone: 605 773-3296

Tennessee Dept. of Env. & Conservation
Div. of USTs
4th Floor, L&C Tower
401 Church St.
Nashville, TN
37243-1541
phone: 615 532-0945

Texas Natural Resources Conservation Comm.
Petroleum Storage Tank Div.
Box 13087
Austin, TX 78711-3087
phone: 512 239-2000

Utah Dept. of Env. Quality
Div. of Env. Response and Remediation
UST Branch
168 North 1950 West,
1st Floor
Salt Lake City, UT 84116
phone: 801 536-4100

Vermont Dept. of Natural Resources
UST Program
103 South Main St.,
West Bldg.
Waterbury, VT 05676
phone: 802 244-8702

[UST contact]
Virginia Dept. Env. Quality
Response Program
4951 Cox Road
Glen Allen, VA 23060
phone: 804 527-5189

[LUST contact]
Virginia Dept. Env. Quality
Groundwater Mgt. Program
4951 Cox Road
Glen Allen, VA 23060
phone: 804 527-5189

☆U.S. GOVERNMENT PRINTING OFFICE: 1995 - 615-003/21020

State UST/LUST Contacts

Washington Dept. of Ecology
Box 47655
Olympia, WA
98504-7655
UST phone:
206 407-7211
LUST phone:
206 407-7218

West Virginia Div. of Env. Protection
Office of Waste Mgt.
UST Section
1356 Hansford St.
Charleston, WV 25301
phone: 304 558-6371

[UST contact]
Wisconsin Dept. of Industry, Labor & Human Relations
Division of Safety and Buildings
Box 7969
Madison, WI 53707
phone: 608 267-7605

[LUST contact]
Wisconsin Dept. of Natural Resources
Bureau of Solid and Hazardous Waste Mgt.
Box 7921
Madison, WI
53707-7921
phone: 608 267-7560

Wyoming Dept. of Env. Quality
Water Quality Div.
Herschler Bldg., 4th Floor
122 West 25th St.
Cheyenne, WY 82002
phone: 307 777-7096

U.S. TERRITORIES

American Samoa Env. Protection Agency
Office of the Governor
American Samoa Government
ATTN: UST Program
Pago Pago, American Samoa 96799
phone: 684 633-2304

Commonwealth of Northern Mariana Islands Div. of Env. Quality
Box 1304
Dr. Torres Hospital
Saipan, MP 96950
phone: 607 234-6984

Guam Env. Protection Agency
IT&E
Harmon Plaza,
Complex Unit D-107
130 Rojas St.
Harmon, Guam 96911
phone: 671 646-8863

Puerto Rico Env. Quality Board
Box 11488
Commonwealth of Puerto Rico
Santurce, Puerto Rico
00910
phone: 809 767-8109

Virgin Islands Div. of Env. Protection
Dept. of Planning and Natural Resources
Nisky Center 231
45A Estate Nisky
Charlotte Amalie
St. Thomas, Virgin Islands 00802
phone: 809 774-3320

■ APPENDIX F.1 Occupational Safety and Health Administration 29 CFR 1910.120 Appendix E: Training Curriculum Guidelines

The following non-mandatory general criteria may be used for assistance in developing site-specific training curriculum used to meet the training requirements of 29 CFR 1910.120(e); 29 CFR 1910.120(p)(7), (p)(8)(iii); and 29 CFR 1910.120(q)(6), (q)(7), and (q)(8). These are generic guidelines and they are not presented as a complete training curriculum for any specific employer. Site-specific training programs must be developed on the basis of a needs assessment of the hazardous waste site, RCRA/TSDF, or emergency response operation in accordance with 29 CFR 1910.120. It is noted that the legal requirements are set forth in the regulatory text of §1910.120. The guidance set forth here presents a highly effective program that in the areas covered would meet or exceed the regulatory requirements. In addition, other approaches could meet the regulatory requirements.

Suggested General Criteria

Definitions:

"Competent" means possessing the skills, knowledge, experience, and judgment to perform assigned tasks or activities satisfactorily as determined by the employer.

"Demonstration" means the showing by actual use of equipment or procedures.

"Hands-on training" means training in a simulated work environment that permits each student to have experience performing tasks, making decisions, or using equipment appropriate to the job assignment for which the training is being conducted.

"Initial training" means training required prior to beginning work.

"Lecture" means an interactive discourse with a class lead by an instructor.

"Proficient" means meeting a stated level of achievement.

"Site-specific" means individual training directed to the operations of a specific job site.

"Training hours" means the number of hours devoted to lecture, learning activities, small group work sessions, demonstration, evaluations, or hands-on experience.

Suggested core criteria:

1. Training facility. The training facility should have available sufficient resources, equipment, and site locations to perform didactic and hands-on training when appropriate. Training facilities should have sufficient organization, support staff, and services to conduct training in each of the courses offered.
2. Training Director. Each training program should be under the direction of a training director who is responsible for the program. The Training Director should have a minimum of two years of employee education experience.
3. Instructors. Instructors should be deemed competent on the basis of previous documented experience in the area of instruction, successful completion of a "train-the-trainer" program specific to the topics they will teach, and an evaluation of instructional competence by the Training Director.

   Instructors should be required to maintain professional competency by participating in continuing education or professional development programs or by completing successfully an annual refresher course and having an annual review by the Training Director.

The annual review by the Training Director should include observation of an instructor's delivery, a review of those observations with the trainer, and an analysis of any instructor or class evaluations completed by the students during the previous year.

4. Course materials. The Training Director should approve all course materials to be used by the training provider. Course materials should be reviewed and updated at least annually. Materials and equipment should be in good working order and maintained properly.

   All written and audio-visual materials in training curricula should be peer reviewed by technically competent outside reviewers or by a standing advisory committee.

   Reviews should possess expertise in the following disciplines where applicable: occupational health, industrial hygiene and safety, chemical/environmental engineering, employee education, or emergency response. One or more of the peer reviewers should be an employee experienced in the work activities to which the training is directed.

5. Students. The program for accepting students should include:
   a. Assurance that the student is or will be involved in work where chemical exposures are likely and that the student possesses the skills necessary to perform the work.
   b. A policy on the necessary medical clearance.

6. Ratios. Student-instructor ratios should not exceed 30 students per instructor. Hands-on activity requiring the use of personal protective equipment should have the following student-instructor ratios. For Level C or Level D personal protective equipment the ratio should be 10 students per instructor. For Level A or Level B personal protective equipment the ratio should be 5 students per instructor.

7. Proficiency assessment. Proficiency should be evaluated and documented by the use of a written assessment and a skill demonstration selected and developed by the Training Director and training staff. The assessment and demonstration should evaluate the knowledge and individual skills developed in the course of training. The level of minimum achievement necessary for proficiency shall be specified in writing by the Training Director.

   If a written test is used, there should be a minimum of 50 questions. If a written test is used in combination with a skills demonstration, a minimum of 25 questions should be used. If a skills demonstration is used, the tasks chosen and the means to rate successful completion should be fully documented by the Training Director.

   The content of the written test or of the skill demonstration shall be relevant to the objectives of the course. The written test and skill demonstration should be updated as necessary to reflect changes in the curriculum and any update should be approved by the Training Director.

   The proficiency assessment methods, regardless of the approach or combination of approaches used, should be justified, documented and approved by the Training Director.

The proficiency of those taking the additional courses for supervisors should be evaluated and documented by using proficiency assessment methods acceptable to the Training Director. These proficiency assessment methods must reflect the additional responsibilities borne by supervisory personnel in hazardous waste operations or emergency response.

8. Course certificate. Written documentation should be provided to each student who satisfactorily completes the training course. The documentation should include:
   a. Student's name.
   b. Course title.
   c. Course date.
   d. Statement that the student has successfully completed the course.
   e. Name and address of the training provider.
   f. An individual identification number for the certificate.
   g. List of the levels of personal protective equipment used by the student to complete the course.

   This documentation may include a certificate and an appropriate wallet-sized laminated card with a photograph of the student and the above information. When such source certificate cards are used, the individual identification number for the training certificate should be shown on the card.
9. Recordkeeping. Training providers should maintain records listing the dates courses were presented, the names of the individual course attenders, the names of those students successfully completing each course, and the number of training certificates issued to each successful student. These records should be maintained for a minimum of five years after the date an individual participated in a training program offered by the training provider. These records should be available and provided upon the student's request or as mandated by law.
10. Program quality control. The Training Director should conduct or direct an annual written audit of the training program. Program modifications to address deficiencies, if any, should be documented, approved, and implemented by the training provider. The audit and the program modification documents should be maintained at the training facility.

Suggested Program Quality Control Criteria

Factors listed here are suggested criteria for determining the quality and appropriateness of employee health and safety training for hazardous waste operations and emergency response.

A. Training Plan

Adequacy and appropriateness of the training program's curriculum development, instructor training, distribution of course materials, and direct student training should be considered, including:

1. The duration of training, course content, and course schedules/agendas;
2. The different training requirements of the various target populations, as specified in the appropriate generic training curriculum;
3. The process for the development of curriculum, which includes appropriate technical input, outside review, evaluation, program pretesting.

4. The adequate and appropriate inclusion of hands-on, demonstration, and instruction methods;
5. Adequate monitoring of student safety, progress, and performance during the training.

B. Program Management, Training Director, Staff and Consultants.

Adequacy and appropriateness of staff performance and delivering an effective training program should be considered, including:

1. Demonstration of the training director's leadership in assuring quality of health and safety training.
2. Demonstration of the competency of the staff to meet the demands of delivering high quality hazardous waste employee health and safety training.
3. Organization charts establishing clear lines of authority.
4. Clearly defined staff duties including the relationship of the training staff to the overall program.
5. Evidence that the training organizational structure suits the needs of the training program.
6. Appropriateness and adequacy of the training methods used by the instructors.
7. Sufficiency of the time committed by the training director and staff to the training program.
8. Adequacy of the ratio of training staff to students.
9. Availability and commitment of the training program of adequate human and equipment resources in the areas of
   a. Health effects,
   b. Safety,
   c. Personal protective equipment (PPE),
   d. Employee protection practices/procedures.
10. Appropriateness of management controls.
11. Adequacy of the organization and appropriate resources assigned to assure appropriate training.
12. In the case of multiple-site training programs, adequacy of satellite centers management.

C. Training facilities and resources.

Adequacy and appropriateness of the facilities and resources for supporting the training program should be considered, including:

1. Space and equipment to conduct the training.
2. Facilities for representative hands-on training.
3. In the case of multiple-site programs, equipment and facilities at the satellite centers.
4. Adequacy and appropriateness of the quality control and evaluations program to account for instructor performance.
5. Adequacy and appropriateness of the quality control and evaluation program to ensure appropriate course evaluation, feedback, updating, and corrective action.
6. Adequacy and appropriateness of disciplines and expertise being used within the quality control and evaluation program.
7. Adequacy and appropriateness of the role of student evaluations to provide feedback for training program improvement.

D. Quality control and evaluation.

Adequacy and appropriateness of quality control and evaluation plans for training programs should be considered, including:

1. A balanced advisory committee and/or competent outside reviewers to give overall policy guidance.
2. Clear and adequate definition of the composition and active programmatic role of the advisory committee or outside reviewers.
3. Adequacy of the minutes or reports of the advisory committee or outside reviewers' meetings or written communication.
4. Adequacy and appropriateness of the quality control and evaluations program to account for instructor performance.
5. Adequacy and appropriateness of the quality control and evaluation program to ensure appropriate course evaluation, feedback, updating, and corrective action.
6. Adequacy and appropriateness of disciplines and expertise being used within the quality control and evaluation program.
7. Adequacy and appropriateness of the role of student evaluations to provide feedback for training program improvement.

E. Students.

Adequacy and appropriateness of the program for accepting students should be considered, including:

1. Assurance that the students already possess the necessary skills for their job, including necessary documentation.
2. Appropriateness of methods the program uses to ensure that recruits are capable of satisfactorily completing training.
3. Review and compliance with any medical clearance policy.

F. Institutional Environment and Administrative Support.

The adequacy and appropriateness of the institutional environment and administrative support system for the training program should be considered, including:

1. Adequacy of the institutional commitment to the employee training program.
2. Adequacy and appropriateness of the administrative structure and administrative support.

G. Summary of Evaluation Questions.

Key questions for evaluating the quality and appropriateness of an overall training program should include the following:

1. Are the program objectives clearly stated?
2. Is the program accomplishing its objectives?
3. Are appropriate facilities and staff available?
4. Is there an appropriate mix of classroom, demonstration, and hands-on training?
5. Is the program providing quality employee health and safety training that fully meets the intent of regulatory requirements?
6. What are the program's main strengths?
7. What are the program's main weaknesses?
8. What is recommended to improve the program?
9. Are instructors instructing according to their training outlines?
10. Is the evaluation tool current and appropriate for the program content?
11. Is the course material current and relevant to the target group?

Suggested Training Curriculum Guidelines

The following training curriculum guidelines are for those operations specifically identified in 29 CFR 1910.120 as requiring training. Issues such as qualifications of instructors, training certification, and similar criteria appropriate to all categories of operations addressed in §1910.120 have been covered in the preceding section and are not re-addressed in each of the generic guidelines. Basic core requirements for training programs that are addressed include

1. General Hazardous Waste Operations.
2. RCRA Operations Training for Treatment, Storage, and Disposal Facilities.
3. Emergency Response.

A. General Hazardous Waste Operations and Site-Specific Training

1. Off-site training. Training course content for hazardous waste operations, required by 29 CFR 1910.120(e), should include the following topics or procedure:
   a. Regulatory knowledge.
      1. A review of 29 CFR 1910.120 and the core elements of an occupational safety and health program.
      2. The content of a medical surveillance program as outlined in 29 CFR 1910.120(f).
      3. The content of an effective site safety and health plan consistent with the requirements of 29 CFR 1910.120(b)(4)(ii).
      4. Emergency response plan and procedures as outlined in 29 CFR 1910.38 and 29 CFR 1910.120(I).
      5. Adequate illumination.
      6. Sanitation recommendation and equipment.
      7. Review and explanation of OSHA's hazard-communication standard (29 CFR 1910.1200) and lock-out/tag-out standard (29 CFR 1910.147).
      8. Review of other applicable standards including but not limited to those in the construction standards (29 CFR Part 1926).
      9. Rights and responsibilities of employers and employees under applicable OSHA and EPA laws.

   b. Technical knowledge.
      1. Type of potential exposures to chemical, biological, and radiological hazards; types of human responses to these hazards and recognition of those responses; principles of toxicology and information about acute and chronic hazards; health and safety considerations of new technology.
      2. Fundamentals of chemical hazards including but not limited to vapor pressure, boiling points, flash points, pH, other physical and chemical properties.
      3. Fire and exposure hazards of chemicals.
      4. General safety hazards such as but not limited to electrical hazards, powered equipment hazards, motor vehicle hazards, walking-working surface hazards, excavation hazards, and hazards associated with working hot and cold temperature extremes.
      5. Review and knowledge of confined space entry procedures in 29 CFR 1910.146.
      6. Work practices to minimize employee risk from site hazards.

7. Safe use of engineering controls, equipment, and any new relevant safety technology or safety procedures.
8. Review and demonstration of competency with air sampling and monitoring equipment that may be used in a site monitoring program.
9. Container sampling procedures and safeguarding; general drum and containing handling procedures including special requirement for laboratory waste packs, shock-sensitive wastes, and radioactive wastes.
10. The elements of a spill control program.
11. Proper use and limitations of material handling equipment.
12. Procedures for safe and healthful preparation of containers for shipping and transport.
13. Methods of communication including those used while wearing respiratory protection.

c. Technical skills.
1. Selection, use, maintenance, and limitations of personal protective equipment including the components and procedures for carrying out a respiratory program to comply with 29 CFR 1910.134.
2. Instruction in decontamination programs including personnel, equipment, and hardware; hands-on training including level A, B, and C ensembles and appropriate decontamination lines; field activities including the donning and doffing of protective equipment to a level commensurate with the employee's anticipated job function and responsibility and to the degree required by potential hazards.
3. Sources for additional hazard information; exercises using relevant manuals and hazard coding systems.

d. Additional suggested items.
1. A laminated, dated card or certificate with photo, denoting limitations and level of protection for which the employee is trained should be issued to those students successfully completing a course.
2. Attendance should be required at all training modules, with successful completion of exercises and a final written or oral examination with at least 50 questions.
3. A minimum of one-third of the program should be devoted to hands-on exercises.
4. A curriculum should be established for the 8-hour refresher training required by 29 CFR 1910.120(e)(8), with delivery of such courses directed toward those areas of previous training that need improvement or reemphasis.
5. A curriculum should be established for the required 8-hour training for supervisors. Demonstrated competency in the skills and knowledge provided in a 40-hour course should be a prerequisite for supervisor training.

2. Refresher training.
The 8-hour annual refresher training required in 29 CFR 1910.120(e)(8) should be conducted by qualified training providers. Refresher training should include at a minimum the following topics and procedures:

a. Review of and retraining on relevant topics covered in the 40-hour program, as appropriate, using reports by the students on their work experiences.
b. Update on developments with respect to material covered in the 40-hour course.

c. Review of changes to pertinent provisions of EPA or OSHA standards or laws.
d. Introduction of additional subject areas as appropriate.
e. Hands-on review of new or altered PPE or decontamination equipment or procedures. Review of new developments in personal protective equipment.
f. Review of newly developed air and contaminant monitoring equipment.

3. On-site training.
   a. The employer should provide employees engaged in hazardous waste site activities with information and training prior to initial assignment into their work area, as follows:
      1. The requirements of the hazard communication program including the location and availability of the written program, required lists of hazardous chemicals, and material safety data sheets.
      2. Activities and locations in their work area where hazardous substance may be present.
      3. Methods and observations that may be used to detect the presence or release of a hazardous chemical in the work area (such as monitoring conducted by the employer, continuous monitoring devices, visual appearances, or other evidence (sight, sound or smell) of hazardous chemicals being released), and applicable alarms from monitoring devices that record chemical releases.
      4. The physical and health hazards of substances known or potentially present in the work area.
      5. The measures employees can take to help protect themselves from work-site hazards including specific procedures the employer has implemented.
      6. An explanation of the labeling system and material safety data sheets and how employees can obtain and use appropriate hazard information.
      7. The elements of the confined space program including special PPE, permits, monitoring requirements, communication procedures, emergency response, and applicable lock-out procedures.
   b. The employer should provide hazardous waste employees information and training and should provide a review and access to the site safety and plan as follows:
      1. Names of personnel and alternate responsible for site safety and health.
      2. Safety and health hazards present on the site.
      3. Selection, use, maintenance, and limitations of personal protective equipment specific to the site.
      4. Work practices by which the employee can minimize risks from hazards.
      5. Safe use of engineering controls and equipment available on site.
      6. Safe decontamination procedures established to minimize employee contact with hazardous substances, including:
         a. Employee decontamination,
         b. Clothing decontamination, and
         c. Equipment decontamination.
      7. Elements of the site emergency response plan, including:
         a. Pre-emergency planning.
         b. Personnel roles and lines of authority and communication.
         c. Emergency recognition and prevention.

d. Safe distances and places of refuge.
e. Site security and control.
f. Evacuation routes and procedures.
g. Decontamination procedures not covered by the site safety and health plan.
h. Emergency medical treatment and first aid.
i. Emergency equipment and procedures for handling emergency incidents.

c. The employer should provide hazardous waste employees information and training on personal protective equipment used at the site, such as the following:
1. PPE to be used based upon known or anticipated site hazards..
2. PPE limitations of materials and construction; limitations during temperature extremes, heat stress, and other appropriate medical considerations; use and limitations of respirator equipment as well as documentation procedures as outlined in 29 CFR 1910.134.
3. PPE inspection procedures prior to, during, and after use.
4. PPE donning and doffing procedures.
5. PPE decontamination and disposal procedures.
6. PPE maintenance and storage.
7. Task duration as related to PPE limitations.

d. The employer should instruct the employees about the site medical surveillance program relative to the particular site, including:
1. Specific medical surveillance programs that have been adapted for the site.
2. Specific signs and symptoms related to exposure to hazardous materials on the site.
3. The frequency and extent of periodic medical examinations that will be used on the site.
4. Maintenance and availability of records.
5. Personnel to be contacted and procedures to be followed when signs and symptoms of exposures are recognized.

e. The employees will review and discuss the site safety plan as part of the training program. The location of the site safety plan and all written programs should be discussed with employees including a discussion of the mechanisms for access, review, and references described.

B. RCRA Operations Training for Treatment, Storage and Disposal Facilities.

1. As a minimum, the training course required in 29 CFR 1910.120(p) should include the following topics:
a. Review of the applicable paragraphs of 29 CFR 1910.120 and the elements of the employer's occupational safety and health plan.
b. Review of relevant hazards such as, but not limited to, chemical, biological, and radiological exposures; fire and explosion hazards; thermal extremes; and physical hazards.
c. General safety hazards including those associated with electrical hazards, powered equipment hazards, lock-out/tag-out procedures, motor vehicle hazards and walking-working surface hazards.

d. Confined-space hazards and procedures.
e. Work practices to minimize employee risk from workplace hazards.
f. Emergency response plan and procedures including first aid meeting the requirements of paragraph (p)(8).
g. A review of procedures to minimize exposure to hazardous waste and various type of waste streams, including the materials handling program and spill containment program.
h. A review of hazard communication programs meeting the requirements of 29 CFR 1910.1200.
i. A review of medical surveillance programs meeting the requirements of 29 CFR 1910.120(p)(3) including the recognition of signs and symptoms of overexposure to hazardous substances including known synergistic interactions.
j. A review of decontamination programs and procedures meeting the requirements of 29 CFR 1910.120(p)(4).
k. A review of an employer's requirements to implement a training program and its elements.
l. A review of the criteria and programs for proper selection and use of personal protective equipment, including respirators.
m. A review of the applicable appendices to 29 CFR 1910.120.
n. Principles of toxicology and biological monitoring as they pertain to occupational health.
o. Rights and responsibilities of employees and employers under applicable OSHA and EPA laws.
p. Hands-on exercises and demonstrations of competency with equipment to illustrate the basic equipment principles that may be used during the performance of work duties, including the donning and doffing of PPE.
q. Sources of reference, efficient use of relevant manuals, and knowledge of hazard coding systems to include information contained in hazardous waste manifests.
r. At least 8 hours of hands-on training.
s. Training the job skills required for an employee's job function and responsibility before they are permitted to participate in or supervise field activities.

2. The individual employer should provide hazardous waste employees with information and training prior to an employee's initial assignment into a work area. The training and information should cover the following topics:
   a. The emergency response plan and procedures including first aid.
   b. A review of the employer's hazardous waste handling procedures including the materials handling program and elements of the spill containment program, location of spill response kits or equipment, and the names of those trained to respond to releases.
   c. The hazardous communication program meeting the requirements of 29 CFR 1910.1200.
   d. A review of the employer's medical surveillance program including the recognition of signs and symptoms of exposure to relevant hazardous substance including known synergistic interactions.
   e. A review of the employer's decontamination program and procedures.
   f. A review of the employer's training program and the parties responsible for that program.

g. A review of the employer's personal protective equipment program including the proper selection and use of PPE based upon specific site hazards.
h. All relevant site-specific procedures addressing potential safety and health hazards. This may include, as appropriate, biological and radiological exposures, fire and explosion hazards, thermal hazards, and physical hazards such as electrical hazards, powered equipment hazards, lock-out/tag-out hazards, motor vehicle hazards, and walking-working surface hazards.
i. Safe use engineering controls and equipment on site.
j. Names of personnel and alternates responsible for safety and health.

C. Emergency Response Training.

Federal OSHA standards in 29 CFR 1910.120(q) are directed toward private sector emergency responders. Therefore, the guidelines provided in this portion of the appendix are directed toward that employee population. However, they also impact indirectly, through state OSHA or U.S. EPA regulations, some public sector emergency responders. Therefore, the guidelines provided in this portion of the appendix may be applied to both employee populations.

States with OSHA state plans must cover their employees with regulations at least as effective as the Federal OSHA standards. Public employees in states without approved state OSHA programs covering hazardous waste operations and emergency response are covered by the U.S. EPA under 40 CFR 311, a regulation virtually identical to §1910.120.

Since this is a non-mandatory appendix and therefore not an enforceable standard, OSHA recommends that those employers, employees or volunteers in public sector emergency response organizations outside Federal OSHA jurisdiction consider the following criteria in developing their own training programs. A unified approach to training at the community level between emergency response organizations covered by Federal OSHA and those not covered directly by Federal OSHA can help ensure an effective community response to the release or potential release of hazardous substances in the community.

a. General Considerations.

Emergency response organizations are required to consider the topics listed in §1910.120(q)(6). Emergency response organizations may use some or all of the following topics to supplement those mandatory topics when developing their response training programs. Many of those topics would require an interaction between the response provider and the individuals responsible for the site where the response would be expected.

1. Hazard recognition, including:
   a. Nature of hazardous substances present.
   b. Practical applications of hazard recognition, including presentations on biology, chemistry, and physics.
2. Principles of toxicology, biological monitoring, and risk assessment.
3. Safe work practices and general site safety.
4. Engineering controls and hazardous waste operations.
5. Site safety plans and standard operating procedures.
6. Decontamination procedures and practices.
7. Emergency procedures, first aid, and self-rescue.
8. Safe use of field equipment.

9. Storage, handling, use and transportation of hazardous substances.
10. Use, care, and limitations of personal protective equipment.
11. Safe sampling techniques.
12. Rights and responsibilities of employees under OSHA and other related laws concerning right-to-know, safety and health, compensation and liability.
13. Medical monitoring requirements.
14. Community relations.

b. Suggested Criteria for Specific Courses.

1. First responder awareness level.
   a. Review of and demonstration of competency in performing the applicable skills of 29 CFR 1910.120(q).
   b. Hands-on experience with the U.S. Department of Transportation's Emergency Response Guidebook (ERG) and familiarization with OSHA standard 29 CFR 1910.1201.
   c. Review of the principles and practices for analyzing an incident to determine both the hazardous substances present and the basic hazard and response information for each hazardous substance present.
   d. Review of procedures for implementing actions consistent with the local emergency response plan, the organization's standard operating procedures, and the current edition of DOT's ERG including emergency notification procedures and follow-up communications.
   e. Review of the expected hazards including fire and explosions hazards, confined space hazards, electrical hazards, powered equipment hazards, motor vehicle hazards, and walking-working surface hazards.
   f. Awareness and knowledge of the competencies for the First Responder at the Awareness Level covered in the National Fire Protection Association's Standard No. 472, Professional Competence of Responders to Hazardous Materials Incidents.
2. First responder operations level.
   a. Review of and demonstration of competency in performing the applicable skills of 29 CFR 1910.120(q).
   b. Hands-on experience with the U.S. Department of Transportation's Emergency Response Guidebook (ERG), manufacturer material safety data sheets, CHEMTREC/CANUTEC, shipper or manufacturer contacts, and other relevant sources of information addressing hazardous substance releases. Familiarization with OSHA standard 29 CFR 1910.1201.
   c. Review of the principles and practices for analyzing an incident to determine the hazardous substances present, the likely behavior of the hazardous substance and its container, the types of hazardous substance transportation containers and vehicles, the types and selection of the appropriate defensive strategy for containing the release.
   d. Review of procedures for implementing continuing response actions consistent with the local emergency response plan, the organization's standard operating procedures, and the current edition of DOT's ERG including extended emergency notification procedures and follow-up communications.
   e. Review of the principles and practice for proper selection and use of personal protective equipment.

f. Review of the principles and practice of personnel and equipment decontamination.
g. Review of the expected hazards including fire and explosions hazards, confined space hazards, electrical hazards, powered equipment hazards, motor vehicle hazards,and walking-working surface hazards.
h. Awareness and knowledge of the competencies for the First Responder at the Operations Level covered in the National Fire Protection Association's Standard No. 472, Professional Competence of Responders to Hazardous Materials Incidents.

3. Hazardous material technician.
   a. Review of and demonstration of competency in performing the applicable skills of 29 CFR 1910.120(q).
   b. Hands-on experience with written and electronic information relative to response decision making including but not limited to the U.S. Department of Transportation's Emergency Response Guidebook (ERG), manufacturer material safety data sheets, CHEMTREC/CANUTEC, shipper or manufacturer contacts, computer data bases and response models, and other relevant sources of information addressing hazardous substance releases. Familiarization with OSHA standard 29 CFR 1910.1201.
   c. Review of the principles and practices for analyzing an incident to determine the hazardous substances present, their physical and chemical properties, the likely behavior of the hazardous substance and its container, the types of hazardous substance transportation containers and vehicles involved in the release, the appropriate strategy for approaching release sites and containing the release.
   d. Review of procedures for implementing continuing response actions consistent with the local emergency response plan, the organization's standard operating procedures, and the current edition of DOT's ERG including extended emergency notification procedures and follow-up communications.
   e. Review of the principles and practice for proper selection and use of personal protective equipment.
   f. Review of the principles and practices of establishing exposure zones, proper decontamination and medical surveillance stations and procedures.
   g. Review of the expected hazards including fire and explosions hazards, confined space hazards, electrical hazards, powered equipment hazards, motor vehicle hazards, and walking-working surface hazards.
   h. Awareness and knowledge of the competencies for the Hazardous Materials Technician covered in the National Fire Protection Association's Standard No. 472, Professional Competence of Responders to Hazardous Materials Incidents.
4. Hazardous materials specialist.
   a. Review of and demonstration of competency in performing the applicable skills of 29 CFR 1910.120(q).
   b. Hands-on experience with retrieval and use of written and electronic information relative to response decision making including but not limited to the U.S. Department of Transportation's Emergency Response Guidebook(ERG), manufacturer material safety data sheets, CHEMTREC/CANUTEC, shipper or manufacturer contacts, computer data bases and response models, and other relevant sources of information addressing hazardous substance releases. Familiarization with OSHA standard 29 CFR 1910.1201.

c. Review of the principles and practices for analyzing an incident to determine the hazardous substances present, their physical and chemical properties, and the likely behavior of the hazardous substance and its container, vessel, or vehicle.
d. Review of the principles and practices for identification of the types of hazardous substance transportation containers, vessels and vehicles involved in the release; selecting and using the various types of equipment available for plugging or patching transportation containers, vessels or vehicles; organizing and directing the use of multiple teams of hazardous material technicians and selecting the appropriate strategy for approaching release sites and containing or stopping the release.
e. Review of procedures for implementing continuing response actions consistent with the local emergency response plan, the organization's standard operating procedures, including knowledge of the available public and private response resources, establishment of an incident command post, direction of hazardous material technician teams, and extended emergency notification procedures and follow-up communications.
f. Review of the principles and practice for proper selection and use of personal protective equipment.
g. Review of the principles and practices of establishing exposure zones and proper decontamination, monitoring and medical surveillance stations and procedures.
h. Review of the expected hazards including fire and explosions hazards, confined space hazards, electrical hazards, powered equipment hazards, motor vehicle hazards, and walking-working surface hazards.
i. Awareness and knowledge of the competencies for the Off-site Specialist Employee covered in the National Fire Protection Association's Standard No. 472, Professional Competence of Responders to Hazardous Materials Incidents.

5. Incident commander.

The incident commander is the individual who, at any one time, is responsible for and in control of the response effort. This individual is the person responsible for the direction and coordination of the response effort. An incident commander's position should be occupied by the most senior, appropriately trained individual present at the response site. Yet, as necessary and appropriate by the level of response provided, the position may be occupied by many individuals during a particular response as the need for greater authority, responsibility, or training increases. It is possible for the first responder at the awareness level to assume the duties of incident commander until a more senior and appropriately trained individual arrives at the response site. Therefore, any emergency responder expected to perform as an incident commander should be trained to fulfill the obligations of the position at the level of response they will be providing including the following:

a. Ability to analyze a hazardous substance incident to determine the magnitude of the response problem.
b. Ability to plan and implement an appropriate response plan within the capabilities of available personnel and equipment.
c. Ability to implement a response to favorably change the outcome of the incident in a manner consistent with the local emergency response plan and the organization's standard operating procedures.

d. Ability to evaluate the progress of the emergency response to ensure that the response objectives are being met safely, effectively, and efficiently.

e. Ability to adjust the response plan to the conditions of the response and to notify higher levels of response when required by the changes to the response plan.

[55 Fed. Reg. 14073, Apr 13, 1990; 56 Fed. Reg. 15832, Apr 18, 1991; 59 Fed. Reg. 43268, Aug 22, 1994; 61 Fed. Reg. 9228, Mar. 07, 1996]

■ APPENDIX G.1 Emergency Response Exercise

Instructions: The following scenarios have been designed for application in a tabletop exercise. They are focused on management issues associated with emergency response. They are more effective when:

- Customized to identify materials and circumstances faced within your organization.
- An answer sheet of possible responses for the organization has been developed.

Scenario 1: It is 1:00 P.M. on a mild spring day. You have just come back from lunch to learn that a forklift has punctured a 55-gallon container of: ____________ . The puncture is midway from the top. The container has been removed to the warehouse floor and laid on its side to stop the flow.

1. What additional information is needed in order to direct emergency response?
2. Whom do you notify about this event? Why? When?
3. What equipment do you have available to help in emergency response?

Scenario 2: While proceeding with the cleanup of ____________, you receive a call from CHEMTREC informing you that a truck transporting hazardous waste from your site has been involved in a "sudden stop" event at a busy intersection. As a result, some containers got damaged and there is liquid leaking from the trailer. The trailer is placarded: ____________. The waste stream is: ____________.

1. What questions should be asked of the CHEMTREC communicator?
2. How would you begin emergency response to this transportation incident?
3. What equipment and other resources can be dispatched to assist in emergency response?

Scenario 3: The driver of the vehicle (yours or the transporter's) has called in and stated that the fire department is evacuating a 1/2-mile area surrounding the vehicle. The driver is being taken to a local hospital for observation.

1. What contact should be made with local government officials about this event? Why? When?
2. What special reporting issues are created because it is a hazardous waste release event?
3. What do you do when the news media call?

Scenario 4: An industrial neighbor has called to ask for details about a spill in your warehouse. An employee learned about it while in line behind two of your organization's employees at a fast-food restaurant. The neighbor asks, "What should be done?"

1. What information can the neighbor be provided with at this time?
2. How will it be communicated to the neighbor?
3. This neighbor is not listed in the organization's emergency response plan. What future action does this require?

Scenario 5: The emergency room physician evaluating the driver has received the material safety data sheet(s) for the virgin product that comprised the waste stream leaking from the truck.

1. What medical surveillance should you request be performed?
   The ____________ container in the warehouse has been patched and is being transferred to another serviceable drum.
2. Describe the decontamination procedure that can be used in this circumstance.
3. Describe the type of decontamination that may be required to address the waste leak.

■ APPENDIX G.2 Domestic Chemical Diversion Control Act of 1993 List I and List II Chemicals (21 CFR 1310.02)

List I Chemicals	DEA Chemical Code Number
Anthranilic acid, its esters and its salts	8530
Benzyl cyanide	8735
Ephedrine, its salts, optical isomers and salts of optimal isomers	8113
Ergonovine and its salts	8675
Ergotamine and its salts	8676
N-acetylanthranilic acid, its esters and its salts	8522
Norpseudoephedrine, its salts, optical isomers and salts of optical isomers	8317
Phenylacetic acid, its esters and its salts	8791
Phenylpropanolamine, its salts, optical isomers and salts of optical isomers	1225
Piperidine and its salts	2704
Pseudoephedrine, its salts, optical isomers and salts of optical isomers	8112
3,4-methylenedioxyphenyl-2-propanone	8502
Methylamine and its salts	8520
Ethylamine and its salts	8678
Propionic anhydride	8328
Insosafrole (isosafrole)	8704
Piperonal	8750
N-methylephedrine, its salts, optical isomers and salts of optical isomers	8119
Hydriotic acide (hydriodic acid)	6695
Benzaldehyde	8256
Nitroethane	6724
List II Chemicals	
Acetic anhydride	8519
Acetone	6532
Benzyl chloride	8570
Ethyl ether	6584
Potassium permanganate	6579
2-butanone (or methyl ethyl ketone or MEK)	6714
Toluene	6594
Hydrochloric acid	6545
Sulfuric acid	6552
Methyl isobutyl ketone (MIBK)	6715

■ APPENDIX G.3 Waste Reduction: Make It Your First Priority

RECYCLING
WORKS
It's Good Business

WASTE REDUCTION: MAKE IT YOUR FIRST PRIORITY

WHY?

- Reduce raw material costs
- Reduce waste disposal costs
- Conserve valuable resources
- Reduce purchasing costs
- Enhance your image with the public and your employees

HOW?

For **Office Areas**:

- Make double-sided copies.
- Reuse single-sided pages for drafts and note paper. Establish a draft paper tray for laser printers so that single-sided sheets can be reused internally.
- Consolidate and reduce the size of office forms.
- Encourage employees to communicate via electronic mail.
- Reduce the number of copies needed by circulating or posting memos. Share reports and publications.
- Use centralized files for hard copies.
- Use outdated letterhead and forms for in-house memos.
- Proofread documents on computer screen before printing.
- Avoid disposable. Instead, recharge laser printers, fax machines, and copiers; reink dot matrix ribbons; use refillable pens and pencils.
- Reuse supplies, such as file folders, envelopes, boxes and containers.
- Use window-less envelopes. When billing customers, use envelopes that can be folded inside-out and returned to you with payment.
- Keep mailing lists current.
- Reduce junk mail: remove your company name from unwanted mailing lists by writing to Direct Marketing Associates, Mail Preference Service, P. 0. Box 9008, Farmingdale, NY 11735-9008. Contact the **Business Recycling Program** (296-8800) for additional ways to reduce junk mail.
- Avoid cover sheets for faxes. Use a rubber stamp especially made for fax transmittal.
- Start a "waste exchange" in your building or office park. (See "Industrial Products & Surplus Materials," below.)

For **Shipping/Receiving Areas**:

- Order merchandise with minimal packaging, in a concentrated form and/or in bulk.
- Ask suppliers to use less packaging in filling your orders.
- Request that deliveries be shipped in returnable containers.
- Return, reuse, and repair wooden pallets and crates.
- Replace cardboard boxes with durable boxes for shipping to branch offices, stores & warehouses.
- Set up a system for returning cardboard boxes and polystyrene "peanuts" to distributors for reuse. Reuse boxes that cannot be returned.
- Use shredded newspapers, tissue, and foam pads from incoming shipments as alternatives to purchasing polystyrene "peanut" packaging.

(over)

King County Solid Waste Division

For **Restaurants** and **Lunch Room Areas**:

- Use cloth items instead of paper towels, tablecloths, napkins, and hand towels.
- Encourage employees to use durable cups and tableware.
- Offer beverages dispensed from tanks or refillable bottles instead of individual packages, provided that you make available reusable glasses for these beverages. If this is not feasible, provide beverages in aluminum cans and glass bottles, which are readily recyclable.
- Invest in a dishwasher and avoid serving food on disposable dishes.

For **Outdoor Areas and Food Waste**:

- Choose a landscape design that needs low maintenance and little water. Eliminate pesticide use: go organic.
- Compost grass clippings and leaves into a valuable soil amendment. Leave grass clippings on lawn using a mulching mower.
- Set up a worm bin to convert non-fatty food wastes into vermicompost.

For **Equipment**:

- Invest in equipment that facilitates waste reduction, such as copiers that automatically make double-sided copies.
- Install reusable furnace and air conditioner filters.
- Purchase high-quality, durable and repairable equipment.
- Use solar calculators; switch to rechargeable batteries for other electronics.
- Don't change your vehicles' motor oil more often than necessary. Newer vehicles require oil changes every 7,000 highway miles or so. Check your manual.

For **Industrial Products** and **Surplus Materials**:

- Join a waste exchange, like "**IMEX**" (Phone: 296-4899) to share industrial byproducts, surplus materials, and "wastes." IMEX matches waste generators with waste users. They are now part of a National Materials Exchange, accessible by modem at 1-800-858-6625. Access is free and available 24 hours each day.

For **Retailers**:

- Implement practices described in the **Preferred Packaging Procurement Guidelines of the Washington Retail Association** (WRA). To obtain a copy, write or phone Patty Schwegman of WRA at PO Box 2227, Olympia, WA 98507-2227; (206) 943-9198. The guidelines, listed in order of preferred implementation, are as follows. *(These have been paraphrased to apply broadly to small as well as large retailers.)*
    - **Eliminate:** Whenever possible, eliminate the package altogether.
    - **Minimize:** Minimize the amount of material used in packaging.
    - **Refill/Reuse:** Design *(or stock)* packages that are either consumable, returnable, or refillable/reusable.
    - **Recyclable/Recycled Content:** Produce *(or stock)* packages that are recyclable and are comprised of post-consumer material.
- Provide financial incentives for customers to support your efforts to eliminate waste. For example, give discounts for bulk purchasers and for those who bring their own bags or drinking mugs.

*If you have questions or would like to share your waste reduction successes with us, please call the **King County Business Recycling Program** at 296-8800.*

■ APPENDIX G.4 Food Lifeline Vermicomposting Project

WASTE PREVENTION STRATEGY
Vermicomposting

RECYCLING WORKS
It's Good Business

Food Lifeline Vermicomposting Project

Food Lifeline is a distribution center for most of the food banks and meal programs in King County. Located on the Fircrest Campus in north King County, it receives, evaluates and recovers donated food stuffs. Through vermicomposting, Food Lifeline diverts as much as 6,720 pounds of spoiled produce and canned goods that would otherwise be bound for the landfill each month.

EXAMPLES OF BUSINESSES AND FACILITIES THAT CAN USE THIS STRATEGY:

- Restaurants
- Hospitals
- Resorts
- Institutions
- Hotels
- Grocery Stores

FOOD LIFELINE PROCESS:

- Volunteers are trained in composting techniques
- Apple crates have been made into 28 worm bins
- A collection system of plastic totes collects food scraps from the receiving center on site
- Cardboard, straw and leaves are shredded for use as bedding
- Food scraps are mixed with bedding materials in a 2:1 ratio by weight (1:3 by volume)
- Cured compost is harvested and used for landscaping

SAVINGS:

- Each worm bin can potentially compost 60 - 150 pounds of food scraps as well as 30 - 75 pounds of bedding per week
- Paid staff time is approximately two hours per day with volunteers providing additional labor as needed
- Diverting food scraps saves hauling fees and landfill space

ADDITIONAL INFORMATION:

- Although the Fircrest campus has extra space for more and larger bins, small individual worm bins are more efficient compost units and can be used in smaller spaces
- Food is donated in bulk, which sometimes requires grinding prior to worm feeding
- Maintaining aerobic conditions varies according to feedstocks
- Vermicomposting requires regular maintenance and a knowledge of composting. A facility wishing to compost their own food scraps must submit a description of proposed composting methods and regular maintenance plan to the local Health Department.

Call the King County Business Recycling Program for more information at 296-4356.

King County Solid Waste Division

Setting Up Onsite Vermicomposting

About 9% of the total waste stream in King County is composed of food scraps. However, for individual businesses like restaurants, food scraps can represent as much as 50% of their garbage. Vermicomposting (composting with worms) has been successfully implemented by a variety of businesses and facilities around the country, including hotels, restaurants, grocery wholesalers and retailers, and recreational facilities.

Vermicomposting can be an efficient way to compost food scraps on your business premises. You won't need much to set up a worm bin on site. Conduct a preliminary waste steam audit for your establishment. Look at what you throw away, on a daily, weekly, and monthly basis. What volume of food scraps is generated? Are there compostable materials such as cardboard, shredded newspaper, or fallen leaves that can be used as worm bedding?

WHAT YOU WILL NEED:

- a worm bin
- bedding for the worms to live in
- red worms
- vegetative food scraps
- a committed person knowledgeable about composting to tend the bin
- a description of proposed composting methods and maintenance plan to provide to the local Health Department before vermicomposting can begin

A worm bin is a sturdy box that can be made from materials like a packing crate, pallet boxes, or built from plywood and two-by-fours. Concrete blocks can be used to build a more permanent structure. Sturdy weather resistant lids are needed for bins and paint improves bin appearance. Worms need to breathe so the active composting layer should be up to one and a half feet deep. Drill 1/4 - 1/2 inch holes in the bottom for drainage. If intensively managed, a worm bin can have about one square foot of surface area for each 2 - 3 pounds of food scraps added per week. The number and size of worm bins needed will depend on the amount of food scraps generated. To obtain free instructions for building an attractive worm bin, call the King County Recycling and Composting Information Line at 296-4466.

Bedding for worm bins provides the worms with a nest to live and work in. Bedding also holds moisture and serves as a place to bury the food scraps. Composting food scraps without bedding can produce bad odors and a slimy mess. Common bedding materials include shredded newspaper or corrugated cardboard, peat moss, coarse sawdust, or fallen leaves. Moisten bedding materials by immersing them in water for several minutes before adding them to the worm bin. When the bedding is thoroughly wet, remove it from the water and wring out excess water. Fill the bin to the top with loose bedding. Pull apart any compacted paper strips before adding them to the bin so that the worms have room to roam.

Red worms, also known as "red wigglers," or "manure worms," are the composting champions. Red worms are not the same as earthworms, or night crawlers, which live in soil. Use about two pounds of worms for every pound of food scraps generated per day. A starter batch of worms can be taken from an existing worm bin or can be purchased at the Seattle Tilth Association (633-0224). Stores that sell fish bait often sell red worms, too.

Red Wrigglers are vegetarians. Do feed the worms:

- vegetable scraps
- grains
- fruit rinds and peels
- breads
- coffee grounds, filters, and tea bags

Don't feed your worms:

- meat
- fish
- cheese, oils or oily foods
- butter

These foods take too long to fully decompose and are most attractive to rats and other scavenging animals.

Feed the worms by burying food scraps in the bedding. Set up a regular collection system for the food scraps and bedding that you will need for the worms. You may need to add shredded paper for bedding or otherwise adjust your ratio of food scraps. Covering food scraps with a few inches of bedding, worm castings, or a sheet of plastic discourages flies and odors.

This information sheet presents basic information to get you started. An excellent source of information on vermicomposting is Mary Appelhof's book, Worms Eat My Garbage. You can find a copy in libraries, for sale at local bookstores or at Seattle Tilth (633-0224). Vermicomposting principles are also covered in King County Master Recycler Composter training. For more information about this training, vermicomposting and compost uses call King County's Recycling and Composting Information Line (296-4466).

VERMICOMPOSTING MUST BE DONE IN ACCORDANCE WITH LOCAL HEALTH DEPARTMENT REGULATIONS Call your local Health Department for details. For businesses located in King County call the Seattle King County Department of Public health at 296-4722. An operating and regular maintenance plan must be submitted to the Health Department for approval before vermicomposting can begin.

revised October 6, 1993

■ APPENDIX G.5 Washington Drycleaners Association

WASTE PREVENTION STRATEGY

Returnable/Reusable Drycleaning Bag Project

RECYCLING WORKS
It's Good Business

Washington Drycleaners Association

Drycleaning establishments individually contribute approximately 22,000 polyethylene bags to the waste stream each year. The Washington Drycleaning Association further estimates that there are 200 drycleaners in King County whose clients dispose of 100 tons of plastic garment bags per year. Regular customers at drycleaners often patronize the same establishment on a weekly basis. As an alternative, a drycleaner can provide reusable, zippered garment bags to customers at cost. Customers can bring their clothing in to the drycleaners in the bag. The bag is then saved and reused when the customers pick up their clothes. It offers incentives for customer loyalty for those customers who are concerned about both environmental degradation and protecting their clothes. For drycleaners, the use of reusable bags offers an opportunity for advertising (via a logo) and a chance to build up repeat business.

WHO CAN USE THIS STRATEGY:

- ◆ Garment retailers
- ◆ Uniform services
- ◆ Commercial laundries
- ◆ Hospitals and institutions with laundries

PROCESS:

- ◆ Several area drycleaners have agreed to participate in the pilot study
- ◆ The pilot project will substitute reusable garment bags for disposable plastic bags for about 20% of customers
- ◆ The drycleaners will track the number of plastic bags used each month, monitor the waste reduction resulting from the use of reusable garment bags and record changes in purchasing

SAVINGS:

- ◆ Approximately 70% of customers are repeat customers
- ◆ Waste will be diverted at the customers' end of the waste cycle; an estimated one-half ton per cleaner may be diverted, with a total potential diversion of 100 tons County-wide
- ◆ The cost of the bag to customers can be quickly paid for with discounts on drycleaning

UNUSUAL FEATURES:

- ◆ Many drycleaners have repeat customers
- ◆ Customers who regularly use drycleaning services often store their garments in bags prior to transit
- ◆ Providing a reusable bag at low cost increases the value of drycleaning services by further protecting clothes in transport

Call the King County Business Recycling Program at 296-4356 for more information.

revised July 15, 1993

King County Solid Waste Division

INDEX

Lightning Source UK Ltd.
Milton Keynes UK
UKOW01n1929310817
308345UK00002B/119/P

9 780471 180159